Algorithms for Intelligent Systems

This book series publishes research on the analysis and development of algorithms for intelligent systems with their applications to various real world problems. It covers research related to autonomous agents, multi-agent systems, behavioral modeling, reinforcement learning, game theory, mechanism design, machine learning, meta-heuristic search, optimization, planning and scheduling, artificial neural networks, evolutionary computation, swarm intelligence and other algorithms for intelligent systems.

The book series includes recent advancements, modification and applications of the artificial neural networks, evolutionary computation, swarm intelligence, artificial immune systems, fuzzy system, autonomous and multi agent systems, machine learning and other intelligent systems related areas. The material will be beneficial for the graduate students, post-graduate students as well as the researchers who want a broader view of advances in algorithms for intelligent systems. The contents will also be useful to the researchers from other fields who have no knowledge of the power of intelligent systems, e.g. the researchers in the field of bioinformatics, biochemists, mechanical and chemical engineers, economists, musicians and medical practitioners.

The series publishes monographs, edited volumes, advanced textbooks and selected proceedings.

More information about this series at http://www.springer.com/series/16171

Kaiser J. Giri · Shabir Ahmad Parah ·
Rumaan Bashir · Khan Muhammad
Editors

Multimedia Security

Algorithm Development, Analysis and Applications

Springer

Editors
Kaiser J. Giri
Department of Computer Science
Islamic University of Science
and Technology
Pulwama, Jammu and Kashmir, India

Shabir Ahmad Parah
Department of Electronics
and Instrumentation Technology
University of Kashmir
Srinagar, Jammu and Kashmir, India

Rumaan Bashir
Department of Computer Science
Islamic University of Science
and Technology
Pulwama, Jammu and Kashmir, India

Khan Muhammad
Department of Software
Sejong University
Seoul, Republic of Korea

ISSN 2524-7565 ISSN 2524-7573 (electronic)
Algorithms for Intelligent Systems
ISBN 978-981-15-8710-8 ISBN 978-981-15-8711-5 (eBook)
https://doi.org/10.1007/978-981-15-8711-5

This Springer imprint is published by the registered company Springer Nature Singapore Pte Ltd.
The registered company address is: 152 Beach Road, #21-01/04 Gateway East, Singapore 189721, Singapore

Preface

The contemporary multimedia communication has converted the world into a global village. The digitization of sensitive multimedia, its storage and transmission over the Internet has increased massively. There is an exponential increase in the use of the Internet for applications in healthcare, commercial, government, and industrial sectors because of improved networked technology and computing power. The advantages of a completely digital system include ease of data management, cost reduction, high speed, and reliability among many others. However, there is a possibility of encountering numerous threats like phishing attacks, malware, ransomware, unauthorized access, and tampering of the digital content. Therefore, it is of utmost importance that digital information is secured and objectives of confidentiality, integrity, availability, authentication, and non-repudiation are fulfilled. One of the most important challenges is to develop algorithms that can protect sensitive data. Sensitive multimedia can include text files, audios, images, and videos. However, images form a large percentage of the transmitted and stored content and therefore need to be secured completely. This book shall serve as a comprehensive reference to multimedia security and copyright protection, and it deliberates on various enabling technologies in the field. It provides a good account of multimedia security, quality metrics that govern security, and algorithm development for multimedia secure transmission.

Multimedia security technologies broadly include cryptography and information hiding. Cryptography transforms data into an observable meaningless form and includes variants like symmetric-key- and asymmetric-key-based algorithms and more recently have included techniques like chaos-based encryption and DNA encryption. The security of the cryptosystem lies in the secrecy of the key and as such requires proper key management techniques. Information hiding methods on the other hand hide the complete existence and possibility of a secret into a cover media and can be divided into major domains like steganography, watermarking, multimedia mining, and digital forensics. Steganography deals with the optimization of algorithms for embedding secret data into cover images such that embedding capacity is increased without affecting the detectability, whereas digital watermarking is concerned with embedding a small number of bits into a document such that it is protected from forgery. The watermark can be robust, fragile, visible, or concealed and is mainly embedded for copyright protection of the data.

This book provides an in-depth explanation of several robust security algorithms aiming to provide confidentiality and authenticity to digital multimedia content. It attempts to include novel research works about several aspects of multimedia security which are generally not much investigated. The book is mainly divided into three main parts. Part I: *Security Using Cryptography and Chaotic Theory: Algorithm Development and Analysis* includes four chapters. In Chap. 1, the medical image encryption is achieved by the extraction of local features like textures and the use of the chaotic framework. Chapter 2 focusses on image encryption using DNA-based computing and chaotic encryption while Chap. 3 elucidates the impact of computational power on cryptography. Furthermore, Chap. 4 discusses the anti-forensic techniques for providing privacy and security to multimedia.

Part II: *Information Embedding for Secure Data Communication and Authentication* includes Chaps. 5–9 and considers algorithms used to embed information into cover media. Chapter 5 is mainly concerned with embedding information using DNA sequences. Chapters 6 and 7 describe the process of data embedding in colour Images utilizing algorithms like multiple-bit substitution patterns and modular arithmetic. Chapter 8 is concerned with the survey of different watermarking techniques for medical images, and Chap. 9 discusses the data privacy and security aspects in the e-health scenario.

Furthermore, Part III: *Security in IoT and Big Data-Driven Enterprise* focusses on developing techniques to ensure security in IoT-based systems. Chapter 10 describes the implementation and optimization of machine learning-based intrusion detection systems for the Internet of Things. Chapter 11 explains an approach to detect phishing using machine learning. Chapters 12 and 13 discuss the use of blockchain to ensure security in the Internet of things (IoT)-based systems. Furthermore, Chap. 14 presents a review of secure distributed range-free hop-based localization algorithms in wireless communication.

This book presents a comprehensive discussion about the aspects of multimedia security in digital medical systems and IoT-based systems. Several encryption algorithms are discussed in detail providing security and confidentiality to general as well as medical images. Besides, it provides an insight into the underpinnings of machine learning algorithms like SVMs and the use of blockchain technology for securing multimedia. Various aspects of cryptography are investigated which include computational power, ease of implementation, efficiency, and time requirement for proper implementation. Furthermore, the book also includes the development and optimization of state-of-the-art localization, recognition, detection, and security techniques.

We hope this book presents a comprehensive insight into various multimedia security aspects to its audience in general and prospective researchers in particular and paves a way for them to design and implement better and efficient multimedia security techniques.

Pulwama, India	Kaiser J. Giri
Srinagar, India	Shabir Ahmad Parah
Pulwama, India	Rumaan Bashir
Sejong, Republic of Korea	Khan Muhammad

Contents

About the Editors

Kaiser J. Giri received his Master's Degree, M.Phil. and Ph.D. in Computer Applications from the University of Kashmir, India, in the year 2004, 2011 & 2016, respectively. He is working as a Sr. Assistant Professor in the Department of Computer Science, Islamic University of Science & Technology, Awantipora, J & K, India, since 2006. He has participated in more than 50 national & international conferences, seminars & workshops. He has more than 20 papers in various journals & conferences of international repute to his credit. His research interests include image processing, signal processing, information security, multimedia security, digital image watermarking and big data analytics.

Shabir Ahmad Parah received M.Sc., M.Phil. and Ph.D. degrees in Electronics from the University of Kashmir, Srinagar, in 2004, 2010 and 2013, respectively. He is currently an Assistant Professor with the Department of Electronics and IT, University of Kashmir. He has authored or co-authored over 100 papers in the journals and conferences of international repute. His fields of interest are multimedia communication and security, image processing, information hiding, electronichealth care, secure communications, steganography and watermarking, biometrics, FPGA-based digital signal processing, digital system design using hardware description languages, nonlinear dynamics and chaos.

Rumaan Bashir is working as a Sr. Assistant Professor in the Department of Computer Science, Islamic University of Science and Technology, Awantipora, J&K, India, since 2006. Her area of research is pattern recognition/document image analysis and information security and is presently working on script identification. The author received her MCA and PGDBA from the University of Kashmir, J&K, India, in the year 2005 & 2009, respectively. She has also done M.Phil. & Ph.D. in Computer Science from the University of Kashmir, J&K, India, in the year 2011 & 2016, respectively. She has more than 20 papers in various journals & conferences of international repute to her credit.

Khan Muhammad is an Assistant Professor in the Department of Software, Sejong University, South Korea. His research interests include information security, video summarization, computer vision and video surveillance. He has authored over 100 papers in peer-reviewed international journals such as IEEE TII, TIE, IoTJ and TSMC Systems and is a reviewer of over 70 SCI/SCIE journals including IEEE Communications Magazine, IEEE Network, IEEE Internet of Things Journal, TIP, TII, TCYB and IEEE Access. He is a member of the ACM.

Part I
Security Using Cryptography and Chaotic Theory: Algorithm Development and Analysis

Chapter 1
Securing Medical Images via a Texture and Chaotic Key Framework

Shoaib Amin Banday, Mohammad Khalid Pandit, and Ab. Rouf Khan

1 Introduction

In the recent past, there has been a massive transformation in the way that data is generated and handled. This can be attributed undoubtedly to advancements and rigorous research in network technologies and system designs. These advanced systems in tandem with the network technologies not only generates the data but also provides with an opportunity to multiply the data during the transmission. Some of the multimedia data is generated by some critical service infrastructures like airports, military, etc., in the form of text documents, images and videos. The necessity of transmitting such multimedia data over a network demands for some measures to be put in place so that authenticity, integrity and confidentiality are not compromised. Therefore, as can be concluded intuitively, a stronger, robust and an efficient mechanisms should be put in place so that the multimedia data flowing over an insecure network systems are safeguarded against likely attacks. To achieve the basic information security goals: confidentiality, integrity and authentication, researchers from multiple domains have collaborated effectively to come up with some acceptable solutions. Multimedia information can be broadly classified into cryptography and information hiding techniques. Ever since the duplication of multimedia was observed which is considered as an illegal activity in the current digital era,

S. A. Banday (✉)
Department of E&C Engineering, School of Engineering and Technology, IUST, Awantipora, Jammu and Kashmir, India
e-mail: shoaibee.a@gmail.com

M. K. Pandit
Department of CS and Engineering, School of Engineering and Technology, IUST, Awantipora, Jammu and Kashmir, India

Ab. R. Khan
Department of CS and Engineering, National Institute of Technology Srinagar, Srinagar, India

K. J. Giri et al. (eds.), *Multimedia Security*, Algorithms for Intelligent Systems,
https://doi.org/10.1007/978-981-15-8711-5_1

researchers have come up with digital watermarking, digital signatures, steganography, etc., which has a wide variety of applications like copyright insurance and secret communication as put up in the work proposed by [1–16].

There has been a many-fold progress in the health care along various fronts from last 2–3 decades. From the diagnostic front, this has been only possible by the inclusion of new medical imaging techniques that have come into existence from collaborated efforts from different professional corners. These medical imaging modalities ensure every possible detail is at hand of a medical healthcare professional so that an accurate diagnosis and prognosis can be made. This diverse detail from a subject may be obtained by multiple imaging modalities. For example, a patient may be recommended to have both CT and MRI done for a certain investigation. This implies that there is a huge amount of image data that can be generated from a particular subject's anatomical site. Even if we consider only a single modality, the image data generated can still be large enough to bring in two potential problems: storage and security during transmission. This image data may be sometimes consumed or utilized by an intermediate healthcare module for a specific and well-defined analysis like image segmentation, image de-noising and image enhancement [17–21]. The telemedicine on the other hand has progressed significantly from the past decade. The services that the modern healthcare systems provide certainly play a central role in the overall efficacy of the detection and diagnostic procedures. The telemedicine on the other hand has progressed significantly from the past decade. The medical image data generated now often needs to be transmitted over an insecure channel. In doing so, medical organizations must ensure that they stay within the guidelines as put by Health Insurance Portability and Accountability Act (HIPAA). Therefore, it becomes critical that medical data images be made prone to different attacks which are intended to compromise on the authentication, integrity and confidentiality [22–24]. There has been plethora of work across various domains to address these issues. The next section gives a glimpse of the previous work that is carried to incorporate the security aspect in the medical images during their transmission.

2 Related Work

To make diagnosis of a disease as effective as possible on an e-health framework, it is important to ensure that no data is lost or compromised during its transmission [24–29]. One of the broader ways of classifying conventional cryptographic systems is whether these systems utilize a common key for both ciphering and deciphering purposes [30]. Along either line, block ciphers have been proposed from time to time. However, there are few such cryptosystems that were considered as a benchmark in early times of data security. These include cryptosystems like AES, IDEA and DES [31]. These standards served for a long time in data securing in general and have caught an eye in the domain of image security from the past decade. These are numerous applications of image processing were security is of prime important, but at the same time, some of these applications are very time critical. This implies that the time to cipher and decipher an image should be done at a high speed. The same notion holds true for medical image security wherein the time becomes even more critical because of the large amount of medical image data that we need to transmit. Cryptosystems suggested above have not proven efficient as there results are unacceptable. Along the lines of continuous effort of improving the cryptosystems for large medical image datasets, chaos-based approaches were proposed [32–35]. Based on the chaotic behavior, such methods came up primarily with an advantage of the key sensitivity which in essence is very important for the success of any cryptosystem [36].

The literature survey in this work is mainly oriented toward the use of chaotic sequences as we explore the same in our methodology [37] presents a diffusion confusion framework by exploring chaotic map (2D) and key streams. The authors in [38] suggested the use of pseudo-random number generation by employing shift registers alongside with the Kolmogorov chaotic maps for a cryptosystem. Similarly, the work reported in [39–41] suggest implementation of permutation strategies via baker map and cat maps. It must be remembered that most of techniques for medical mage encryption are purely based on the permutation–substitution framework or one of these methodologies as did by [42] wherein only the permutation has been used. As a step forward, the work proposed in [43] specifically addresses the issue of periodicity by employing sorting of chaotic sequence. This results in this work depict that the encryption and decryption times have reduces considerably. The authors [44] use a cascaded type of cryptosystem which essentially are driven by 1D map each at every step. Similarly, a multi-level scheme is introduced by [45] which use 4 rounds of symmetric cryptosystems. These results suggest the use of logistic map and the standard maps to incorporate diffusion and confusion, respectively. The methodologies introduced by [46] and [47] use two image ciphers separately; pixel by pixel shuffling of the image is done initially, and then, it is followed by a confusion strategy by hyper-chaotic system. The work suggested in [48] presents a spatial chaos cipher which encrypts image pixels one by one. The chaos-based techniques have been widely used in the pursuit of image security over an insecure media. There have however been other directions which have been experimented for providing medical image security. For example, the use of a uniquely extracted part of an image to encrypt that very image has been one of the major directions in the domain of medical image security. These techniques, which we call as selective encryption techniques reduce the encryption and decryption times [49–56]. The success of these techniques primarily depends on how well the portion of image to be encrypted has been calculated. Since the cipher does not utilize an external component for generating the encryption key, these techniques can be well thought of as a potential candidate for real-time medical image encryption techniques if other parameters are carefully looked at as well.

The work introduced here is fundamentally a fusion of two strategies: chaotic maps and texture maps. As depicted from the literature survey, chaos-based cryptosystems have been extensively used either separately or in a cascaded form. However, the texture driven maps for medical image encryption has not been explored fully. To enhance the basic medical image encryption parameters like resilience to common attacks, the length of the key and sensitivity to the initial condition, we present this methodology for secure image transmission.

3 Materials and Methods

Like most of the image cryptosystems, the methodology presented in this work is again based on a diffusion–confusion framework. The overall flow of this methodology is presented in Fig. 1. This methodology is mainly compromised of three components. Initially, the permutation of original medical image is done by employing Arnold cat map. Alongside this permutation, we generate two keys from (1) using multiple chaotic maps and (2) maps generated from Gabor filters. The Gabor filters are the texture characterization of an image along different orientations and scales.

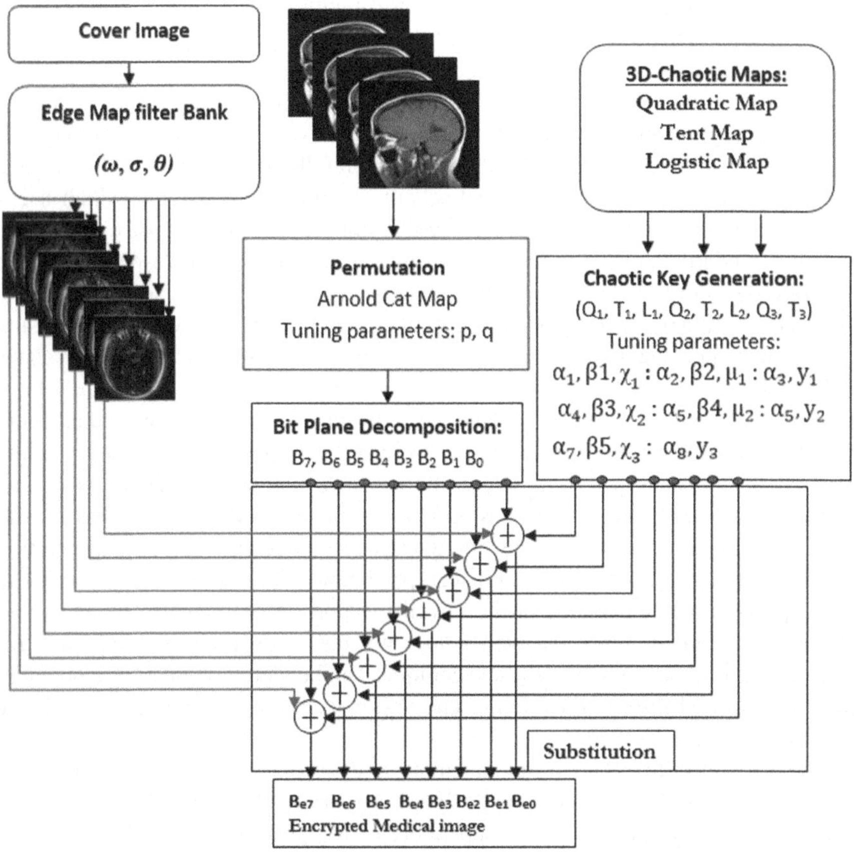

Fig. 1 Proposed schematic of texture and chaos map encryption framework

3.1 Initial Pixel Shuffling via ACM

The Arnold cat map is a chaotic map based on permutation with inevitability property [57, 58]. Consider a medical image of size $n \times m$. The Arnold transformation can be simply put as:

$$R(x, y) = \begin{bmatrix} 1 & r \\ s & rs + 1 \end{bmatrix} \begin{bmatrix} x \\ y \end{bmatrix} \bmod n \tag{1}$$

where *(x, y) is the initial index of the pixel to be permuted and variables r and s are the transformation control parameters which implies that the values of r and s would control the amount of image squeeze and stretch. To illustrate this amount of squeeze and stretch, we assume that we have an image of size 256 × 256. Also, we assume that the pixel that we want to transform is currently indexed as (5,5). We also assume that the initial seed value for r and s is 1 and 1, respectively. After the first transformational iteration, we get:*

$$R(5, 5) = \begin{bmatrix} 1 & 1 \\ 1 & 2 \end{bmatrix} \begin{bmatrix} 5 \\ 5 \end{bmatrix} \bmod (256) = (10, 15)$$

Permuting with the same transformational matrix, we obtain the next two indexes of the pixel at positions of (25, 40) and (65, 105). The permutation is carried simultaneously for all other pixels using the same Arnold transformational matrix. The same is depicted in Fig. 2, wherein the original can be easily seen to shuffle. We can easily choose the values of the ***r*** and ***s*** to generate a unique shuffling pattern of the image. The expression below can be executed to generate the inverse transformation of the shuffling as:

$$R'(x, y) = \begin{bmatrix} rs + 1 & -r \\ -s & 1 \end{bmatrix} \begin{bmatrix} x \\ y \end{bmatrix} \bmod n \tag{2}$$

$R'(x, y)$ signifies the inverse transformation of a pixel.

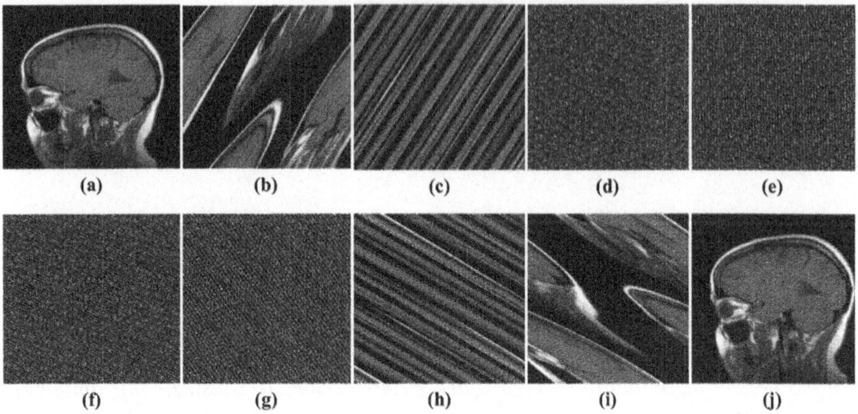

Fig. 2 **a** Sagittal Slice of Brain MR Image undergoing Arnold pixel permutation via Arnold transformation and instances are shown after iterations number(s): **b** 1, **c** 3, **d** 10, **e** 15, **f** 40, **g** 54, **h** 65, **i** 67 and **j** 68

Figure 2 shows an Arnold cat map transformation of normal brain MR image. Figure 2b shows the permuted image after three iterations of the original MR brain image as per Eq. 1. This image Fig. 2b appears skewed after three iterations indicating the shuffling of the pixels. Continuing with more iterations of the original image as shown through Fig. 2c–e, the original image pixels are completely shuffled. It must be observed through Fig. 2f–i that the shuffled image starts to take the order of the original image again. Figure 2j shows the original image back after 68 transformational iterations.

3.2 Low-/Mid-Frequency Maps via Gabor Filters

The texture is a perceptual property that every surface of an object possesses. This property is quantified by estimating a set of parameters from the given image. Texture is a spatial property of an image which varies significantly from one image to another. The spatial property of an image gives an insight into the pattern with which different intensities, or color is distributed from within the image. One formal way of defining texture is: "an entity consisting of mutually related pixels and group of pixels." There is a plethora of work inclined toward the use of texture for applications like detection of objects from an image, image segmentation and classification and security [59–61]. Since the texture is a unique property possessed by an object in an image, the same can be effectively used for the generation of the key for encrypting an image to attain a secure image transmission. The only challenge is that, how effectively can this texture be quantified. In this work, we have successfully employed the use of Gabor filters for texture quantification for the generation of encryption key.

Gabor filters are a type of linear filters used for the analysis of the texture in an image. It captures the low- and mid-frequency details in an image along different directions with varying scales. These filters can be thought of Gaussian envelops modulated by complex sinusoid in the spatial domain [62, 63]. Consider an even function (a cosine) multiplied with a Gaussian window as:

$$h_{\text{even}}(x) = \frac{1}{\sqrt{2\pi\sigma}} e^{\frac{-x^2}{2\sigma^2}} \cos(2\pi\omega_0 x) \tag{3}$$

Similarly, an odd function (a sine) multiplied with a Gaussian window can be expressed as:

$$h_{\text{odd}}(x) = \frac{1}{\sqrt{2\pi\sigma}} e^{\frac{-x^2}{2\sigma^2}} \sin(2\pi\omega_0 x) \tag{4}$$

Based on the above two functions, Gabor mathematically defined a complex one-dimensional filter as:

$$h_{\text{gabor}}(x) = \frac{1}{\sqrt{2\pi\sigma}} e^{\frac{-x^2}{2\sigma^2}} e^{j(2\pi\omega_0 x)} \tag{5}$$

A 2D Gabor filter may similarly be represented in terms of its even and odd components as:

$$h_{\text{even}}(x, y) = \frac{1}{\sqrt{2\pi\sigma_x\sigma_y}} e^{-\frac{1}{2}\left[\frac{x^2}{\sigma x}+\frac{y^2}{\sigma y}\right]} \cos(2\pi\omega_0 x + 2\pi\omega_0 y) \tag{6}$$

$$h_{\text{odd}}(x, y) = \frac{1}{\sqrt{2\pi\sigma_x\sigma_y}} e^{-\frac{1}{2}\left[\frac{x^2}{\sigma x}+\frac{y^2}{\sigma y}\right]} \sin(2\pi\omega_0 x + 2\pi\omega_0 y) \tag{7}$$

which yields a 2D complex Gabor filter function h_{gabor} as:

$$h_{\text{gabor}}(x, y) = \frac{1}{\sqrt{2\pi\sigma_x\sigma_y}} e^{-\frac{1}{2}\left[\frac{x^2}{\sigma x}+\frac{y^2}{\sigma y}\right]} e^{j\omega_0(x\cos\theta + y\sin\theta)} \tag{9}$$

where ω_0 is the frequency, σ is the spatial spread of the Gaussian window to be modulated and θ is the orientation of the window. This implies that the window can be used to extract a wide range of frequency details from an image by properly choosing the values of ω, σ and θ. Figure 3 shows some Gaussian kernels with different values of ω, σ and θ.

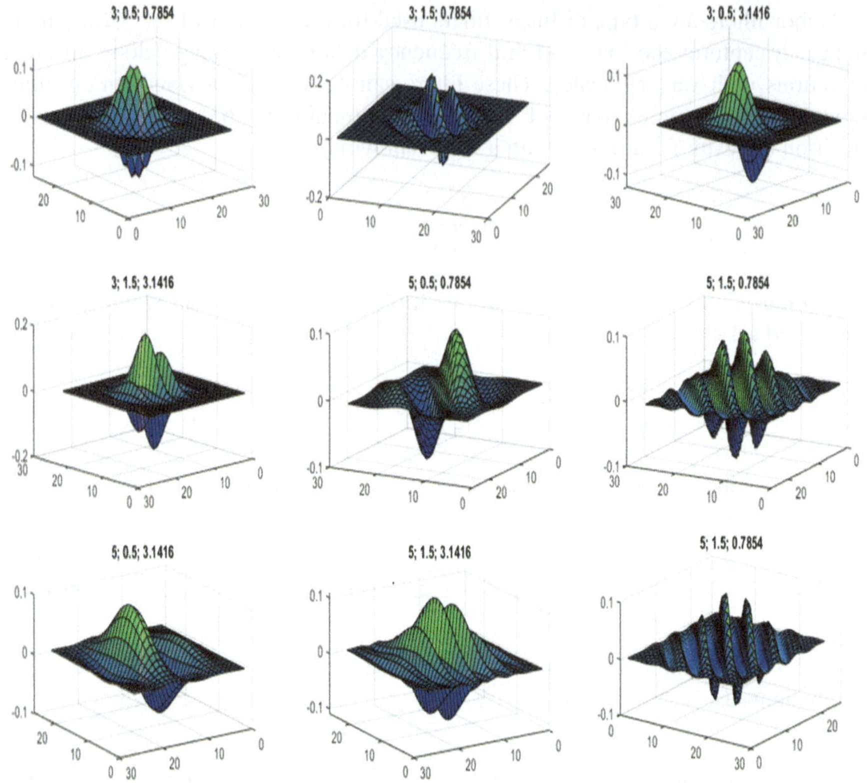

Fig. 3 A bank of Gabor filter kernals with $\sigma = 3$,$\omega = 0.5$, $\theta = \pi/4$ (top middle) $\sigma = 3$, $\omega = 1.5$, $\theta = \pi/4$ (top right) $\sigma = 3$, $\omega = 0.5$, $\theta = \pi$ (middle left) $\sigma = 3$, $\omega = 1.5$, $\theta = \pi$ (center) $\sigma = 5$, $\omega = 0.5$, $\theta = \pi/4$ (middle right) $\sigma = 5$, $\omega = 1.5$, $\theta = \pi/4$ (bottom left) $\sigma = 5$, $\omega = 0.5$, $\theta = \pi$ (top middle) $\sigma = 5$, $\omega = 1.5$, $\theta = \pi$ (Right bottom corner, window slightly rotated) $\sigma = 5$, $\omega = 1.5$, $\theta = \pi/4$

Based on the Gabor kernels bank that is set up, we obtain the magnitude and phase responses using the random values of σ,ω and θ as shown in Figs. 4 and 5, respectively, for an MR brain image that acts an cover image for the generation of first texture-based encryption key. Each of these Gabor-based texture maps are obtained from an image known to both the parties. Therefore, we can generate as many keys as we want considering three tuneable parameters. This provides us with the first key for our medical image cryptosystem. The pseudocode for the encryption process is given as algorithm one later in the article.

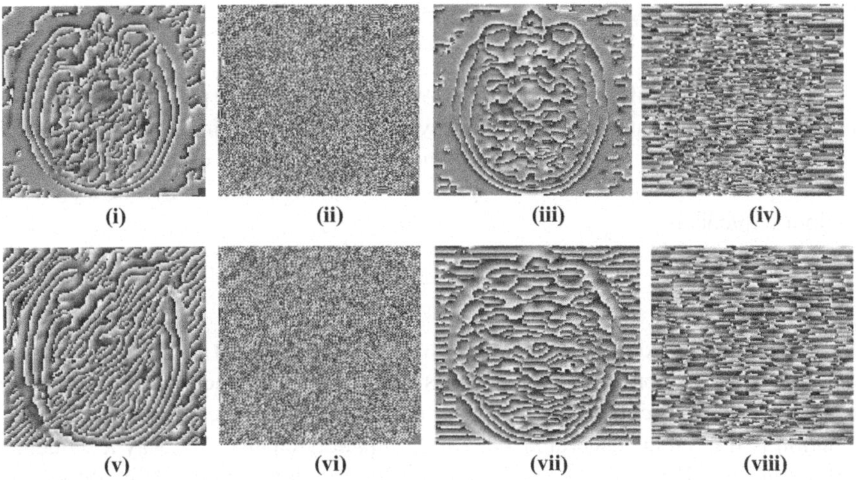

Fig. 4 Phase Responses Corresponding to kernels bank of Fig. 4 at: (i) $\sigma = 3$, $\omega = 0.5$, $\theta = \pi/4$ (ii) $\sigma = 3$, $\omega = 1.5$, $\theta = \pi/4$ (iii) $\sigma = 3$, $\omega = 0.5$, $\theta = \pi$ (iv) $\sigma = 3$, $\omega = 1.5$, $\theta = \pi$ (v) $\sigma = 5$, $\omega = 0.5$, $\theta = \pi/4$ (vi) $\sigma = 5$, $\omega = 1.5$, $\theta = \pi/4$ (vii) $\sigma = 5$, $\omega = 0.5$, $\theta = \pi$ (viii) $\sigma = 5$, $\omega = 1.5$, $\theta = \pi$

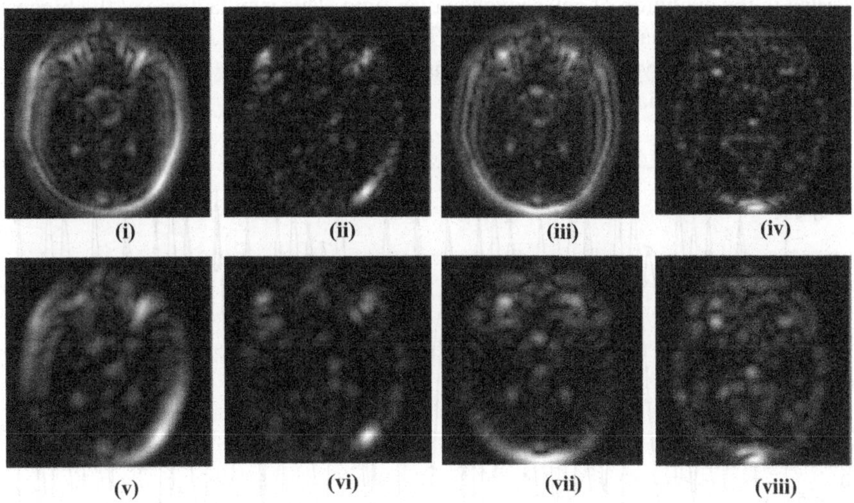

Fig. 5 Magnitude Responses Corresponding to kernels bank of Fig. 4 at: (i) $\sigma = 3$, $\omega = 0.5$, $\theta = \pi/4$ (ii) $\sigma = 3$, $\omega = 1.5$, $\theta = \pi/4$ (iii) $\sigma = 3$, $\omega = 0.5$, $\theta = \pi$ (iv) $\sigma = 3$, $\omega = 1.5$, $\theta = \pi$ (v) $\sigma = 5$, $\omega = 0.5$, $\theta = \pi/4$ (vi) $\sigma = 5$, $\omega = 1.5$, $\theta = \pi/4$ (vii) $\sigma = 5$, $\omega = 0.5$, $\theta = \pi$ (viii) $\sigma = 5$, $\omega = 1.5$, $\theta = \pi$

3.3 Key Generation via Chaotic Maps

Pixel confusion and diffusion strategies have been continuously proposed and implemented via key generation from chaotic keys [64–69]. Logistic maps are the classical ways of implementing the chaos theory and can effectively be used for the purpose of encryption. Equation (10) below gives a logistic map equation which yields a random sequence:

$$x(n+1) = rx_n(1 - x_n) \tag{10}$$

The above equation does generate a random sequence that can be used for the purpose of encryption. This equation has two tunable parameters; each uniquely defines how the generated sequence will evolve. The real strength from with this equation is the initial seed value x_n. This completely changes the way a sequence evolves in time. Figure 6 below shows three different logistic maps. The three start with the same chaotic value of $r = 3.99$ but different values of x_n. The values of x_n are chosen to be S_0, S_1 and S_2 as shown in the figure. The value of S_0, S_1 and S_2 are 0.2, 0.002 and 0.00002, respectively. As evident, the three sequences evolve completely differently even though S_0, S_1 and S_2 differ subtly. There is hardly any overlap in the evolution of any sequence over time. This is in fact one of the biggest advantages that a chaotic map presents. These sequences generated by these maps are extremely sensitive to initial conditions, thus making it very hard for any attacker to launch a differential attack.

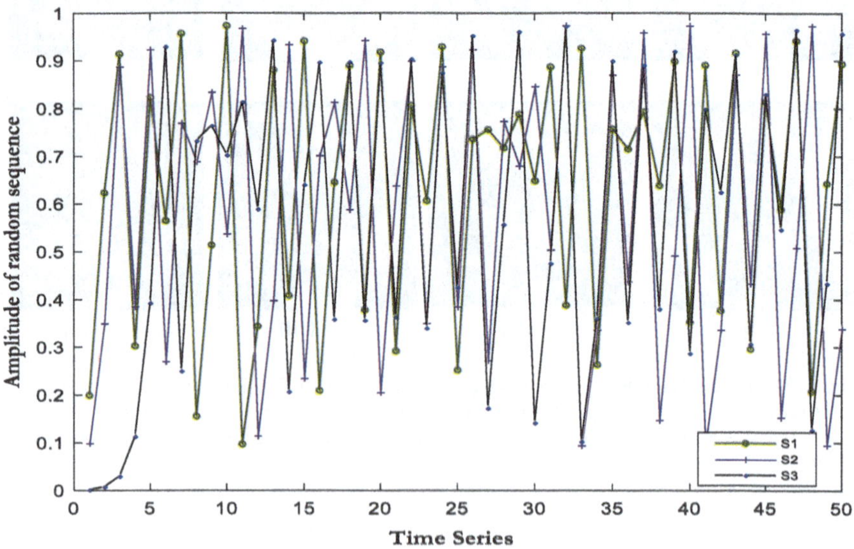

Fig. 6 Three Chaotic logistic map sequences evolving over time with initial conditions set to **a** 0.2 (green plot), **b** 0.002 (purple plot) and **c** 0.00002 (blue plot)

In this work, three chaotic map functions, viz: logistic, quadratic and tent, have been employed to obtain eight encryption keys to encipher an original image. These eight chaotic map keys are fused to the Gabor map-based keys. Table 1 summarizes the governing map functions with their respective tunable parameters. Table 1 underlines a fact that the key space provided by these three chaotic map equations is large with 21 parameters. For any image medical image that is being encrypted, it is very hard to be deciphered by brute force attack by any attacker. Also, as mentioned above, sensitivity to initial conditions makes chaotic maps a good candidate for key generation for encrypting medical images. The randomness property of chaotic sequences is illustrated in Fig. 7. It shows a high correlation value of a sequence with itself as shown by a spike in the graph. The correlation of a sequence with a slightly

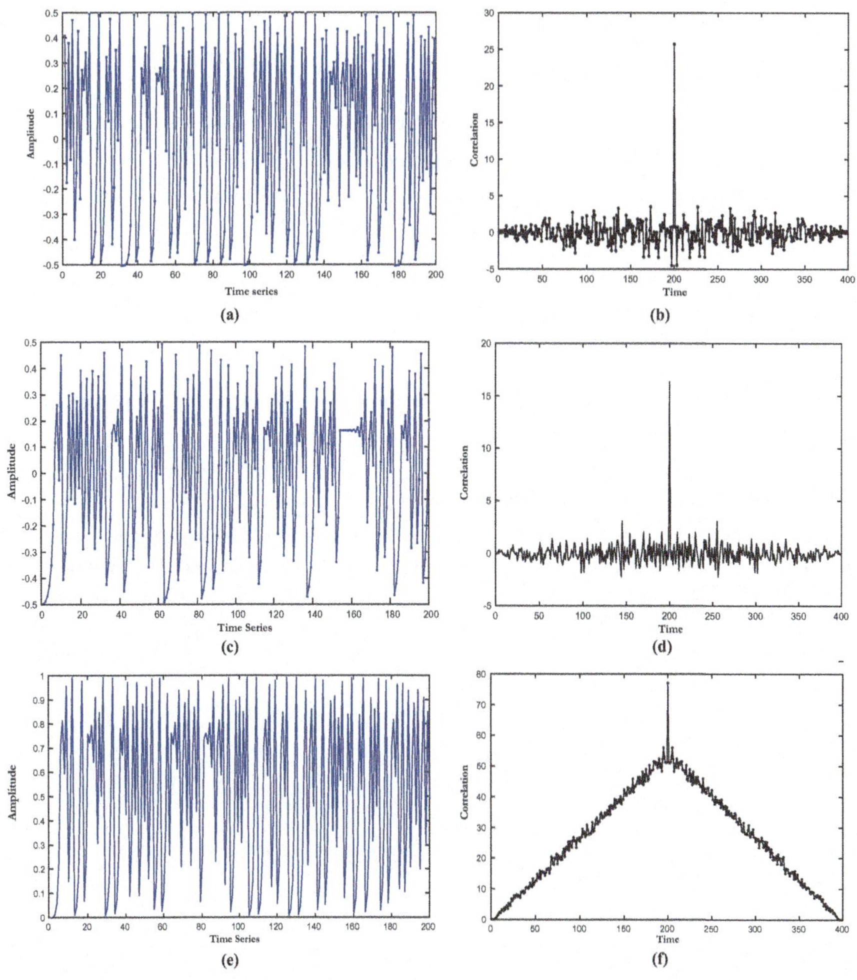

Fig. 7 Illustration of randomness of three chaotic maps used in this work **a** quadratic chaotic sequence, **b** auto-correlation of quadratic chaotic sequence, **c** tent chaotic sequence, **d** auto-correlation of tent chaotic sequence, **e** logistic map sequence and **f** auto-correlation of logistic map sequence

Table 1 Chaotic sequence equations and corresponding tuneable parameters

Keys	Mapping functions	Mapping Equations	Parameters that can be tuned
1	Quadratic	$\chi_1(n+1) = \beta 1 - (\alpha 1 * \chi_1(n-1)^2)$ $\chi_1(1) = \beta 1 - \alpha 1 * \chi_1\hat{}2$	$\alpha_1, \beta 1$ and χ_1
2	Tent	$\mu_1(n+1) = \beta 2 - (\alpha 2 * \lvert\mu_1(n-1)\rvert)$ $(\mu_1(1) = \beta 2 - \alpha 2 * \mu_1)$	$\alpha_2, \beta 2$ and μ_1
3	Logistic	$y_1(n+1) = \alpha_3 y_1(1-y_1)$	α_3 and y_1
4	Quadratic	$\chi_2(n+1) = \beta 3 - (\alpha 4 * \chi_2(n-1)^2)$ $\chi_2(1) = \beta 3 - \alpha 4 * \chi_2\hat{}2$	$\alpha_4, \beta 3$ and χ_2
5	Tent	$\mu_2(n+1) = \beta 4 - (\alpha 5 * \lvert\mu_2(n-1)\rvert)$ $(\mu_2(1) = \beta 4 - \alpha 5 * \mu_2)$	$\alpha_5, \beta 4$ and μ_2
6	Logistic	$y_2(n+1) = \alpha_6 y_2(1-y_2)$	α_5 and y_2
7	Quadratic	$\chi_3(n+1) = \beta 5 - (\alpha 7 * (\chi_3(n-1)^2)$ $\chi_3(1) = \beta 5 - \alpha 8 * \chi_3^2$	$\alpha_7, \beta 5$ and χ_3
8	Logistic	$y_3(n+1) = \alpha_8 y_3(1-y_3)$	α_8 and y_3

delayed or shifted sequence is very low as shown in the Fig. 7. This is one of the important properties of random sequences.

Algorithm 1: Texture and Chaos-driven Encryption

Input: Original Medical Image (m), Reference Texture Image (r),preset chaos initial conditions ($\alpha_1, \beta 1, \chi_1, \alpha_2, \beta 2, \mu_1, \alpha_3, y_1, \alpha_4, \beta 3, \chi_2, \alpha_5, \beta 4, \mu_2, \alpha_5, y_2, \alpha_7, \beta 5, \chi_3, \alpha_8, y_3$), Number of Gabor kernals (g_i)
Outputπ Encrypted Image (e_i)

(continued)

(continued)

Algorithm 1: Texture and Chaos-driven Encryption

1. Apply $\boldsymbol{g}_i$ gaborkernals with an optimally chosen values of σ,ω and θ to $\boldsymbol{r}$to obtain$\boldsymbol{t}_i$ texture gabor maps ($\boldsymbol{g}_i$=$\boldsymbol{t}_i$)
2. Apply chaotic functions using ($\alpha_1, \beta 1, \chi_1, \alpha_2, \beta 2, \mu_1, \alpha_3, y_1,\alpha_4, \beta 3, \chi_2, \alpha_5, \beta 4, \mu_2, \alpha_5, y_2\alpha_7, \beta 5, \chi_3,\alpha_8, y_3$) to obtain $\boldsymbol{c}_i$ chaotic sequences
3. **for** each m**do**
4. bm_i=extbp (m_i)
5. bt_i=extbp (t_i)
6. bc_i=extbp (c_i)
7. **for** i=0 to n-1 **do**
8. **for** i=0 to n-1 **do**
9. be1{i,j}=bc{i,j} **xor**bt{i,j}
10. be2{i,j}=be{i,j} **xor**bm{i,j}
11. **end**
12. **end**
13. e_i=appbp (be2)
14. **end**

By choosing the proper values of $\boldsymbol{\alpha}_1, \boldsymbol{\beta}1$ and $\boldsymbol{\chi}_1$, the generated key from the chaotic maps ensures that the key is random and highly sensitive to initial conditions. It must be reiterated that both randomness and key sensitivity are very important properties with regard to image encryption. So, as per the proposed schema, we obtain two sets of random keys in the substitution stage of this cryptosystem: one coming from the Gabor texture maps and the other coming from the chaotic maps. The two maps are fused to obtain a resultant key which encrypts the original medical image to be transmitted. This is graphically illustrated in Fig. 8.

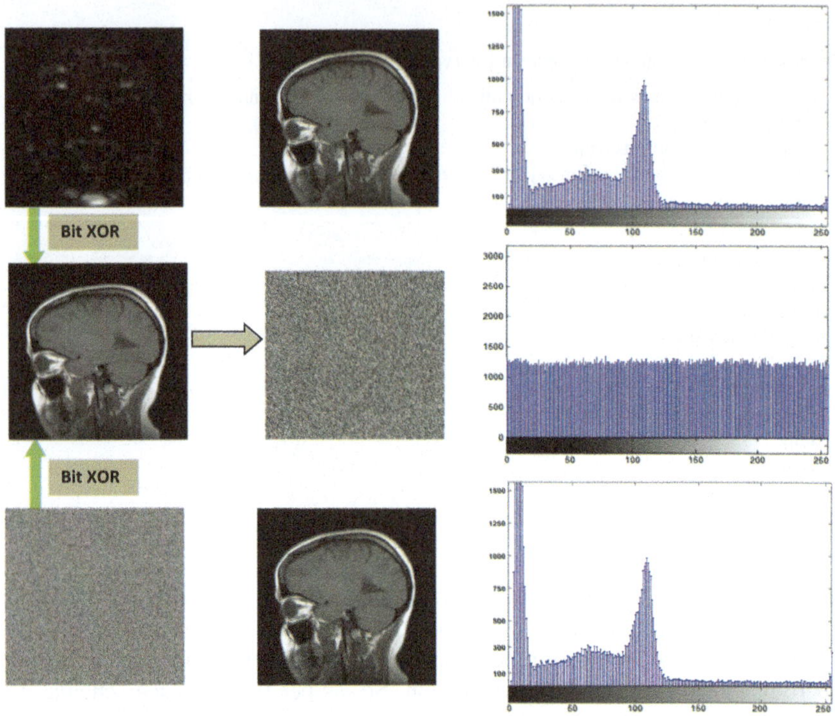

Fig. 8 Histograms of original (2nd and 3rd column of top row), encrypted (2nd and 3rd column of middle row) and decrypted images (2nd and 3rd column of bottom row). The first column illustrates the generation of the fused key for this medical image cryptosystem

4 Results and Analysis

The experimentation and analysis of this work has been carried out using MATLAB. The magnetic resonance images of brain obtained manually from some diagnostic centers have been used in this work. The dataset pertains to 233 subjects which together represent brain abnormalities like meningioma, glioma, pituitary adenomas. The next sub-sections illustrate the results and analysis of the proposed work.

4.1 Histogram Analysis

Although histogram is not a strong indicator that can reflect the similarity of the images efficiently, it has widely been used to determine the difference between the two images in various cryptosystems. The histogram gives a subjective comparison of the two images. Figure 6 shows histograms of the original image on the top row strongly pointing that this histogram pertains to a bimodal image. The next row is

Table 2 Average entropies of original, encrypted and decrypted medical images

Image	Original image	Encrypted image	Decrypted image
256 × 256	6.152	7.509	6.152
400 × 400	5.914	7.782	5.914
512 × 512	5.862	7.910	5.862

the histogram of the encrypted image, and as evident, no idea can be extracted by an attacker as to what image this histogram corresponds to. This histogram indicates that all intensity values of the grayscale image are equally present in the image. The last row is the histogram of the decrypted image, and as required, this should have been precisely similar to the histogram of the original image.

4.2 *Image Entropy*

For a strong encryption, the entropy of the encrypted image should be high. The average entropy difference between the original image and the encrypted image is shown in Table 2. It can be observed that entropy of the encrypted image is significantly higher than the original image.

4.3 *Line Profile*

The same is illustrated in Fig. 9. The purple graph shown in Fig. 9 is a line profile of an original image corresponding to 100th row. The blue graph of Fig. 7 shows a line profile of encrypted brain image on the same row. It can be clearly observed that the two plots are completely different. Similarly, line profiles of the original image and the decrypted image are exactly the same and are overlapped in Fig. 9.

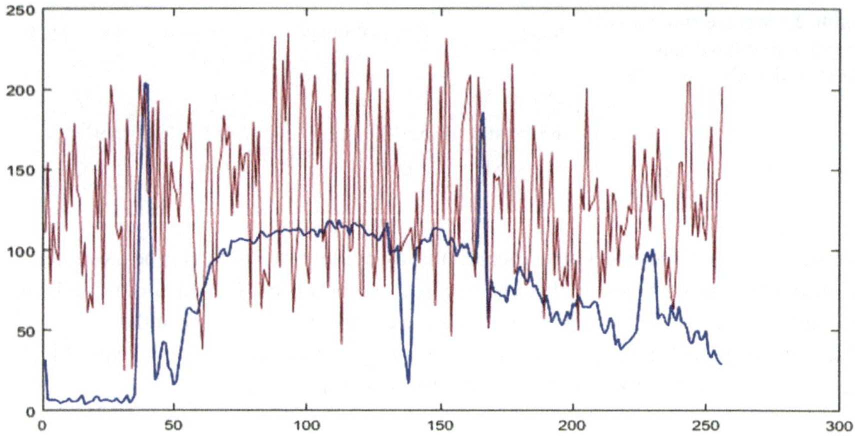

Fig. 9 Line profile of 100th row pertaining to original image (purple graph), encrypted (blue graph) and decrypted image (overlapped with purple)

4.4 Correlation Coefficient

Correlation coefficient is the fourth parameter that is used in the analysis of the results. It is a statistical measure that condenses the comparison of two 2D images down to a single scalar value. From Fig. 10, it can be observed that there is a high correlation between the original and the decrypted image (shown in blue). This is signified by the clustered points along the diagonal of the plot. On the other hand, the correlation between the original and the encrypted version of the image show a weak correlation and the same is signified by the scattered points on the plot (shown in brown).

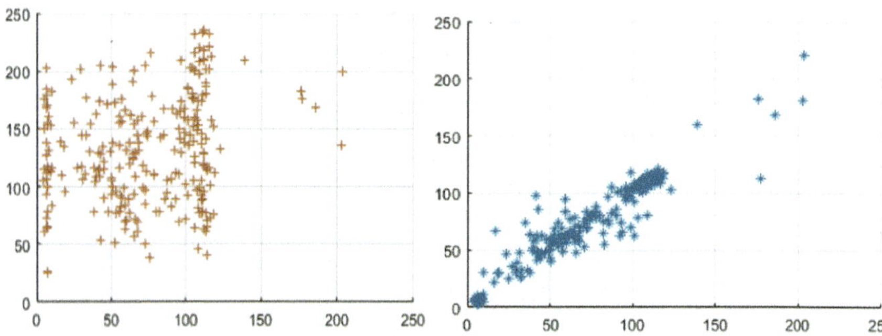

Fig. 10 Correlation plots of original image (blue) showing high correlation and the encrypted image (brown) showing nearly no correlation among the pixels of image

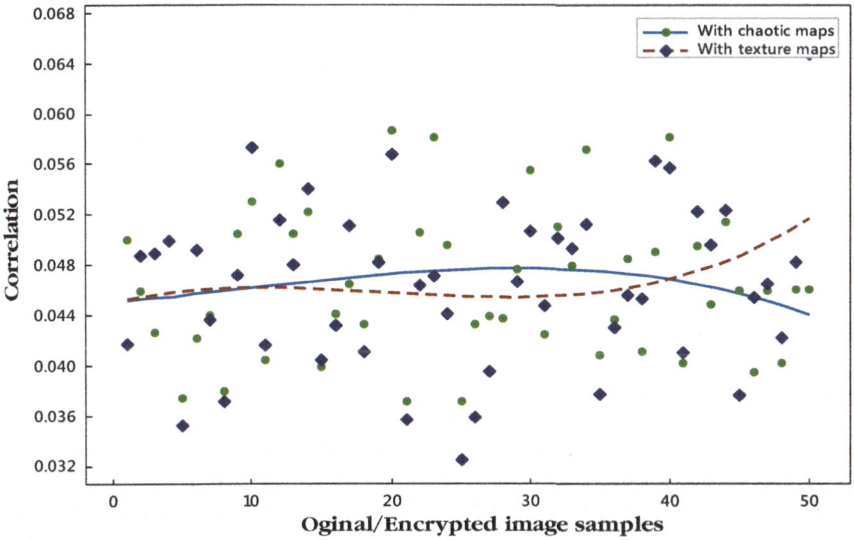

Fig. 11 Correlation coefficient of original and encrypted images with chaotic maps and without chaotic maps

Figure 11 is a plot of correlation for original and encrypted images when (1) Arnold cat map is followed with only the chaotic map-based key for encryption and (2) Arnold cat map is followed by only the texture-based encryption key. The correlation values obtained in both these cases are very close to 0. This again emphasis the fact that fusing both the texture and the chaotic keys offers us a large key space and hence makes the proposed system robust to attacks like brute force.

4.5 *Key Space of the Proposed Cryptosystem*

Since the proposed medical image cryptosystem uses three key spaces, one coming from the use of the Arnold cat transformation, second comes from the use of chaotic maps and the third is obtained from texture maps. The Arnold cat map transformation uses two tunable parameters (p *and* q) and chaotic maps employs $\alpha_1, \beta 1, \chi_1, \alpha_2, \beta 2, \mu_1, \alpha_3, y_1, \alpha_4, \beta 3, \chi_2, \alpha_5, \beta 4, \mu_2, \alpha_5, y_2 \alpha_7, \beta 5, \chi_3, \alpha_8, y_3$ as tunable parameters. In addition to this, we have $\omega \times \sigma \times \theta$ tunable parameters with which we can define a Gabor filter for obtaining the texture map.Therefore, we have a total of 26 tunable parameters that determine the key space, and it is evident that this cryptosystem offers a significantly large key space to avoid brute force attack.

Table 3 Average image quality evaluation values

Evaluation parameters	Image size	Original and encrypted
PSNR (dB)	256 × 256	+8.15
UIQI	256 × 256	−0.00038
MSE	256 × 256	10,029.48
RMSE	256 × 256	100.14
SNR (dB)	256 × 256	−17.45
MAE	256 × 256	84.88

4.6 Image Quality Performance

The image quality of original image, encrypted and decrypted versions is evaluated by employing parameters like peak signal-to-noise ratio (PSNR), signal-to-noise ratio (SNR), mean absolute error (MAE), mean square error (MSE), root mean square error (RMSE), Universal Image Quality Index (UIQI). The average values are summarized in Table 3. It depicts that there is no loss in information in this cryptosystem. The decrypted image information is identical to that of the encrypted version.

5 Conclusion

The work proposed herein successfully attempted to utilize the random and sensitivity advantage of chaotic maps and the unique, random and large key space advantage of Gabor edge maps computed from a cover image. This makes the transmission of the medical images secure with a high level of reliability. The technique introduced in this work initially permuted the image pixel as per the pre-selected Arnold cat parameters. This is followed by bit plane extraction of this image. The key is generated from multiple chaotic maps and the texture maps. These maps are fused and used as a key for encrypting the medical image. The experimental results and their evaluation using statistical analysis, key sensitivity analysis and the key space analysis show the efficacy of the proposed work. This work assumes that the key is shared with the receiver at the onset of the communication; thus, the overhead of the key has to be tolerated just once. During the course of the transmission, only one of just few parameters can be changed and that overhead is manageable. Based on the results and their analysis, it can be concluded that the proposed work has both potential and suitability to be used in real-time medical image transmissions. One of the possible directions of this work can be the estimation of the less number of parameters from the set of proposed parameters. This in essence would imply to applying a kind of reversible parameter reduction technique.

References

1. Hurrah N, Parah S, Loan N, Sheikh J (2019) Dual watermarking framework for privacy protection and content authentication of multimedia. Future Gener Comput Syst 94:654–673
2. Gull S, Loan NA, Parah SA et al (2020) An efficient watermarking technique for tamper detection and localization of medical images. J Ambient Intell Human Comput 11:1799–1808. https://doi.org/10.1007/s12652-018-1158-8
3. Gull S, Parah SA (2019) Color image authentication using dual watermarks. In: 2019 fifth international conference on image information processing (ICIIP), Shimla, India, pp 240–245. https://doi.org/10.1109/ICIIP47207.2019.8985760
4. Loan NA, Parah SA, Sheikh JA, Akhoon JA, Bhat GM (2017) Hiding electronic patient record (EPR) in medical images: a high capacity and computationally efficient technique for e-healthcare applications. J Biomed Inf 73:125–136. ISSN 1532-0464. https://doi.org/10.1016/j.jbi.2017.08.002
5. Parah SA, Loan NA, Shah AA et al (2018) A new secure and robust watermarking technique based on logistic map and modification of DC coefficient. Nonlinear Dyn 93:1933–1951. https://doi.org/10.1007/s11071-018-4299-6
6. Parah S, Sheikh J, Ahad F, Loan N, Bhat G Information hiding in medical images: a robust medical image watermarking system for E-healthcare. Multimedia Tools Appl 76(8):10599–10633
7. Parah SA, Ahad F, Sheikh JA, Loan NA, Bhat GM (2017) A new reversible and high capacity data hiding technique for E-healthcare applications. Multimedia Tools Appl 76:3:3943–3975. https://doi.org/10.1007/s11042-016-4196-2
8. Loan NA, Parah SA, Sheikh JA, Bhat GM (2018) Secure and robust digital image watermarking using coefficient differencing and chaotic encryption. IEEE Access 6(1):19876–19897
9. Parah SA, Sheikh JA, Hafiz AM, Bhat GM (2014) Data hiding in scrambled images: a new double layer security data hiding technique. Comput Electr Eng 40(1):70–82. https://doi.org/10.1016/j.compeleceng.2013.11.006
10. Parah SA, Sheikh JA, Hafiz AM, Bhat GM (2015) A secure and robust information hiding technique for covert communication. Int J Electron 102(8):1253–1266. https://doi.org/10.1080/00207217.2014.954635
11. Parah SA, Sheikh JA, Bhat GM (2014) A secure and efficient spatial domain data hiding technique based on pixel adjustment
12. Giri KJ, Peer MA, Nagabhushan P (2014) A channel wise color image watermarking scheme based on discrete wavelet transformation. In: 2014 international conference on computing for sustainable global development (INDIACom), New Delhi, pp 758–762. https://doi.org/10.1109/IndiaCom.2014.6828064
13. Giri KJ, Peer MA, Nagabhushan P (2015) A robust color image watermarking scheme using discrete wavelet transformation. IJ Image, Graph Sig Process
14. Giri KJ, Bashir R (2017) Digital watermarking: a potential solution for multimedia authentication. In: Intelligent techniques in signal processing for multimedia security. Studies in computational intelligence, vol 660. Springer, Berlin
15. Giri KJ, Bashir R (2018) A block based watermarking approach for color images using discrete wavelet transformation. Int J Inf Tecnol 10:139–146. https://doi.org/10.1007/s41870-017-0075-y
16. Giri KJ, Bashir R, Bhat JI (2019) A discrete wavelet based watermarking scheme for authentication of medical images. Int J E-Health Med Commun (IJEHMC) 10(4)
17. Mitra S, Uma Shankar B (2015) Medical image analysis for cancer management in natural computing framework. Inf Sci 306:111–131
18. Phophalia A, Rajwade A, Mitra SK (2014) Rough set based image denoising for brain MR images. Signal Process 103:24–35
19. Ji Z, Xia Y, Sun Q, Cao G, Chen Q (2015) Active contours driven by local likelihood image fitting energy for image segmentation. Inf Sci 301:285–304

20. Wu H, Huang J, Shi Y (2015) A reversible data hiding method with contrast enhancement for medical images. J Vis Commun Image Represent 31:146–153
21. Jung H, Sung K, Nayak KS, Kim EY, Ye JC (2009) k-t focus: a general compressed sensing framework for high resolution dynamic MRI. Magnet Reson 61:103–116
22. Lou LDC, Hua MC, Liua JL (2009) Multiple layer data hiding scheme for medical images. Comput Stand Interfaces 31(2):329–335
23. Hu JK, Han FL (2009) A pixel-based scrambling scheme for digital medical images protection. J Netw Comput Appl 32(4):788–794
24. Lin CF, Chung CH, Lin JH (2009) A chaos-based visual encryption mechanism for clinical EEG signals. Med Biol Eng Comput 47(7):757–762
25. Mousavi SM, Naghsh A, Abu-Bakar SAR (2014) Watermarking techniques used in medical images: a survey. J Digit Imaging 27(6):714–729
26. Das S, Kundu MK (2012) Effective management of medical information through a novel blind watermarking technique. J Med Syst 36(5):3339–3351
27. Giakoumaki A, Pavlopoulos S, Koutsouris D (2006) Multiple image watermarking applied to health information management. IEEE Trans Inf Technol Biomed 10(4):722–732
28. LiXW KimST (2013) Optical 3D watermark based digital image watermarking for telemedicine. Opt Lasers Eng 51(12):1310–1320
29. Arsalan M, Malik SA, Khan A (2012) Intelligent reversible watermarking in integer wavelet domain for medical images. J Syst Softw 85(4):883–894
30. Deng X, Chen Z, Zeng F, Zhang Y, Mao Y (2013) Authentication and recovery of medical diagnostic image using dual reversible digital watermarking. J Nanosci Nanotechnol 13(3):2099–2107
31. Hong J, Abdurahman K, Yangling L (2016) Asymmetric color pathological image encryption scheme based on complex hyper chaotic system. Optik 127:5812–5819
32. Menezes AJ, van Oorschot PC, Vanstone SA (1996) Handbook of applied cryptography. CRC Press
33. Diaconu AV (2016) Circular inter-intra pixels bit-level permutation and chaos-based image encryption. Inf Sci 355:314–327
34. Abdmouleh AKMK, Bouhlel MS (2013) Dynamic chaotic look-up table for MRI medical image encryption. In: Proceedings of the 2013 international conference on systems, control, signal processing and informatics, pp 241–246
35. Ashtiyani M, Birgani PM, Hosseini HM (2008) Chaosbased medical image encryption using symmetric cryptography. In: 3rd international conference on information and communication technologies: from theory to applications, ICTTA. IEEE, pp 1–5
36. Fu C et al (2013) An efficient and secure medical image protection scheme based on chaotic maps. Comput Biol Med 43(8):1000–1010
37. Sarker MZH, Parvez MS (2005) A cost effective symmetric key crypto-graphic algorithm for small amount of data. In: Proceedings of the 9th IEEE international multi topic conference, pp 1–6
38. Fridrich J (1998) Symmetric ciphers based on two-dimensional chaotic maps. Int J Bifurcation Chaos 8(6):1259–1284
39. Scharinger J (1998) Fast encryption of image data using chaotic Kolmogorov flows. J Electron Imaging 7(2):318–325
40. Chen GR, Mao YB, Chui CK (2004) Asymmetric image encryption scheme basedon 3D chaotic cat maps. Chaos, Solitons Fractals 21(3):749–761
41. Mao YB, Chen GR, Lian SG (2004) A novel fast image encryption scheme based on 3D chaotic baker maps. Int J Bifurcation Chaos 14(10):3613–3624
42. Belkhouche F, Gokcen I, Qidwai U (2005) Chaotic gray-level image transformation. J Electron Imaging 14(4):043001
43. Behnia S, Akhshani A, Mahmodi H (2008) A novel algorithm for image encryption based on mixture of chaotic maps. Chaos, Solitons Fractals 35(2):408–419
44. Patidar V, Pareek NK, Sud KK (2009) A new substitution–diffusion based image cipher using chaotic standard and logistic maps. Commun Non-Linear Sci Numer Simulation 14(7):3056–3075

45. Gao TG, Chen ZQ (2008) A new image encryption algorithm based on hyper-chaos. Phys Lett A 372(4):394–400
46. Gao TG, Chen ZQ (2008) Image encryption based on a new total shuffling algorithm. Chaos, Solitons Fractals 38(1):213–220
47. Sun FY, Liu ST, Li ZQ (2008) A novel image encryption scheme based on spatial chaos map. Chaos, Solitons Fractals 38(3):631–640
48. Puech W, Rodrigues JM (2005) Crypto-compression of medical images by selective encryption of DCT. In: EUSIPCO'05: European signal processing conference, Antalya, Turkey
49. Zhou Y, Panetta K, Agaian S (2009) A lossless encryption method for medical images using edge maps. In: Engineering in medicine and biology society. Annual international conference of the IEEE. IEEE, pp 3707–3710
50. Rao YS, Mitra A, Prasanna SM (2006) A partial image encryption method with pseudo random sequences. In: Information systems security. Springer, Berlin, Heidelberg, pp. 315–325
51. Panduranga HT, Naveen Kumar SK (2013) Selective image encryption for medical and satellite images. Int J Eng Sci Tech 5(2)
52. Panduranga HT, Naveenkumar SK (2013) Partial image encryption using block wise shuffling and chaotic maps. In: 2013 international conference on optical imaging sensor and security (ICOSS). IEEE, pp 1–5
53. Ou Y, Sur C, Rhee KH (2007) Region-based selective encryption for medical imaging. In: Proceedings of the 1st annual international conference on frontiers in algorithmics. Springer, Berlin Heidelberg, pp 62–73
54. Brahimi Z, Bessalah H, Tarabet A, Kholladi MK (2008) Selective encryption techniques of JPEG2000 codestream for medical images transmission. WSEAS Trans Circuits Syst 7(7):718–727
55. Xiang T, Wong KW, Liao X (2007) Selective image encryption using a spatiotemporal chaotic system. Chaos: Interdiscip J Nonlinear Sci 17(2)
56. Liu JL (2006) Efficient selective encryption for JPEG 2000 images using private initial table. Pattern Recogn 39(8):1509–1517
57. Som S, Sen S (2013) A non-adaptive partial encryption of grayscale images based on chaos. Procedia Tech 10:663–671
58. Shang Z, Ren H, Zhang J (2008) A block location scrambling algorithm of digital image based on Arnold transformation. In: 9th IEEE international conference for young computer scientists, 978-0-7695-3398-8/08/$25.00
59. Zhu L, Li W, Liao L, Li H (2006) A novel algorithm for scrambling digital image based on cat chaotic mapping. In: Proceedings of the 2006 international conference on intelligent information hiding and multimedia signal processing (IIH-MSP'06), 0-7695-2745-0/2006© IEEE
60. Banday SA, Mir AH (2017) Statistical textural feature and deformable model based brain tumor segmentation and volume estimation. Multimed Tools Appl 76:3809–3828. https://doi.org/10.1007/s11042-016-3979-9
61. Banday, SA, Ajaz HM (2016) Statistical textural feature and deformable model based MR brain tumor segmentation. In: 2016 international conference on advances in computing, communications and informatics (ICACCI), pp 657–663
62. Materka A, Strzelecki M (1998) Texture analysis methods–A review. Technical University of Lodz, Institute of Electronics, COST B11 Report, Brussels, pp 9–11
63. Daugman JG (1985) Uncertainty relation for resolution in space, spatial frequency, and orientation optimized by 2D visual cortical filters. J Opt Soc A2(7):1160–1169
64. Lee TS (1996) Image representation using 2D gabor wavelets. IEEE Trans Pattern Anal Mach Intell 18(10):959–971
65. Alvarez E, Fernandez A, Jimenez GP, Marcano A (1999) New approach to chaotic encryption. Phys Lett A 263:373–375
66. Wong WK, Lee LP, Wong KW (2000) A modified chaotic cryptographic method. Comput Phys Commun 138:23–236

67. Wong KW (2002) A fast chaotic cryptography scheme with dynamic look-up table. Phys Lett A 298:238–242
68. Khan MK, Zhang J (2008) Multimodal face and fingerprint biometrics authentication on space-limited tokens. Neurocomputing 71(13–15):3026–3031
69. Pareek NK, Patidar V, Sud KK (2003) Discrete chaotic cryptography using external key. Phys Lett A 309:75–82

Chapter 2
Image Encryption Based on DNA Substitution and Chaotic Theory

Nimara Habib, Solihah Gull, and Shabir A. Parah

1 Introduction

The utilization of the Internet has increased massively since its advent. Sending and receiving of multimedia data have increased to a large extent. The data is being sent over insecure channels and hence demands security. This security can be provided by various techniques including watermarking [1–3] and encryption techniques that not only ensure security but also help in stopping unauthorized access of data and provides privacy [4–7]. The images frequently consist of private or confidential data, and such data can be protected before sending it by encrypting [8]. Various encryption techniques are present in the literature, yet none of them satisfies all image types. In the early days, most of the algorithms were developed for textual data, but these algorithms are not frequently suitable for images because of their large size [9]. Various algorithms including triple-data encryption standard (T-DES) [10] and international data encryption algorithm (IDEA) [11] have been providing high security; however, they may not be well suited for images. Other encryption techniques including Data Encryption Standard (DES) [12], International Data Encryption Standard (IDEA), and Advanced Encryption Standard (AES) [13] have also been implemented for textual data. These traditional algorithms of encrypting text can be taken further to be implemented for images, but it may not prove to be a good idea because image data is larger than the text data. Hence, it may take too long for these algorithms to encrypt the images. Also, in the case of text encryption, the decrypted text should be the same as that of encrypted one, but it is not true for images. Images contain redundant data, and the Human Visual System (HVS) is not able to perceive small

N. Habib · S. Gull · S. A. Parah (✉)
PG Department of Electronics and Instrumentation Technology, University of Kashmir, Hazratbal, Srinagar 190006, India
e-mail: shabireltr@gmail.com

N. Habib
e-mail: shabireltr@gmail.com

K. J. Giri et al. (eds.), *Multimedia Security*, Algorithms for Intelligent Systems,
https://doi.org/10.1007/978-981-15-8711-5_3

distortions present in images [14, 15]. Thus, a decrypted image having a small amount of distortion can usually be considered. Comprehending the information within an image is possible because of the correlation between various elements. Once this correlation is reduced, the image is completely diffracted. Further, to reduce this correlation, various transformation techniques are implemented. Cryptography along with chaotic encryption techniques is the transformations that have acquired huge impetus and are used widely [16]. Encryption using chaotic image transformation diffracts the image to an extent that no one could be able to peep what information the image is carrying [17]. Using both cryptography and chaotic image transformation together, we doubly secure the image information. Another encryption technique, first proposed by Adleman [18], is DNA computing. This field is new to encryption and is being used for image cryptography [19]. DNA computing makes use of molecular biology, possesses massive parallelism and higher storage capacity with lower energy consumption [20, 21]. Thus, DNA computing provides an additional advantage to traditional methods of cryptography. However, it can be said that DNA encryption cannot be used alone for making a system secure. It can be combined with other encryption schemes to obtain better encryption results. In this chapter, we present a scheme using chaotic encryption along with DNA-based image encryption to solve the problems in existing systems. Chaotic encryption is used to confuse the pixels of the image. Further, the confused pixels are diffused using DNA encoding.

2 Introduction to DNA

All organisms (except viruses), including humans, contain DNA as the genetic material. Figure 1 shows the actual DNA structure. The location of DNA is not restricted only to the nucleus of organisms (also known as nuclear DNA). A certain amount of DNA is also present in mitochondria, where it is referred to as the mitochondrial DNA (mt DNA). In prokaryotes, apart from the nucleotide, DNA also occurs in the extrachromosomal form where it is referred to as plasmid. In the case of plants, DNA is also found inside chloroplasts, and there it is referred to as cpDNA. DNA, in itself, is made up of four nitrogenous bases viz. Adenine (A), guanine (G), cytosine (C), and thymine (T). These bases form a specific DNA sequence that plays an essential role in the determination of outcomes in the processes of replication, translation, transcription, and other cellular processes. The structure of these molecules has been shown in Fig. 2.

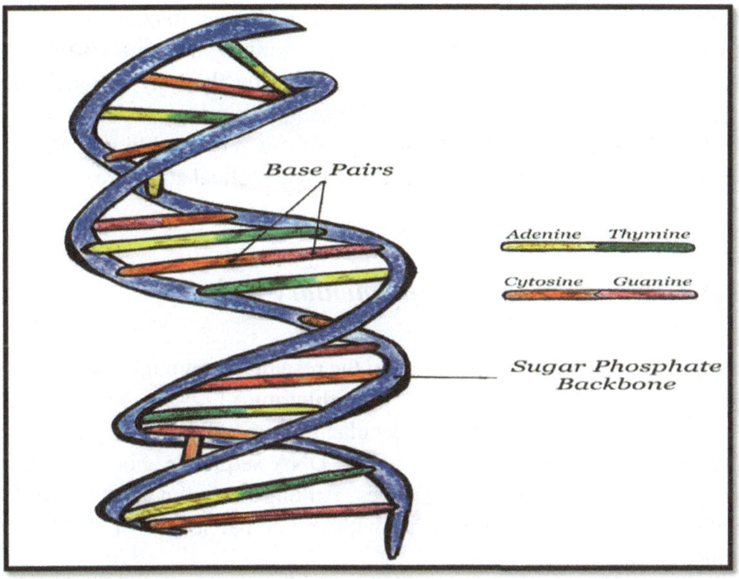

Fig. 1 Structure of DNA

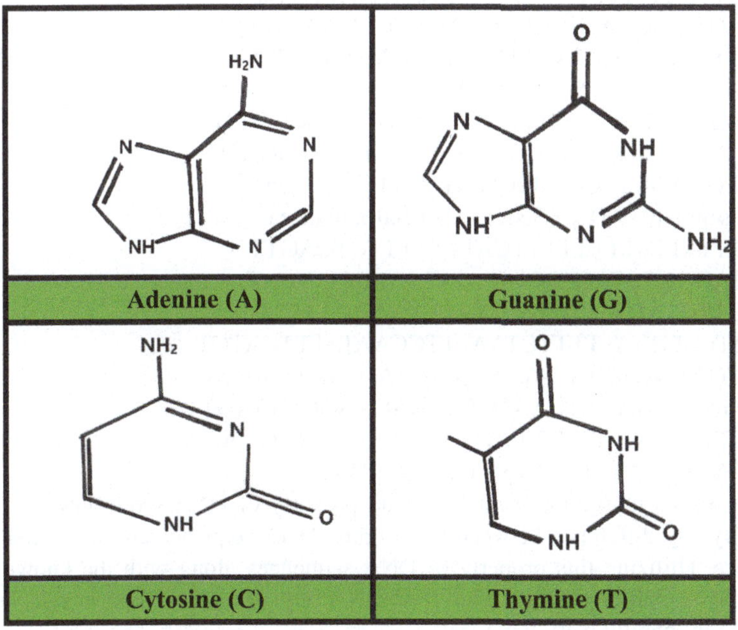

Fig. 2 Structure of DNA nucleotide base pairs

The nitrogenous bases of DNA lie opposite to each other to undergo pairing. This pairing occurs in a particular fashion wherein adenine pairs with thymine by two hydrogen bonds while guanine pairs with cytosine by three hydrogen bonds. The latter is, therefore, harder to breakdown. These base pairings occur fixedly so that the overall width of the DNA helix remains uniform. The double-ringed bases pair with single ringed ones to provide uniformity to the helical structure.

2.1 DNA Encoding and Complementary Rule

DNA computing gained huge interest from the research community, and many DNA encryption-based methods are present in the literature [22–24]. Most of the DNA encryption-based methods use the molecular biology of DNA sequences. In this work, we use the following property of the DNA sequence. For example, let us consider segments of two DNA sequences. Among them, one belongs to Litmus, and another belongs to Balsaminaccae with the actual length of 2858 and 2283 nucleotides, respectively.

The segment of DNA Sequence of Litmus:
ATCGAATTCGCGCTGAGTCACAATTCGCGCTG
AGTCACAATTCGCGCTGAGTCACAATTGTGAC
TCAGCCGCGAATTCCTGCAGCCCCGAATTCCG
CATTGCAGAGATAATTGTATTTAAGTGCCTAGC
TCGATACAATAAACGCCATTTGACCATTCACCAC
ATTGGTGTGCACCTCCAAGCTCGCGCACCGTAC
CGTCTCGAGGAATTCCTGCAGGATATCTGGATC
CACGAAGCTTCCCATGGTGACGTCACC [25].

The segment of DNA Sequence of Balsaminaccae:
TTTTTATTATTTTTTTTCATTTTTTTCTCAGTT
TTTAGCACATATCATTACATTTTATTTTTTCAT
TACTTCTATCATTCTATCTATAAAATCGATTATTT
TTATCACTTATTTTTCTAATTTCCAATATTTCATCT
AATGATTATATTACATTAAAGAAATCGGTTAAAA
GCGACTAAAAATCAATCTGGAACAAGGCTTAGTT
TATTTAATATATTATTTTATGTAATTTCTATTGAAAA
ATTAGTTAAAAGGCAAGTATTTGAGAT [25].

This makes us see the most important property of DNA sequences, i.e., there is merely any difference between the original DNA sequence and the faked DNA sequence. Utilizing this property of DNA sequences along with the known facts about the availability of over 55 million DNA sequences over Web sites makes the guessing probability very low [25].

The basic genetic code is formed by four nucleotide bases namely adenine (A), guanine (G), thymine (T), and cytosine (C) where 'A' binds with 'T' and 'G' binds with 'C' [22]. Further, we know that '0' is complementary to '1' and similarly '00' complements to '11', '10' complements to '01'. Now looking at the base pairs 'A',

Table 1 DNA encoding rules

1	2	3	4	5	6	7	8
00 A	00 A	00 C	00 C	00 G	00 G	00 T	00 T
01 C	01 G	01 A	01 T	01 A	01 T	01 C	01 G
10 G	10 C	10 T	10 A	10 T	10 A	10 G	10 C
11 T	11 T	11 G	11 G	11 C	11 C	11 A	11 A

'T', 'G', and 'C', since they are complementary, they can be represented as '00', '11', '01', and '10', respectively. And hence can be used to encode each pixel of the image because each gray pixel can be represented by an 8-bit binary value. For example, let us consider a gray level pixel value as 178; it can be represented in binary form as '10110010'. Now, it can be converted to DNA representation using the above complementary rule as 'CTAC'. Twenty four different types of combinations shown in Table 1 are complementary.

In this chapter, we have selected the rules from Table 1 to encode grayscale image 'Io' of M × N dimensions. Firstly, the original image is converted to binary form, then this binary data is encoded using one of the above rules to obtain encoded image 'Ic'. This is expressed as one-dimensional sequence '*Y*' and is obtained as under:

$$Y = \{y1,\ y2,\ y3,\ \ldots,\ y4MN\},\ yi \in \{A,\ T,\ G,\ C\} \tag{1}$$

According to the principle of the complementary base, we set the nucleotide string of the encoding nucleotides as follows:

$$yi \neq Q(yi) \neq Q(Q(yi)) \neq Q(Q(Q(yi))) \tag{2}$$

$$yi = Q(Q(Q(Q(yi)))) \tag{3}$$

where $Q(yi)$ and yi are complementary; i.e., they form a set of base pairs. The pixel diffusion is carried out using complementary rules for substitution.

2.2 DNA-Based Image Encryption

With the advancement of various technologies and since the advent of the Internet, the transmission of data via the Internet has increased. However, the transmission of data over an insecure channel makes it vulnerable to interceptions by hackers. Therefore, the security of the data over insecure channels is highly important. Most often, information hiding [22] has been used as a tool to protect data during transmission over insecure channels. One of the potent tools for information hiding is image encryption [26, 27]. Encryption is a technique in which the sender encrypts the original data (plaintext) into ciphertext (encrypted data) [28]. This ciphertext can be decoded back to plaintext by key(s) which is only provided to the receiver or the authorized user. The basic image encryption methods fall into three categories viz., pixel replacement, pixel scrambling, or combination of the two. Pixel replacement can be applied to grayscale values by altering those using operations like XOR, but it shows low resistance to plain text attacks. Pixel scrambling, as the name suggests, only scrambles the pixels without actually altering them. The last one is obtained by combining the two. Various encryption techniques include chaos techniques, modern cryptography mechanisms, compression methodology, DNA techniques, etc. Although many cryptography methods have been developed, yet if the hacker possesses infinite computational power, these methods can be easily broken. Thus, the need of the hour is to build safer and more efficient methods for encryption. This is being achieved using DNA encryption which evolved from genetic engineering [29]. DNA is found to have exceptional properties including high energy efficiency, massive parallelism, and inherent information density. Therefore, they have attained attention to be applied in the field of cryptography. They are used for authentication, encryption, digital signatures, and many more. DNA cryptography gets its advantage from DNA computing. There lies a difference between the two, one is used for solving various complex computational problems, while the other is used to provide security to the information using the biological properties of DNA. The security in the case of DNA-based encryption is limited only by biotechnology and not by computational power. Since the information is carried by DNA itself and the availability of an enormous number of DNA sequences helps to enhance the security of the cryptosystem to a large extent. Other than high security, it provides higher storage capacity making it a more suitable candidate for encryption. There are three basic methods of DNA encryption which include DNA chip-based encryption, DNA strand-based encryption, and DNA model-based encryption. Further DNA strand-based encryption is categorized into two types [30, 31]. First is a method in which secret data is converted to a DNA message. Then, this DNA message is concealed using another DNA sequence using certain encoding rules. Second is a method in which storage and transmission of the secret message are done by directly inserting it into the living organism. In DNA chip-based encryption [32], a DNA chip is used as a carrier of information. The fabrication of a DNA chip is carried using the in-situ synthesized process of Oligo nucleic acids. DNA model-based encryption [32] is another encryption technique that utilizes molecular biology for encrypting and decrypting the data. The field of DNA cryptography can be utilized further to find solutions for various problems.

3 Chaos and Logistic Map

Most of the systems including natural and physical systems tend to show nonlinear behavior [33–35]. These systems have various inherent characteristics including nonlinear interdependence, sensitive dependence, determinism, and order. Chaotic systems show random and unpredictable behavior and are highly sensitive to change in initial conditions. In this chapter, the following logistic map equation has been used

$$x_{k+1} = \mu x_k(1 - x_k) \tag{4}$$

where $\mu \in [0, 4]$, $x_k \in (0, 1)$, $k = 1, 2, 3, 4, 5, \ldots, m$.

The equation can be solved with a pocket calculator, at least in principle. For example, suppose that $\mu = 2.8$ and $x_1 = 0.4$ and that we want to predict the value for $n = 20$. We time-step from x_1 to x_2 to x_3, etc., in the following way:

$x_1 = 0.4$ at $n = 1$;

$x_2 = 2.8 \times 0.4 \times (1 - 0.4) = 0.6720$ at $n = 2$;

$x_3 = 2.8 \times 0.6720 \times (1 - 0.6720) = 0.6172$ at $n = 3$;

etc.

Figure 3 shows the graph obtained for $\mu = 2.8$ and $n = 1{:}20$ with an initial value of $x_1 = 0.4$. The graph also depicts that, for large n, the solution x_n is the same, no matter what is the initial condition.

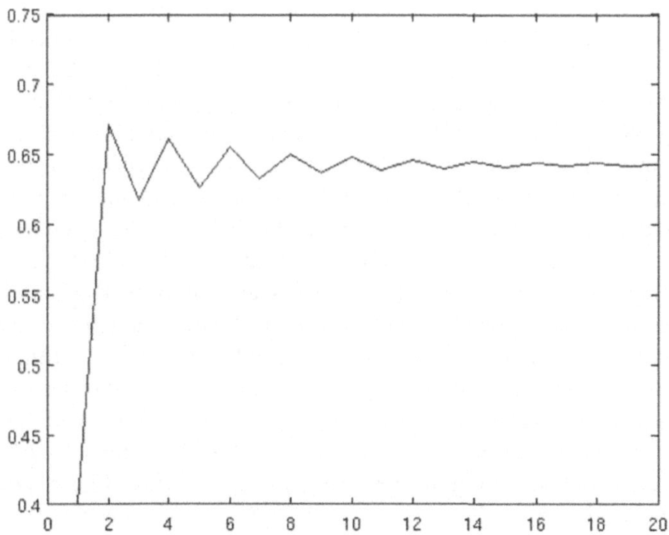

Fig. 3 Solution $\boldsymbol{x_n}$ of the logistic map versus $n = 1, 2, \ldots, 20$ for $\mu = 2.8$ and initial condition $\boldsymbol{x} = 0.4$ at $\boldsymbol{n} = 1$

4 Literature Review

Data security is an essential aspect in today's world to prevent the data over the insecure channel from unauthorized access. These unauthorized users misuse the information in various ways including disclosure, modification, etc. The issues of privacy, integrity, and confidentiality are the major issues in the current framework of internetworked connections. To ensure the reliable and dependable exchange of information, various security approaches have been proposed. One such approach that provides high security is DNA-based cryptography. It utilizes various ways of securing data using DNA-based cryptography. It converts the data into DNA nucleotide sequences and then rearranges these sequences based on certain rules to prevent data from being compromised. Hence, data security gives a platform to secure data from unauthorized access and keeps valuable information safe. In 2011, Prabhu and Adimoolam [36] presented a scheme that makes use of an encryption algorithm, working on the data obtained after its number conversion. This scheme uses PCR two primer pairs as the key to enhancing the security of the system. The use of DNA with its inherent biological properties and difficulty of DNA makes the scheme doubly secure. The scheme is seen to have high confidentiality when evaluated using a security analysis. Borda and Tornea [37] have reported two principles that are adopted in combination, these are bimolecular computation (BMC) and DNA cryptography. DNA cryptography is done using XOR operation and one-time passwords or pads (OTPs). The randomness of DNA and its high efficiency to provide high security in hybrid models is also discussed in this scheme. Another cryptography technique (pseudo cryptography) was developed by Kang [38] in 2009. This scheme first converts the actual data into a DNA sequence and then splits this sequence into two parts, viz. spliced form and protein form. Further, this technique makes use of the actual biological properties of DNA for encryption and as such is called pseudo cryptography. Gehani et al. [39] proposed a scheme that can be used for encoding both natural as well as artificial DNA by converting it to binary form. It makes use of one-time pad encryption via the substitution method and the XOR method. The substitution method includes various libraries containing distinct pads generated randomly. The XOR scheme is based on molecular computation. In 2012, Angeline and Thiruthuvadoss [40] have presented a scheme that makes use of the Triple DES (T-DES) algorithm and DNA cryptography. The evaluation of the scheme has been done by comparison among various parameters including key size, execution time, the complexity of the algorithm, and key analysis. The confidentiality of the proposed scheme has been improved via OTP as the key. Hamed et al. [41] have given a comparative analysis along with future scope for various steganography techniques based on encryption. Bahig and Nassr [42] have reported the scheme for performing various steps of AES while using encoding and decoding of data in a DNA sequence. The summary of a few papers along with their strengths and weaknesses has been provided in Table 2.

Table 2 Comparison between various approaches in the DNA cryptography domain

S. No.	Authors	Strengths	Security Approach adopted	Weaknesses
1	D. Prabhu, M. Adimoolam [35]	The use of two primer pairs makes the scheme more secure	Makes use of number conversion, DNA encoding, and PCR amplification	Because of the use of two primers using the PCR amplification, the computational complexity is higher
2	Monica Borda, Olga Tornea [36]	The cryptanalysis is hard because of chromosome indexing and the use of OTP	One-time pad (OTP) is used as a key, XOR operation for DNA encoding along with DNA chromosome indexing	Use of steganography technique to encrypt the data once the attacker knows the presence of data
3	Ning Kang [37]	The theoretical analysis results that this scheme shows resistance to certain attacks, including brute force attack and also through the translation and transcription ciphertext is more complex	The division of DNA sequence into two parts and using the pseudorandom technique makes the system secure	The ciphertext and plaintext size are not the same, and the availability of partial information can easily compromise data
4	Ashish Gehani, Thomas LaBean, and John Reif [38]	OTPs are used as a key for the XOR-based method. DNA substitution is used to make scheme unbreakable	DNA substitution and one-time pads for XOR operations	This technique assumes that the adversary is not aware of the data being hidden in the media
5	Angeline, Priyadarshini, Thiruthuvadoss [39]	This technique is computationally efficient because of the use of Triple DES. Also, the operation is less complex	One-time pad (OTP) with Triple DES	Consumption of more memory along with the use of a random generator which may repeat the sequence after a certain period

5 Proposed Method

A new image encryption technique is presented in this chapter. The technique makes use of the chaotic binary sequence and the DNA sequences to encrypt the image. Since chaos and DNA have been discussed in the previous sections, in this section, we will first describe the way to form the binary sequence from the chaotic equation, and then, an encryption hiding technique will be discussed which is used to encrypt the image.

5.1 Formation of Chaotic Binary Sequence

As discussed earlier, the mathematical model of the logistic map used in this work is as follows:

$$x_{k+1} = \mu x_k(1 - x_k) \tag{5}$$

Here, $\mu \in [0, 4]$, $x_k \in (0, 1)$, $k = 1, 2, 3, 4, 5, \ldots, m$.

The chaotic binary sequence used in the suggested algorithm is generated as follows:

1. Firstly, the real value sequence is calculated through some particular initial values, μ and x_k, where $k = 1, 2, 3, \ldots, m$, where m represents the total number of bits present in the RGB color image.
2. Binary sequence: The real value sequence obtained in the above step is further converted into the binary sequence using a threshold function $f(k)$, which is calculated as under:

$$\overline{f(x)} = \begin{cases} 0\ 0 < x_k \leq 0.5 \\ 0\ 0.5 < x_k \leq 1 \end{cases} \tag{6}$$

In this way, a chaotic binary sequence is obtained which depends upon the initial conditions.

5.2 DNA Substitution

This method utilizes a DNA sequence, R, obtained from the RGB image and a binary sequence of a reference DNA sequence, B, for data hiding or encryption. To explain the method, let R = TTGGACTGCCAGTACA and B = 10111011. The length of R is 16, which is larger than the length of B, q, which is 8 in this example. In this method, the complementary rule is used that ensures in a DNA sequence, all x, $C(x)$, $C(C(x))$, and $C(C(C(x)))$ should not be equal, where $C(x)$ is the complement. For an example, the following rule can be applied:

(AT)(TG)(GC)(CA).

Let DNA sequence of RGB image represented as $R = r1, r2, \ldots, rq$ and the binary sequence $B = b1, b2, \ldots, bq$. The scheme is summarized as under:

Step 1 Since the length of the actual DNA sequence, R, is 16; so m well-defined numbers are selected randomly from 1 to 16. Assuming that $m = 5$ and these chosen m numbers are 2, 3, 5, 10, and 12, then a point matrix, $A = \{A1, A2, \ldots, Am\} = \{2, 3, 5, 10, 12\}$ is formed.

Step 2 Transformation of R into R' by the rules as under:
For i varying from 1 to 16:
if $i = Aj$ and bj is 1, $1 \leq j \leq m$, set ri to be $C(ri)$;
else if $i = Aj$ and bj is 0, do not change ri;
else if $i \neq Aj$, set ri to be equal $C(C(ri))$;
Thus, $R' =$ CGGATTCATAGCCGTG.

Step 3 Sending to the receiver.

We must be confident that the receiver has the binary DNA sequence B, the point matrix and knows the complementary rule used in the method to decrypt the data correctly. The decryption algorithm will be the same as that of the encryption except for the complementary rules. The complementary rule used on the decryption side will be the inverted version of the complementary rules used for encryption. Since the complementary rule used for encryption is (AT) (TG) (GC) (CA), therefore the complementary rule for decryption will be (AC) (TA) (GT) (CG).

For an attacker to decrypt the data correctly, the information required is: (a) the binary DNA sequence, (b) the complementary rules, and (c) the point matrix.

$$P1 = \frac{1}{1.63 \times 10^8} \tag{7}$$

It is because there are 163 million DNA sequences present openly.

As for (b): The attacker should be aware of the legal complementary used in the algorithm. The probability of the correct guess is 1/6.

As for (c): He must be able to guess the point matrix correctly and the probability of the correct guess is 1.

As for (a): The probability of an intruder predicting a correct guess (P1) is

$$\frac{1}{{}^{n}C_2 {}^{n}C_3 \ldots {}^{n}C_n} \tag{8}$$

Finally, the total probability of guessing correctly is as follows:

$$\frac{1}{1.63 \times 10^8 \times 6} \times \frac{1}{{}^{n}C_2 {}^{n}C_3 \ldots {}^{n}C_n} \tag{9}$$

6 Algorithm Description of the Proposed Technique

This section presents the procedure for image encryption using the substitution method and chaos in detail. The new image encryption algorithm used in this chapter is presented in Fig. 4.

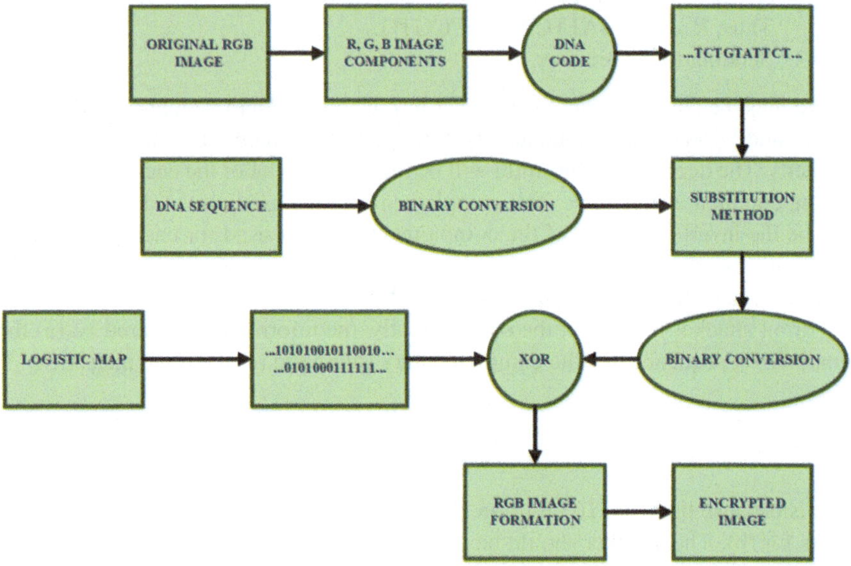

Fig. 4 Flowchart for image encryption

The procedure involved in the proposed algorithm is as under in Algorithm 1:

The Algorithm1:*Image Encryption*

Input:*Logistic Chaos Map Parameters $g_0$and μ_0,DNA sequence, Input Image, Coding/Complementary Rules, Point Matrix*

Step1*. Logistic chaos map parameters $g_0$and $\mu_0$are calculated which serve as the key for this encryption method*
Step2*. A binary chaotic sequence, C, is generated from the above parameters*
Step3*. Then, a DNA sequence is taken out of 163 million DNA's which serve as the second key; it is further converted to binary sequence, M*
Step4*. Further, the image is split into R, G, B planes and is transformed into DNA code using a particular code out of 24 combinations*
Step5*. Then, these planes are concatenated to form a long DNA sequence, S*
Step 6*. Some random points, less than the DNA sequence S's nucleotides are chosen. This serves as the third key*
Step7*. Then, the binary sequence (M) that is obtained is hidden in DNA sequence (S) by the substitution method to produce an encrypted DNA sequence S'*
Step8*. DNA sequence S' is converted back into the binary sequence, B*
Step9*. The chaotic sequence, C, is XORed with the binary sequence, B to produce an encrypted sequence, O*
Step10*. The encrypted sequence, O, is then converted into an RGB image, which will be the final encrypted image*

Output:*Encrypted Image*

To break the algorithm, the attacker must have logistic chaos map parameters x_k and μ, DNA sequence, the coding scheme of the four nucleic acid bases, the complementary rule, and the point matrix, *A*, of the substitution method.

7 Simulation Results

This algorithm is simulated in MATLAB environment and is tested for the standard 256 × 256 RGB Lena image. The chaotic parameters used in the encryption are $x_k = 0.2$ and $\mu = 3.9$. Besides, the DNA sequence used is BN000065 obtained from [20]. The point matrix used is {2, 3, 5, 10, 12}. Original Figure is shown in Fig. 5a. Figure 5b depicts the encrypted image. The decrypted image using the wrong secret key is depicted in Fig. 5c. The image that is decrypted using the correct key is depicted in Fig. 5d. It can be seen that it is hard to identify the original image from Fig. 5b.

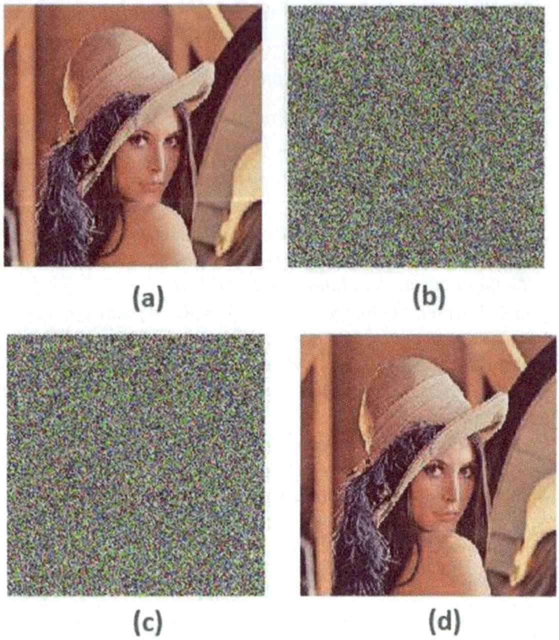

Fig. 5 **a** Original image, **b** encrypted image, **c** decrypted Image with error key, **d** decrypted image with the correct key

7.1 Security Analysis

The security analysis of an encrypted image can be done by various methods, e.g., secret key sensitivity analysis, secret key space analysis, and histogram analysis. Secret key sensitivity analysis specifies how the output changes with the change in the key. For a system to be efficient, the key should be very sensitive. Histogram analysis specifies the intensity of the pixels. For a system to be efficient, there should be uniformity in the histogram of the encrypted images so that the probability of the intensity of the pixels should be nearly equal. Secret key space analysis specifies how large the key is. Further, to repel the brute force attack, the secret key space should be large enough.

7.2 Secret Key's Sensitivity Analysis

Chaotic sequences are very sensitive to the initial value, during decryption of the image, a small change in the value of the secret key will affect the outcome of the overall process. To verify the scheme for the sensitivity of the secret key, we use a different key (having a slight difference) for decryption. Figure 6a shows the image obtained on decryption after using the correct key, $x_k = 0.2$, and Fig. 6b shows its corresponding histogram. Figure 6c shows the image obtained on decryption with $x_k = 0.2000001$ without changing the other keys, and Fig. 6d shows its corresponding histogram.

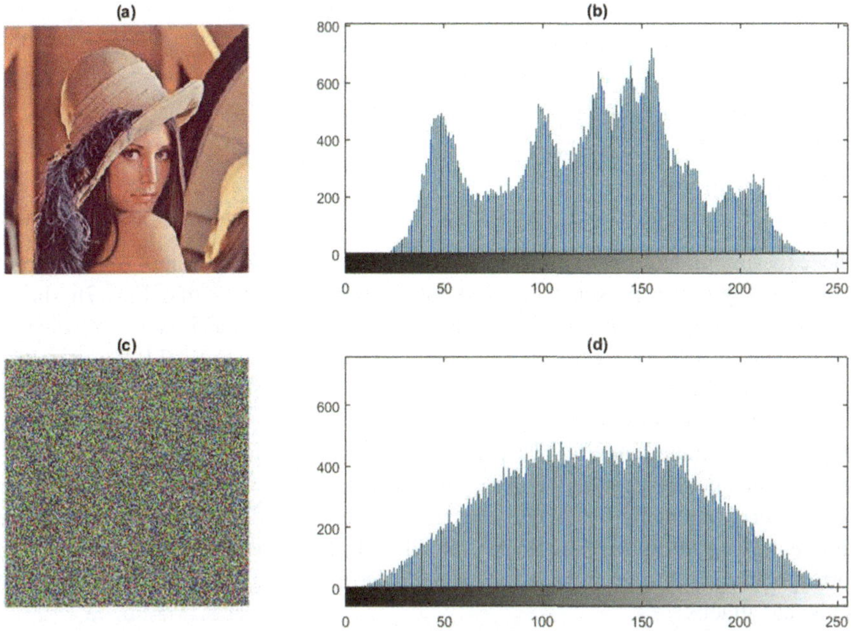

Fig. 6 **a** Decrypted Image with $x_n = 0.2$, **b** corresponding histogram, **c** decrypted image with $x_n = 0.2000001$, **d** corresponding histogram

7.3 Secret Key's Space Analysis

Our algorithm contains six key parameters, i.e., x_k and μ, binary DNA sequence, the coding scheme of the four nucleic acid bases, the complementary rule, and the point matrix, *A*, of the substitution method.

The probability of guessing the correct coding scheme will be 1/24.

The probability of guessing the correct complementary rule will be 1/6.

The probability of estimating the correct point matrix is

$$\frac{1}{{}^{n}C_2{}^{n}C_3\ldots{}^{n}C_n}$$

The probability of solving the correct DNA

$$P1 = \frac{1}{1.63 \times 10^8}$$

Thus, it is concluded that the secret key's space is very large, and it is very difficult to guess the correct keys.

Therefore, the total probability of guessing the keys is

$$\frac{1}{1.63 \times 10^8 \times 6} \times \frac{1}{{}^nC_2{}^nC_3 \ldots {}^nC_n}$$

7.4 Histogram Analysis

The experiment has been performed on three test images: Lena, Baboon, and Airplane. Figure 7a presents the original RGB Lena image, and Fig. 7b shows its corresponding histogram. Figure 7c presents the encrypted image, and Fig. 7d presents its corresponding histogram. Figure 8a presents the original Baboon image, and Fig. 8b presents its corresponding histogram. Figure 8c presents the encrypted image, and Fig. 8d presents its corresponding histogram. Figure 9a presents the original airplane image, and Fig. 9b presents its corresponding histogram. Figure 9c presents the encrypted image, and Fig. 9d presents its corresponding histogram. As can be seen from the figures, the histogram of the encrypted images is uniform in the middle making the statistical attack very difficult. It can be seen that the histogram corresponding to the encrypted image is uniform making the scheme resistant to statistical attacks.

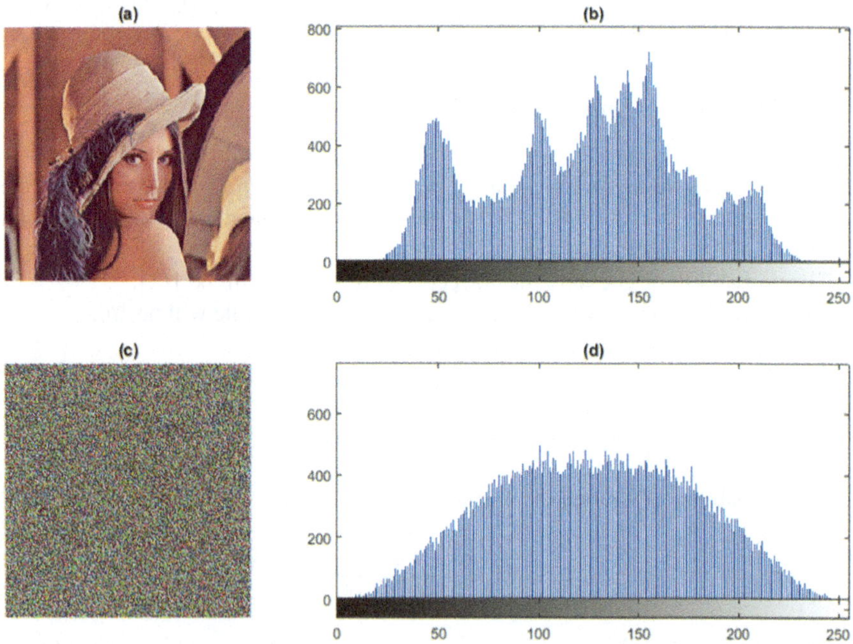

Fig. 7 **a** Original image, **b** corresponding histogram, **c** encrypted image, **d** corresponding histogram

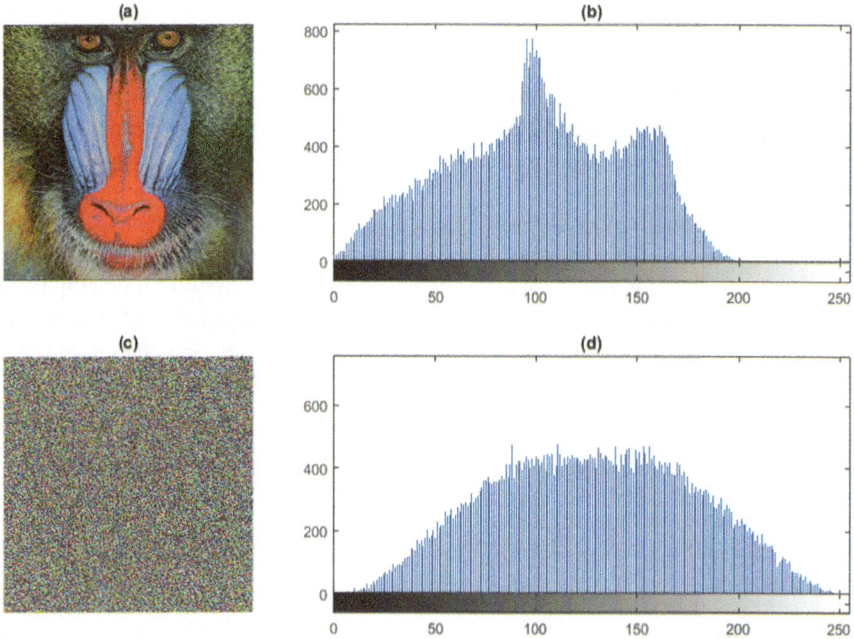

Fig. 8 **a** Original image, **b** corresponding histogram, **c** encrypted image, **d** corresponding histogram

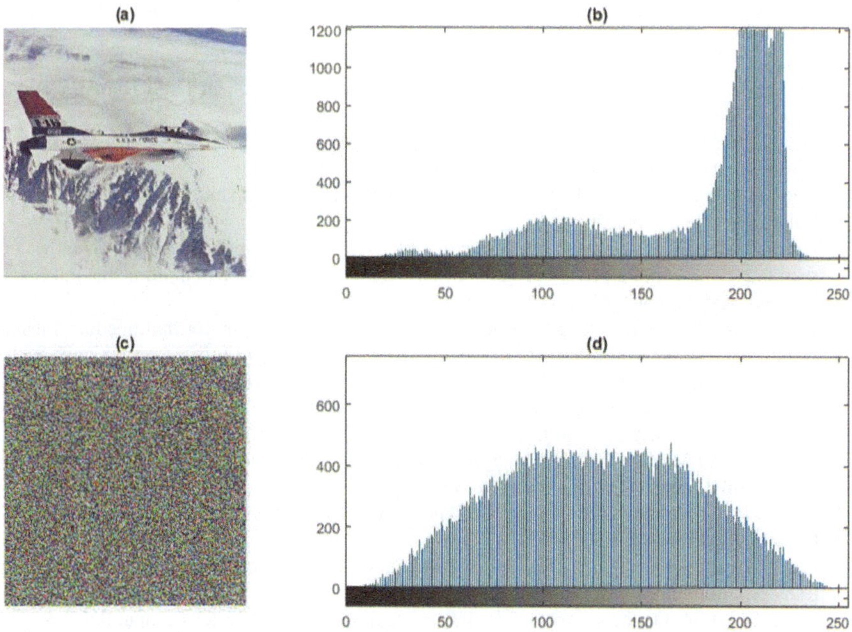

Fig. 9 **a** Original image, **b** corresponding histogram, **c** encrypted image, **d** corresponding histogram

8 Conclusion

In this chapter, a highly secure image encryption scheme has been proposed. After converting an RGB image into a DNA sequence, the encryption has been performed by using another DNA sequence and chaotic map technique. The proposed algorithm results in efficient image encryption using the DNA substitution method. The obtained histogram for the encrypted image is seen to be uniform and hence, making it robust against statistical attacks. The six keys used in encryption make the algorithm very efficient for various attacks and hence making it difficult for unauthorized users or adversaries to crack the proposed algorithm. Encryption and data hiding are done together in this algorithm. Further, using more than one chaotic map for encryption may improve the overall security of the algorithm.

Acknowledgements The authors would thank the Department of Science and Technology (DST) New Delhi, Government of India, for providing financial support under the DST Inspire Fellowship Scheme.

References

1. Giri KJ, Bashir R, Bhat JI (2019) A discrete wavelet based watermarking scheme for authentication of medical images. Int J E-Health Med Commun. https://doi.org/10.4018/IJEHMC.2019100103
2. Gull S, Loan NA, Parah SA, Sheikh JA, Bhat GM (2018) An efficient watermarking technique for tamper detection and localization of medical images. J Ambient Intell Human Comput 11:1799–1808. https://doi.org/10.1007/s12652-018-1158-8
3. Giri KJ, Bashir R (2018) A block based watermarking approach for color images using discrete wavelet transformation. Int J Inf Tecnol 10:139–146
4. Gao TG, Chen ZQ (2008) Image encryption based on a new total shuffling algorithm. Chaos, Solitons Fractals 38(1):213–220
5. Hurrah NN, Parah SA, Loan NA, Sheikh JA, Elhoseny M, Muhammad K (2019) Dual watermarking framework for privacy protection and content authentication of multimedia. Fut Gener Comput Syst 94:654–673
6. Giri KJ, Bashir R (2017) Digital watermarking: a potential solution for multimedia authentication. In: Dey N, Santhi V (eds) Intelligent techniques in signal processing for multimedia security. Studies in computational intelligence, vol 660. Springer, Cham
7. Parah SA, Sheikh JA, Hafiz AM, Bhat GM (2014) Data hiding in scrambled images: a new double-layer security data hiding technique. Comput Electr Eng 40(1):70–82
8. Tong XJ, Cui MJ (2008) Image encryption with compound chaotic sequence cipher shifting dynamically. Image Vis Comput 26(6):843–850
9. Dang PR, Chau PM (2002) Image encryption for secure internet multimedia applications. Consum Electron 46(3):395–403
10. Liu ZJ, Dai JM, Sun XG, Liu ST (2009) Triple image encryption scheme in fractional Fourier transform domains. Opt Commun 282:518–522
11. Meier W (1994) On the security of the IDEA Block Cipher. Lect Notes Comput Sci 765:371–385
12. Li HJ, Wang YR (2008) Double-image encryption by iterative phase retrieval algorithm in fractional Fourier domain. J Mod Opt 55(21):3601–3609

13. Zeghid M, Machout M, Khriji L, Baganne A, Tourkil R (2007) A modified AES based algorithm for image encryption. Int J Comput Sci Eng 1(1):70–75
14. Parah SA, Sheikh JA, Bhat GM (2012) High capacity data embedding using joint intermediate significant bit (ISB) and least significant bit (LSB) technique. J Inf Eng Appl 2224–5782
15. Chang CC, Hwang MS, Chen TS (2001) A new encryption algorithm for image cryptosystems. J Syst Softw 58(7):83–91
16. Loan NA, Hurrah NN, Parah SA, Lee JW, Sheikh JA, Bhat GM (2018) Secure and robust digital image watermarking using coefficient differencing and chaotic encryption. IEEE Access 6:19876–19897
17. Giri KJ, Peer MA, Nagabhushan P (2014) A channel wise color image watermarking scheme based on discrete wavelet transformation. In: 2014 international conference on computing for sustainable global development (INDIACom). https://doi.org/10.1109/IndiaCom.2014.6828064
18. Zhang Q, Guo Y, Wei X (2010) Image encryption using DNA addition combining with chaotic maps. Math Comput Model 52(11–12):2028–2035
19. Zhang Q, Wang Q, Wei X (2010) A novel image encryption scheme based on DNA coding and multi-chaotic maps. Adv Sci Lett 3(4):447–451
20. Jiao SH, Goutte R (2008) Code for encryption hiding data into genomic DNA of living organisms. In: Proceedings of the 9th international conference on signal processing (ICSP '08), pp 2166–2169
21. Kanso A, Ghebleh M (2012) A novel image encryption algorithm based on a 3D chaotic map. Commun Nonlinear Sci Numer Simul 17(7):2943–2959
22. Clelland CT, Risca V, Bancroft C (1999) Hiding messages in DNA microdots. Nature 399:533–534
23. Leier A, Richter C, Banzhaf W, Rauhe H (2000) Cryptography with DNA binary strands. BioSystems 57:13–22
24. Shimanovsky B, Feng J, Potkonjak M (2002) "Hiding data in DNA", Revised papers from the 5th international workshop on information hiding. Lect Notes Comput Sci 2578:373–386
25. European Bioinformatics Institute. https://www.ebi.ac.uk/
26. Ye GD (2010) Image scrambling encryption algorithm of pixel bit based on chaos map. Pattern Recogn Lett 31(5):347–354
27. Zhang YW, Wang YM, Shen XB (2007) A chaos-based image encryption algorithm using alternate structure. Sci China Ser F: Inf Sci 50(3):334–341
28. Parah AP, Javaid AS, Bhat GM (2015) Hiding in encrypted images: a three-tier security data hiding system. Multidim Syst Sig Process 9
29. Zhao L, Liao XF, Xiang T, Xiao D (2009) Multidimensional Z-matrix with control parameters and its applications in image encryption. Chin Phys Lett 26(6):060502:1–4
30. Xiao GZ, Lu MX (2006) New field of cryptography: DNA cryptography. Chin Sci Bull 51(12):1413–1420
31. Gehani A, LeBean T, Reif J (2000) DNA-based cryptography. Dimacs Scries in Disc Math Theor Comput Sci 54:233–249
32. Wong PC, Wong KK, Harlan F (2003) Organic data memory using the DNA approach. Commun ACM 46(1):95–98
33. Lu MX, Lai XJ, Xiao GZ, Qin L (2007) The symmetrical encryption method based on DNA techniques. Sci China Ser E: Technol Sci 37:175–182
34. Matthews R (1989) On the derivation of a "chaotic" encryption algorithm. Cryptologia 13(1):29–42
35. Kanso A, Ghebleh M (2012) A novel image encryption algorithm based on a 3D chaotic map, Commun Nonlinear Sci 17(7):2943–2959
36. Prabhu D, Adimoolam M (2011) Bi-serial DNA encryption algorithm (BDEA). arXiv:1101.2577
37. Borda ME, Tornea O (2010) DNA secret writing Techniques. In: IEEE conferences
38. Ning K (2009) A pseudo-DNA cryptography method. https://arxiv.org/abs/0903.2693

39. Gehani A, LaBean T, Reif J (2004) DNA-based cryptography. In: Lecture notes in computer science, Springer, Berlin
40. Priyadharshini A, Thiruthuvadoss (2012) Comparison and performance evaluation of modern cryptography and DNA cryptography. Department of System on Chip Design Masters of Science
41. Hamed G et al. (2016) Comparative study for various DNA based steganography techniques with the essential conclusions about future research. In: 11th international conference computer engineering and systems (ICCES)
42. Bahig HM, Nassr DI (2019) DNA-Based AES with silent mutations. Arab J Sci Eng 44:3389–3403. https://doi.org/10.1007/s13369-018-3520-8

Chapter 3
Impact of Computational Power on Cryptography

M. Iqbal Bhat and Kaiser J. Giri

1 Introduction

From the time immemorial, security and privacy have been the main concerns of humans. Great and concrete efforts have always been diverted towards the fulfillment of these objectives. These efforts ultimately gave birth to much important field of cryptography—the art of writing or solving codes as defined by Concise Oxford English Dictionary. Though classical cryptography was considered as an art with its primary consumers as military organizations and governments, has radically changed in the modern era. The demands of modern cryptography are beyond the capabilities of an art [1], which include:

(i) Privacy/confidentiality
(ii) Authentication
(iii) Integrity
(iv) Non-repudiation and
(v) Key exchange.

Beyond these fundamental and primary functionalities, modern cryptography is securing e-commerce, e-governance, digital cash, electronic auctions and much more. Therefore modern cryptography has evolved as a science and a mathematical discipline.

M. I. Bhat
Department of Higher Education, J&K, Government Degree College Pulwama, Pulwama, Jammu and Kashmir, India
e-mail: iqbal.profhghredu@jk.gov.in

K. J. Giri (✉)
Department of Computer Science, School of Technology, Islamic University of Science and Technology Awantipora, Pulwama, Jammu and Kashmir, India
e-mail: kaiser.giri@islamicuniversity.edu.in

K. J. Giri et al. (eds.), *Multimedia Security*, Algorithms for Intelligent Systems,
https://doi.org/10.1007/978-981-15-8711-5_4

Putting in simple words, "cryptography has gone from a heuristic set of tools concerned with ensuring secret communication for the military to a science that helps secure systems for ordinary people all across the globe [2]."

The advancements in multimedia field in past few decades have enabled its application in much important fields like digital forensic, e-government, e-health care, social networking, education, military, industry etc. The demand of multimedia security has thus increased. The nature of multimedia data demands some special security schemes like watermarking, in addition to the traditional cryptographic schemes. Since digital platforms convert various kinds of multimedia content like text, audio, video, images etc., into common format of zero's and one's which are indistinguishable at storage and transportation level, the general cryptographic schemes discussed in this paper are applicable for its protection.

The advancements in computational power have made technology available at the doorstep of every individual. With the result, the dependency on cryptography has not only increased exponentially but has become inevitable. According to a report by Statista Research Department, there are about 4.57 billion active Internet users (Fig. 1) with 60% population already online and about 49.7% of the private households worldwide (Fig. 2) are estimated to have a computer at home. In case of developed countries, the share exceeds 80%.

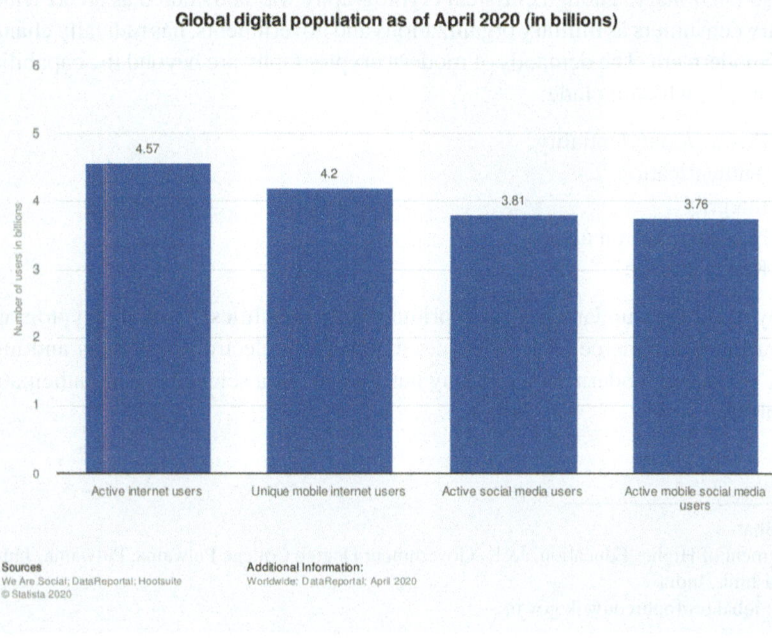

Fig. 1 Global digital population as of April 2020. *Source* https://www.statista.com/statistics/617136/digital-population-worldwide/

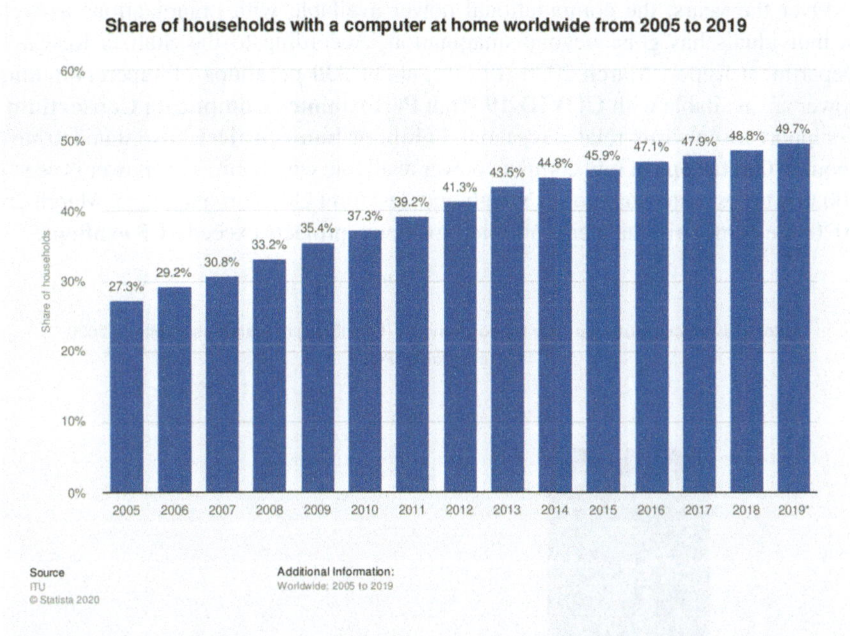

Fig. 2 Share of households with a computer at home worldwide from 2005 to 2019. *Source* https://www.statista.com/statistics/748551/worldwide-households-with-computer/

In such a scenario where almost the entire world population lives within the reach of mobile networks (Fig. 3), the only tool to secure them digitally is the cryptographic techniques.

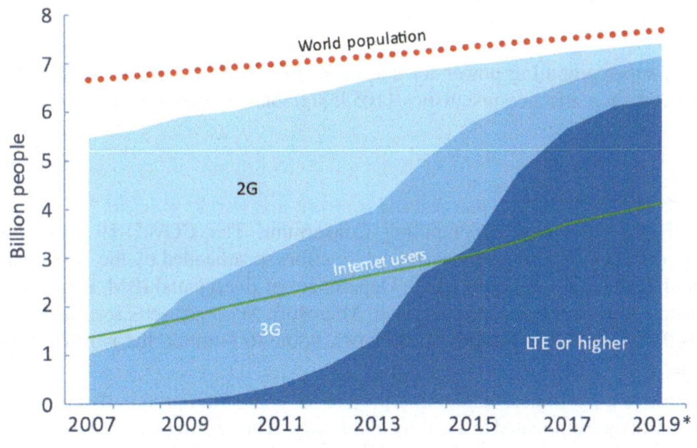

Fig. 3 Reach of mobile networks. *Source* https://itu.foleon.com/itu/measuring-digital-development/network-coverage/

Over the years, the computational power available with organizations as well as individuals has gone beyond imagination. According to the Statista Research Department Report March 2020 (Fig. 4), about **330 petaflops** of supercomputing power are available with **COVID-19 High Performance Computing Consortium**[1] to support coronavirus-related research. **Folding@home project**[2] also stated to have about **470 petaflops** of raw compute power available with them. This power is nearly 500,000 times faster than today's high-end desktop PCs. Moreover, as of March 26, 2020, the cumulative capacity of Folding@home projects exceeds **1.5 exaflops.**

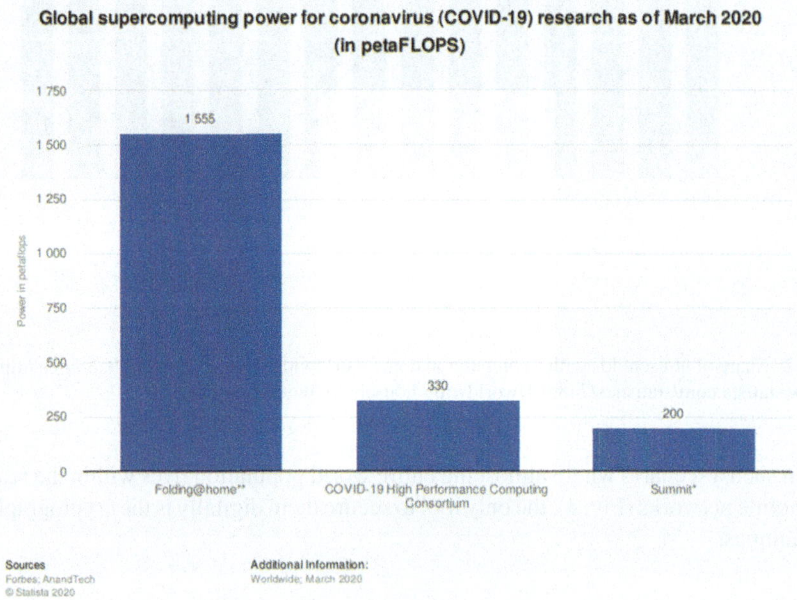

Fig. 4 Global supercomputing power for coronavirus (COVID-19) research as of March 2020. *Source* https://www.statista.com/statistics/1105858/global-supercomputing-power-for-covid-19-research/

[1]COVID-19 High Performance Computing Consortium: The COVID-19 High Performance Computing Consortium is a unique private–public effort spearheaded by the White House Office of Science and Technology Policy, the U.S. Department of Energy and IBM. Consortium partners include Amazon Web Services, Google Cloud, Microsoft, Massachusetts Institute of Technology (MIT), as well as Oak Ridge National Laboratory, home to Summit, the world's most powerful supercomputer.

[2]Folding@home is a distributed computing project for simulating protein dynamics, notably the movements of proteins implicated in a variety of diseases. It brings together scientists who volunteer to run simulations of protein dynamics on their personal computers. Insights from this data are helping scientists to better understand biology and providing new opportunities for developing therapeutics.

This is just tip of an iceberg and gives a glimpse of the computational power available in the world. Emerging technologies like cloud computing, mobile cloud computing (MCC), quantum computing, parallel computing etc., are expected to totally change the computational power scenario of the world. Advancements in GPU's and Compute Unified Device Architecture (CUDA)[3] have brought supercomputing techniques available even for desktop and laptop computers.

These advancements in computational power have registered great impact on the cryptographic world. On one side, favoring cryptanalysts, this enormous computational power proves destructive for most of the modern symmetric and asymmetric cryptographic algorithms. Even elliptic curve cryptography which is considered presently the most secure and efficient scheme is weak against quantum computers [3]. On the other hand, this computational power can be utilized to develop robust cryptographic algorithms. Blockchain is one such example, where extraordinary security has been achieved by utilizing the computational power of miners.

In this paper, we will analyze the impact of computational power on cryptography. Starting from classical cryptography to modern cryptography, first will study the procedures for various cryptosystems and then study the impact of computational power on their security. Our main focus will be on historically important and most widely used cryptosystems.

1.1 Literature Review

The history of cryptography is as old as the importance of data security and data privacy among humans. Though there is no fixed timeline about its beginning, Greeks and Romans had practiced different forms of cryptography from ancient times [4]. Cryptography evolved as an important branch of computer science with the invention of commercial computing machines. Computing machines have given birth to new directions of generating, processing and communicating data. Technological advancements have connected almost entire human population which has further strengthened the importance of security and cryptographic schemes. Paweł Czarnul in "Parallel Programming for Modern High Performance Computing Systems" [5] has given an in-depth coverage of modern parallel computing. There has been tremendous technological and literature progress in both the worlds, i.e., the world of computation and the world of cryptography. Ceruzzi [6] has discussed the history of modern computing from advent of commercial computers in 1945 to the Internet ages in 2001. On the other hand, Dooley [4] discusses the history of cryptography and cryptanalysis from ancient times to cryptography in the era of Internet of things and quantum computing. The groundbreaking work in the field of cryptography was

---

[3]CUDA™ is a revolutionary parallel computing architecture from NVIDIA. As an enabling hardware and software technology, CUDA makes it possible to use many computing cores in a graphics processor to perform general-purpose mathematical calculations, achieving dramatic speedups in computing performance.

done by David Kahn's code breakers [7] and F. L. Bauer's decrypted secrets [8]. A historical and mathematical perspective in this direction was added by Craig P. Bauer in his Secret History: The Story of Cryptology [9]. A recent work in this direction is done by Jonathan Katz and Yehuda Lindell in their book Introduction to Modern Cryptography [2] in which they have presented the basic paradigms and principles of modern cryptography.

Advancements in computational power have resulted in more powerful cryptanalyst tools which have impacted the world of cryptography. Diffie and Hellman in [10] discuss the exhaustive cryptanalysis of DES, and Eli Biham and Shamir have given the differential cryptanalysis of DES like cryptosystem [11]. A comprehensive analysis of cryptanalyst attacks on Feistel ciphers is given by Nachef et al. in their book Feistel Ciphers: Security Proofs and Cryptanalysis [12]. Various cryptanalyst attacks on different variants of AES are given in [13–18]. Ferguson et al. presented a simple algebraic representation of AES cipher [19] which if solved will be end of AES cipher. Mathematical attacks on RSA are given in [20]. A comprehensive analysis of cryptanalysis attacks on RSA and its variants is given by M. Jason Hinek in [21]. A detailed coverage of elliptic curves is given in [22, 23]. A comparative analysis of RSA and ECC is given in [24]. An important algorithm for solving ECDLP known as Pollard's rho algorithm is presented in [25].

Advancements in quantum computing are arising as a great security threat for cryptography. The impact of quantum computing on present cryptography is given in [3]. Quantum attacks against symmetric key schemes are given in [26]. A detailed report of NIST on post-quantum cryptograph is given in [27]. Moreover, an important polynomial time quantum algorithm known as Shor's quantum algorithm is presented in [28].

1.2 Organization of Chapter

The rest of this paper is organized as follows. Section 2 introduces cryptography and advancements in computational technology. Section 3 discusses classical cryptography and impact of computational power on some famous classical cryptosystems. Section 4 covers the modern cryptography and impact of computational power on some famous and most widely used modern cryptosystems. Finally, the conclusion is presented in Sect. 5.

2 Introduction to Cryptography

Cryptography is the science of creating cipher systems or in simple words it is the science of keeping secrets secret. Etymologically, it comes from two Greek words **κρυπτός (krypto)**, meaning "**hidden**," and **γραφία (graphene)**, meaning "**writing**." It is the method of protecting information (numerical, textual, graphical or even executable program) from adversaries by way of hiding or disguising them using secret codes so that only the intended recipients can read and process it.

The sender of the message converts the plaintext message (m) into cipher text (c) by using a special key (k_e). This process is called **encryption** (E) and is given as:

$$E(m,\ k_e) = c$$

This encrypted message is then sent to the intended recipient over the unsecured communication channel in the presence of eavesdropper. Even if the eavesdropper is able to intercept the message, he will not be able to get anything from the cipher text as long as he is not able to get hold of the decryption key (k_d).

At the receiver end, the intended recipient converts the cipher text (c) into original message (m) by using the decryption key (k_d). This process is called **decryption** (D) and is given by:

$$D(c,\ k_d) = m$$

Since decryption process should generate same plaintext message used during encryption process, it can be mathematically written as:

$$D(E(m, k_e), k_d) = m$$

If encryption key (k_e) and decryption key (k_d) are same, as is the case in classical cryptography (before 80s), then it is called private key cryptography or symmetric key cryptography. But if the key used for encryption (k_e) is different from the decryption key (k_d), then it called public key cryptography or asymmetric cryptography and was first time introduced by W. Diffie and M. E. Hellman in their famous paper, New Directions in Cryptography [29].

So, key algorithms of a cryptosystem include [2]:

(a) **The key generation algorithm**: A probabilistic algorithm that outputs a key k chosen according to some distribution.
(b) **The encryption algorithm**: An algorithm $E(m, k_e)$ that accepts a message m and a key k_e as input and outputs a cipher text c.
(c) **The decryption algorithm**: An algorithm $D(c, k_d)$ that accepts cipher text c and a key k_d as input and outputs a plaintext m.

According to Kirchhoff's principle [7]: "*The secrecy should be in the key used for decryption, not in the decryption or encryption algorithm.*" Modern cryptography is based on this principle. So according to this principal, the security of a cryptosystem should not rely on the encryption algorithm but on the secrecy of the key used for encryption. Hence, keys and key generation algorithms have central role in the security of a cryptosystem.

The main thing that distinguishes modern cryptography from classical cryptography is its emphasis on definitions, precise assumptions and rigorous proofs of security [2]. The security of modern cryptosystems is based on some widely believed, unproven mathematical and technological assumption(s). For example, de facto standard for public key encryption throughout the world—RSA public cryptosystem—is based on the assumption of intractability of the integer factorization problem (IFP). If an adversary can factor the public modulus, N into its two prime factors, p and q, he can efficiently calculate the private exponent [20]. In [21], Hinek discusses in detail the best-known algebraic attacks on RSA and its main variants, including rebalanced RSA, multi-prime RSA and multi-factor RSA. According to Kirsch [30], RSA public cryptosystem is theoretically vulnerable if a fast factorizing algorithm is introduced or huge increase in computation power can exist. So, mathematical and technological advancements pose a great threat to the security of modern cryptosystem. The advancements in computational power are sure to change the landscape of the cryptographic world and we need to think of cryptographic solutions that are safe under and post those circumstances. The recent advancements in quantum computing field have raised the question of quantum safe cryptography; i.e., cryptosystems that are safe on quantum computers as research is showing most of the widely used cryptosystems are no longer safe on large-scale quantum computers [27]:

From Table 1, it is evident that the AES cryptosystem is quantum safe but with a large key size and has been adapted by National Security Agency (NSA) to secure secrets and top secrets with key sizes of 192 and 256 bits [31].

Table 1 Impact analysis of quantum computing on encryption schemes

Cryptographic algorithm	Type	Purpose	Impact from large-scale quantum computer
AES	Symmetric key	Encryption	Large key size needed
SHA-2, SHA-3	–	Hash functions	Larger output needed
RSA	Public key	Signatures, key establishment	No longer secure
ECDSA, ECDH (elliptic curve cryptography)	Public key	Signatures, key exchange	No longer secure
DSA (finite field cryptography)	Public key	Signatures, key exchange	No longer secure

Source [27]

Table 2 Comparison of classical and quantum security level for the most used cryptographic schemes

Cryptographic scheme	Key size	Effective key strength/security level (in bits)	
		Classical computing	Quantum computing
RSA-1024	1024	80	0
RSA-2048	2048	112	0
ECC-256	256	128	0
ECC-384	384	256	0
AES-128	128	128	64
AES-256	256	256	128

Source [27]

Comparison of classical and quantum security levels for most used cryptographic schemes is given by Mavroeidis et al. [3], which clearly shows that key size is not going to guard famous and most adopted public cryptosystems (Table 2).

The advancements in computational power can also result in robust and secure cryptosystem by exploiting the growing computation power. It was shown by Dong et al. [32] that by fully exploiting the floating-point computing power of GPUs using designs like floating-point based Montgomery multiplication/exponentiation algorithm and Chinese Remainder Theorem (CRT) the performance may be promoted by about 5%. They have achieved a remarkable 13 times decryption performance of RSA-2048/3072/4096 decryption on NVIDIA GeForce GTX TITAN over the previous fastest floating-point-based implementation.

Cheng et al. [33] proposed a high-performance symmetric cryptography server using a high-speed SM4 kernel with a GPU card as an SM4 algorithm accelerator. The server which can be adopted for mass data encryption can deliver 15.96 Gbps data encryption through network which is about 1.23 times of the existing fastest symmetric cryptographic server.

The SHA256 hashing algorithm is playing an important role in cyber security by ensuring the data integrity and authenticity. In Bitcoin protocol, a huge computational power is diverted in calculation of large number of SHA256 hashes for "proof of work." To improve the performance of SHA256 computation, Zhang et al. [34] proposed a high-performance parallel hardware architecture in AISC.[4] The result shows (Table 3) that it can have 3 times speed improvement at 2.90 times area cost and gain 50.7% improvement in power.

The underlying technology for Bitcoin cryptocurrency [35] is the blockchain which is a distributed ledger used to store, validate, authorize and move digital transactions across the Internet. This blockchain has an extreme potential to change not just the world of digital currencies but a vast number of organizational contexts. Even some predict it as future of Internet [36], and anyone who wants to understand the future of Internet must understand blockchain. Bitcoin mining is an important process by which transactions are verified and added to the blockchain. It is also a means through which new Bitcoins are generated and released. This mining process is computation intensive used to verify Bitcoin transactions for profit. Miners are putting in their computational power to solve mathematical puzzles required to add new blocks to the blockchain, and in return they are rewarded with some Bitcoins. Without miners, blockchain will halt to function.

Table 3 Hardware comparison

	Clock (ps)	Area (μm^2)	Power (mW)
Standard	1959	4916.7	3.3794
Proposed	653	14,272.6	6.855

Source [34]

[4]ASIC: An application-specific integrated circuit (ASIC/ˈeɪsɪk/) is an integrated circuit (IC) chip not intended for general-purpose rather customized for a particular use.

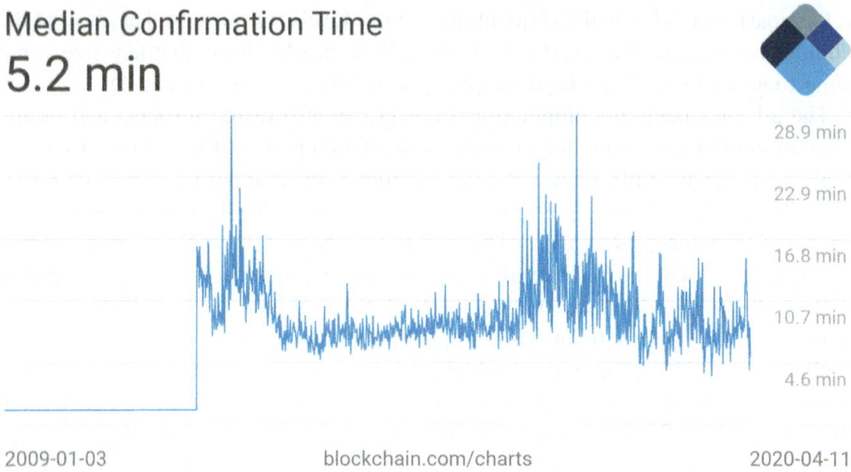

Fig. 5 Blockchain median confirmation time. *Source* https://www.blockchain.com/charts

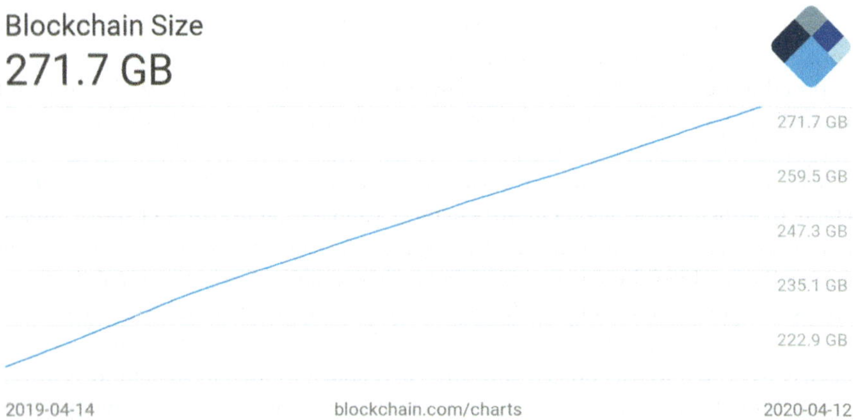

Fig. 6 Block size as on 12–04–2020. *Source* https://www.blockchain.com/charts

The requirement of huge computational power for adding new blocks to the blockchain is the main hurdle in wide applicability of blockchain technology. The median confirmation time to add a transaction block to the blockchain is 5.2 min which is much higher than that of centralized banking (Figs. 5 and 6).

Moreover, to handle the largely growing blockchain which has already exceeded 250 GBs is also rising as a technical challenge.

In addition, there are a number of challenges before Bitcoin as discussed in [37–40]. The recent advancements in computational technology may solve these hard challenges of blockchain and a number of efforts are being put in to optimize blockchain technology. In [41], a parallel proof of work has been proposed with results showing improvement of 34% compared to the current system which is subject to change based on the processing power allocated to the miners.

3 Pre-multimedia Era cryptography—The Classical Cryptography

The multimedia era started around 1973 with the introduction of interactive laser disks and later Atari Video Games in 1977. This was the time when cryptography was also going to shift its era from classical cryptography to modern cryptography. The cryptography before the invention of public key cryptography in 1977 [29] is considered as classical cryptography. There is no fixed timeline about beginning of classical cryptography. Greeks and Romans had practiced different forms of cryptography from ancient times [4]. Julius Caesar, one of the greatest Roman Generals, had used the oldest recorded ciphers, know after his name as Caesar's cipher as described in De Vita Caesarum, Divus Iulius ("The Lives of the Caesars, the Deified Julius"), written in approximately 110 CE [2]:

> There are also letters of his to Cicero, as well as to his intimates on private affairs, and in the latter, if he had anything confidential to say, he wrote it in cipher, that is, by so changing the order of the letters of the alphabet, that not a word could be made out....

3.1 Caesar Cipher

The Caesar cipher is based on changing the order of letters of alphabets and encrypted messages by replacing each letter by the third letter to follow it alphabetically and at the end wrap around by replacing z with C and so on. The cipher text of word "cryptology" using Caesar's cipher looks like:

Plain: *Cryptology*
Cipher: *FUBSWRORJB.*

Due to the fixed nature of cipher and absence of encryption key, the Caesar cipher is insecure and to decipher above message, the cryptanalyst just needs to substitute for each alphabet of cipher text an alphabet three places before it.

3.2 Shift Ciphers

The above problems of the Caesar ciphers are removed to some extent in shift ciphers where a key k is used which varies from 0 to 25. In shift ciphers, letters are shifted by k places and can be written in general form as

$$c = p + (k \bmod 25)$$

where c is the cipher text, p is the plaintext and k is the key. The key space for k ranges from 1 to 25 as the message remains unchanged for values 0 and 26. Due to this small key space, shift ciphers are susceptible to brute-force attacks.

3.3 Impact of Computational Power on Caesar and Shift Ciphers

No computational power is required to break the Caesar cipher as it is trivial to find the plaintext from cipher text by shifting each cipher text character three places before it. In case of shift ciphers, there are only 26 possible keys and a brute-force attack or an exhaustive search attack can easily find out the list of 26 candidate plaintexts.

3.4 Monoalphabetic Substitution Ciphers (MASC)

The monoalphabetic substitution cipher (MASC) is another category of classical ciphers where instead of shift, substitution of a particular character with same cipher text letter is used for encryption. The key is defined as a one-to-one mapping from the set of plaintext alphabets to the set of cipher text. In case of 26 letters of English alphabet, there are 26 possible mappings for letter A, 25 for letter B, 24 for letter C and so on. For last letter Z, there is only one mapping possibility. Using multiplication rule of combinatorics, size of the key space is $26 * 25 * 24 * \cdots * 1 = 26! = 4.0329146112661\text{E}{+}26 \approx 2^{88}$. In spite of having such a large key space, the monoalphabetic substitution ciphers are not secure and can be easily attacked by frequency distribution attack [2]. Since in monoalphabetic substitution ciphers, each plaintext

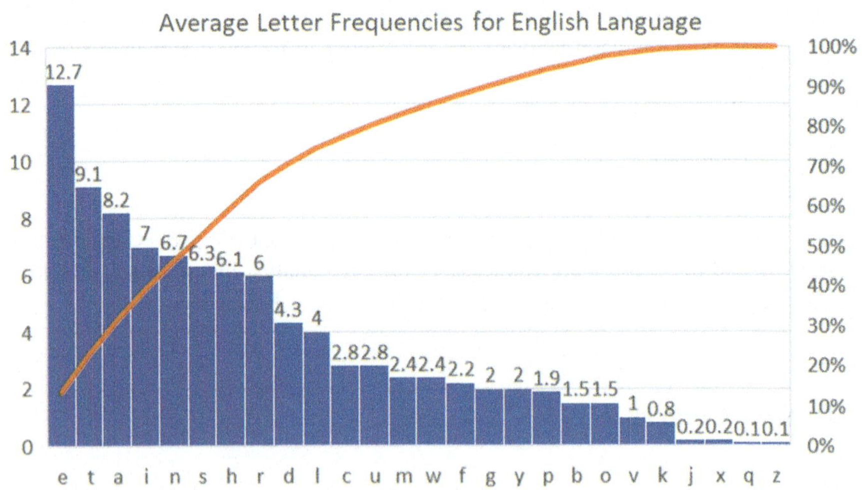

Fig. 7 Average letter frequencies for English language

letter maps to the same cipher text wherever it appears, the average frequency of letters in English language (Fig. 7) can be used to break the cipher.

The mapping of the key can be deduced or guessed from the frequencies of the characters in the cipher text. For examples, since **e** is the most frequently used character in the English language, then the most frequently used character in the cipher text can be guessed as **e**. This way the key can be guessed very quickly.

The monoalphabetic substitution ciphers were used in West for centuries and were considered as strongest cipher systems. The cryptographic world got a setback in West after the fall of Western Roman Empire in 476 C.E. [4]. During the Islamic Golden Age in ninth century C.E, Abu Yūsuf Ya-qūb ibn Isāq as-Sabbāh al-Kindi (801–873 C.E.) first described the technique of **frequency analysis** in his book "Manuscript on Deciphering Cryptographic Messages." This was the first cryptanalytic tool.

3.5 Vigenère Cipher

Blaise de Vigenère (1523–1596) is credited for a **polyalphabetic substitution cipher** know after his name as "Vigenère cipher." Vigenère cipher was a progression toward unbreakable ciphers. In Vigenère cipher, each character in plaintext is not mapped to a fixed cipher character. This feature guards the Vigenère cipher against statistical attack. Vigenère cipher can be considered as use of several Caesar ciphers in sequence

	a	b	c	d	e	f	g	h	i	j	k	l	m	n	o	p	q	r	s	t	u	v	w	x	y	z
A	A	B	C	D	E	F	G	H	I	J	K	L	M	N	O	P	Q	R	S	T	U	V	W	X	Y	Z
B	B	C	D	E	F	G	H	I	J	K	L	M	N	O	P	Q	R	S	T	U	V	W	X	Y	Z	A
C	C	D	E	F	G	H	I	J	K	L	M	N	O	P	Q	R	S	T	U	V	W	X	Y	Z	A	B
D	D	E	F	G	H	I	J	K	L	M	N	O	P	Q	R	S	T	U	V	W	X	Y	Z	A	B	C
E	E	F	G	H	I	J	K	L	M	N	O	P	Q	R	S	T	U	V	W	X	Y	Z	A	B	C	D
F	F	G	H	I	J	K	L	M	N	O	P	Q	R	S	T	U	V	W	X	Y	Z	A	B	C	D	E
G	G	H	I	J	K	L	M	N	O	P	Q	R	S	T	U	V	W	X	Y	Z	A	B	C	D	E	F
H	H	I	J	K	L	M	N	O	P	Q	R	S	T	U	V	W	X	Y	Z	A	B	C	D	E	F	G
I	I	J	K	L	M	N	O	P	Q	R	S	T	U	V	W	X	Y	Z	A	B	C	D	E	F	G	H
J	J	K	L	M	N	O	P	Q	R	S	T	U	V	W	X	Y	Z	A	B	C	D	E	F	G	H	I
K	K	L	M	N	O	P	Q	R	S	T	U	V	W	X	Y	Z	A	B	C	D	E	F	G	H	I	J
L	L	M	N	O	P	Q	R	S	T	U	V	W	X	Y	Z	A	B	C	D	E	F	G	H	I	J	K
M	M	N	O	P	Q	R	S	T	U	V	W	X	Y	Z	A	B	C	D	E	F	G	H	I	J	K	L
N	N	O	P	Q	R	S	T	U	V	W	X	Y	Z	A	B	C	D	E	F	G	H	I	J	K	L	M
O	O	P	Q	R	S	T	U	V	W	X	Y	Z	A	B	C	D	E	F	G	H	I	J	K	L	M	N
P	P	Q	R	S	T	U	V	W	X	Y	Z	A	B	C	D	E	F	G	H	I	J	K	L	M	N	O
Q	Q	R	S	T	U	V	W	X	Y	Z	A	B	C	D	E	F	G	H	I	J	K	L	M	N	O	P
R	R	S	T	U	V	W	X	Y	Z	A	B	C	D	E	F	G	H	I	J	K	L	M	N	O	P	Q
S	S	T	U	V	W	X	Y	Z	A	B	C	D	E	F	G	H	I	J	K	L	M	N	O	P	Q	R
T	T	U	V	W	X	Y	Z	A	B	C	D	E	F	G	H	I	J	K	L	M	N	O	P	Q	R	S
U	U	V	W	X	Y	Z	A	B	C	D	E	F	G	H	I	J	K	L	M	N	O	P	Q	R	S	T
V	V	W	X	Y	Z	A	B	C	D	E	F	G	H	I	J	K	L	M	N	O	P	Q	R	S	T	U
W	W	X	Y	Z	A	B	C	D	E	F	G	H	I	J	K	L	M	N	O	P	Q	R	S	T	U	V
X	X	Y	Z	A	B	C	D	E	F	G	H	I	J	K	L	M	N	O	P	Q	R	S	T	U	V	W
Y	Y	Z	A	B	C	D	E	F	G	H	I	J	K	L	M	N	O	P	Q	R	S	T	U	V	W	X
Z	Z	A	B	C	D	E	F	G	H	I	J	K	L	M	N	O	P	Q	R	S	T	U	V	W	X	Y

Fig. 8 Tabula recta

but with different shift values. Vigenère cipher encryption makes use of table known as **Vigenère tabulae** or **tabula recta** (Fig. 8) in which alphabets are written in 26 rows.

In each row, the alphabets are shifted cyclically to the left compared to the contents of row above it. The left most column is the key alphabet, and the top most row represents the plaintext alphabet. The cipher text for each character is the intersection of the plaintext alphabet column and the key character column. If the size of the key is less than plan text, then it is repeated. For example, to encrypt the string "Kashmir is beautiful" with the key "PARADISE" using Vigenère cipher, we get:

Key	P	A	R	A	D	I	S	E	P	A	R	A	D	I	S	E	P	A
Plaintext	K	A	s	h	m	I	R	i	s	b	e	a	u	t	i	f	u	l
Cipher text	Z	A	J	H	P	Q	J	M	H	B	V	A	X	B	A	J	J	L

During the decryption, the cipher text character is selected from the key character row in the table and the corresponding column heading is taken as plaintext.

If English alphabets are represented by numbers with 0 corresponding to A, 1 to B and so on, then Vigenère cipher encryption and decryption can be represented in modular arithmetic as:

Encryption: $C_i = (P_i + K_j) \bmod 26$
Decryption: $P_i = (C_i - K_i) \bmod 26$

where C_i is the ith cipher text character, P_i is the ith plaintext character and k_j is the jth key character.

Vigenère cipher remained secure and unbroken for a longer period. Casanova was the first to break the cipher, but he did not disclose his procedure [7]. Later cryptanalysts attacked Vigenère cipher by trying to get the length of the key—the main weakness of the cipher. Once the length of the key is determined, then it is easy to break the cipher as it makes use of repeated key where each repetition can be viewed as having been encrypted by shift cipher. The first successful attack on Vigenère cipher was published by Friedrich Kasiski and is based on finding repeated patterns of length 2 or 3 in the cipher text [42]. Another attack based on calculation of index of coincidence (IC) is called Friedman test and was published in 1920 by Friedman [43].

3.6 *Impact of Computational Power on Vigenère Cipher*

If the size of the key is t and we try to use brute-force method to guess the key, then we need to search through all the 26^t different possibilities which is infeasible for larger key sizes. But making use of letter frequency analysis for analysis of independent streams, the attack will require time of $26 * t$ instead of 26^t [2].

3.7 The One-Time Pad or Vernam Cipher

In 1917, Gilbert Vernam and Major Joseph O. Mauborgne at AT&T developed a perfectly secret cipher which is the only cipher that is theoretically unbreakable. It is known as Vernam cipher or a one-time pad. The one-time pad uses a random key or pad equivalent to the size of the message for encryption. It makes use of the bitwise operator exclusive-OR ($\oplus$) for encryption and decryption. If k is the key from some key space K, m is the message from some message space M, then encryption and decryption process in one-time pad is given as:

Encryption: $c = m \oplus k$.
Decryption: $m = c \oplus k$.

To apply above encryption and decryption procedures, the message and key must be in the binary form.

In spite of achieving the level of security as defined by Claude Shannon for perfect secrecy, the one-time pad is rarely used due to the following drawbacks:

(a) The length of the key must match the length of the message. Practically, it is not always possible to determine the length of the message in advance so as to generate the key accordingly. It also limits the applicability of one-time pad for encryption of large messages. For large messages, generation and distribution of large keys is a serious problem. This drawback affects practicability of the cipher but not its security.
(b) Another drawback that affects the security of one-time pad is the usage of same key twice. Using the same key twice reveals enough information to break the cipher [2]. Suppose we are encrypting two messages m_1 and m_2 with same key k generating the cipher text c_1 and c_2 so that:

$$c_1 = k \oplus m_1$$
$$c_2 = k \oplus m_2$$

The advisory can perform the following operation to get the difference of the two messages which complemented with frequency analysis can be used to break the cipher:

$$c_1 \oplus c_2 = (k \oplus m_1) \oplus (k \oplus m_2) = m_1 \oplus m_2$$

3.8 Impact of Computational Power on One-Time Pad

It has been proven that one-time pad is perfectly secret [2]. So, no matter how much computational power is available, the adversary will not be able to exactly retrieve the key except under the advantage of second drawback mentioned above.

3.9 Transposition or Permutation Ciphers

The transposition ciphers are another class of cipher that have been used even during early twentieth century. In the 1920s, transposition ciphers were used by the Irish Republican Army [9]. Unlike substitution ciphers, transposition ciphers change position of the plaintext characters for encryption. For example, in case of **rectangular transposition**, plaintext is arranged in rectangular fashion and the cipher text is taken from columns of the rectangle. The message "KASHMIR IS A PARADISE ON THE EARTH" is encrypted as:

KASHMIR
ISAPARA
DISEONT
HEEARTH

Cipher text = KIDH ASIE SASE HPEA MAOR IRNT RATH.

Another transposition cipher called **rail-fence transposition** distributes the text in a number of lines in some zigzag manner where each line from top to bottom is taken as cipher text. Applying rail-fence transposition on our previous message, we get:

K				M				S				R				S				E				H
	A		H		I		I		A		A		A		I		E		N		A		T	
		S				R				P				D				O				R		

Cipher text = KMSRS EHAHI IAAAI ENATS RPDOR.

The diffusion will increase with increasing number of fence levels. There are more transposition methods like keyword-based columnar transposition, double transposition and more. The methodology in each transposition method is to change the position of the text to obtain cipher text. Substitution and transposition are an important part of the modern ciphers to add confusion and diffusion. Diffusion distributes the influence of each plaintext letter over several cipher text letters and is achieved with the help of transposition. Confusion means to make the relationship between the plaintext and cipher text complexes which is achieved with the help of substitution.

3.10 Impact of Computational Power on Transposition Ciphers

The simplest approach to break transposition ciphers is to try out all possible short keywords staring from a size of 2. This approach is called brute-force approach and requires $n!$ computations to break a key of length n. This approach is not feasible for higher values of n as factorial grows very fast for higher values of n. For example,

Table 4 Time to break transposition cipher with 100 letters for keys of different sizes

Key size	10	11	12	13	14	15
Time (s)	6.45	22.55	29.31	50	72	100

Source [45]

the computations required for a key of length 26 are $26! = 4.0329146112661E + 26 \approx 2^{88}$.

A number of other approaches have been used to reduce the brute-force complexity to a feasible range. Genetic algorithms have been used for cryptanalysis of transposition ciphers with remarkable results [44–46] and even for longer key sizes [47]. As per the results of [45], required time to break the cipher with 1000 letters for keys of different sizes is given in Table 4.

The brute-force approach for a key size of 15 will require about 405 s on a machine with core frequency of 3 GHz assuming one search is performed per cycle. So, above results are better than brute-force approach.

In [48], a cuckoo search algorithm has been adopted for automated cryptanalysis of transposition ciphers. One more automated cryptanalysis of transportation cipher using simulated annealing is presented in [49]. Two more approaches using ant evolutionary computation [50] and particle swarm optimization [51] have been used to break the transportation cipher.

4 Modern Cryptography

The concept of perfect secrecy as given by Claude Elwood Shannon (1916–2001) has some limitations. A perfectly secret encryption scheme must use key only once, and the key space must be at least as large as the message space. Moreover, a perfectly secret encryption scheme must not leak any information regarding encrypted message which is not always the case. These limitations decide a shift from classical cryptography to modern cryptography. The modern cryptography uses the approach of computational secrecy instead of perfect secrecy. To achieve practical encryption schemes, computational secrecy allows two relaxations against the perfect secrecy [2]:

1. Security is only guaranteed against efficient adversaries that run for some feasible amount of time.
2. Adversaries can potentially succeed (i.e., security can potentially fail) with some very small probability.

As per first relaxation, modern cryptosystem need not guarantee security for eternal and unbounded time periods but only for feasible amount of time and computational power. Under unlimited time or unbounded computational power, if the security of the cryptosystem is compromised, it does not affect overall security of the cryptosystem. This relaxation is necessary as it would not be possible to achieve a secure cryptosystem with small key size without limiting the running time of the adversary.

As per second relaxation, if the probability of success of an adversary is extremely small, still the cryptosystem will remain secure. This is again necessary to achieve a secure cryptosystem with small key size by reducing the success probability of the adversary.

The pioneering work toward modern cryptography was done by IBM with the development of a practical and secure encryption algorithm called Data Encryption Standard (DES) in the 1970s. Even before that, IBM had delivered its first of a series of mainframe computers called the System/360 in April 1964 to set a remarkable path for computational world. So, era of modern cryptography started at a time when the industry of computational power also kick-started its journey by release of Intel 4004 microprocessor on November 15, 1971. It had a clock speed of 108 KHz (just over one-tenth of megahertz) and contained 2300 transistors. It was followed by release of 8008 processor in 1972 with a clock speed of 200 KHz containing 3500 transistors. Then in April, 1974, Intel released 8080 processor with a clock speed of 2 MHz having 6000 transistors that was used in first personal computer, the Altair 8800 which eventually gave birth to PC revolution. It was a time when two converging forces were shaping the revolution in personal computing. The semiconductor industry, on one side, was striving for ever-more-powerful microprocessors and ever-more-capacious memory chips. The time-sharing systems like PDP-10 or XDS 940 on other side were setting a base for public access to computing [6]. This is the year which is considered as *annus mirabilis* of personal computing.

Over the decades, both the worlds—the world of modern cryptography and the world of modern computing, have gone beyond imagination in progress and utilization, affecting one another both positively and negatively. In security perspective, modern computing is highly depended on cryptography both when information is at rest and when it is in transit. On the other hand, modern cryptography has been largely affected by the progress in computing power. Many secure cryptographic algorithms like DES have lost their security before increasing computing power. This is due to the fact that increasing computational power favors brute-force solutions. On the other hand, many security means like revolutionary blockchain is existing due to enormous computational power.

Let us now discuss some remarkable modern cryptographic algorithms along with the impact of computational power on their security.

4.1 Data Encryption Standard (DES)

We will start with Data Encryption Standard (DES) which served as a standard for decades and is the most probed cryptosystem by cryptographic community. It was developed at IBM in the 1970s. The main creator of DES was **Horst Feistel** who was an employee at IBM. He wanted to name it Dataseal, but was named by IBM as Demonstration Cipher or in short Demon and finally Lucifer [9]. On May 15,

1973, National Bureau of Standards (NBS) issued a call for an open competition for a cipher system for civilian needs in the Federal Register. IBM submitted Lucifer as its solutions which got accepted and adopted in July 15, 1977, as a Federal Information Processing Standard for the USA. As per the competition conditions, Lucifer was renamed as Data Encryption Standard (DES).

DES is a block cipher with a block size of 64 bits and a key length of 56 bits. Actually, the size of the key in DES is 64 bits but 8 bits at positions 8, 16, 24, 32, 40, 48, 56 and 64 are used for error detection. DES is a 16-round Feistel network where each round is a substitution–permutation network consisting of two self-inverse operations as depicted in Fig. 9.

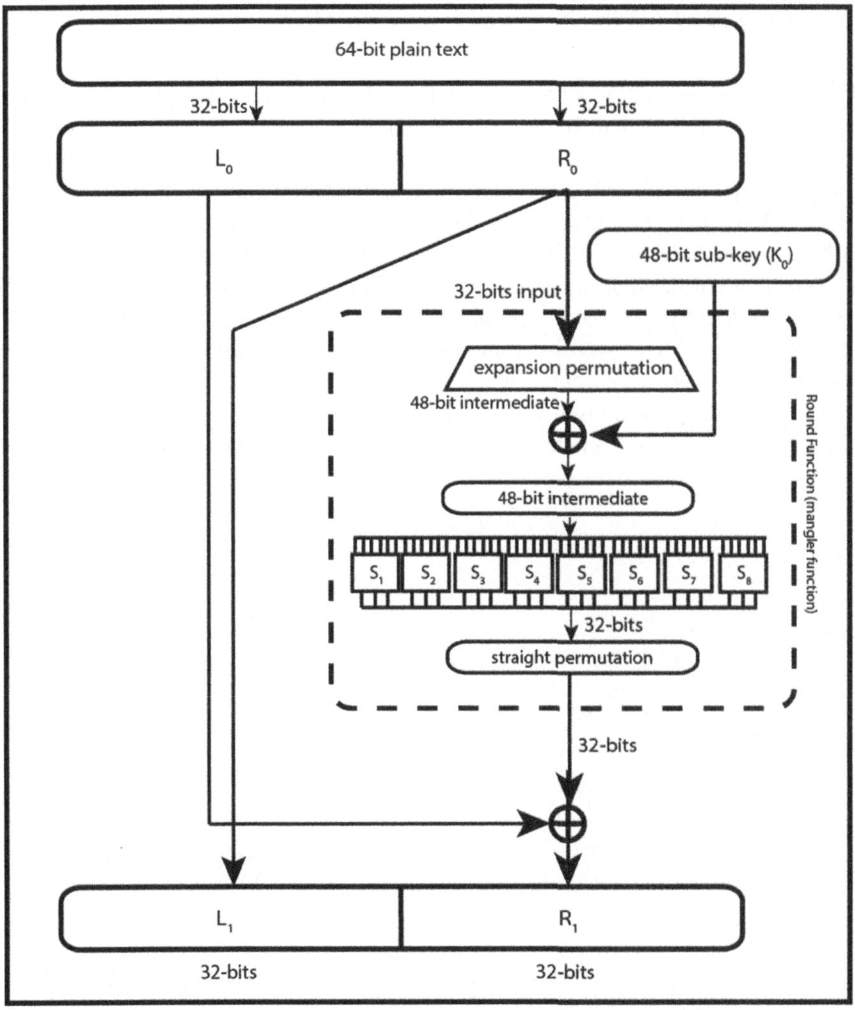

Fig. 9 DES round

The same DES round function (or mangler function) shown in dashed rectangle is applied in each of the 16 rounds. In each round, a 48-bit key generated by the DES schedule from a 56-bit master key is XOR-ed with the 48-bit expansion of the right half of the plaintext. The resulting 48-bit data is divided into 8 blocks of 6 bits each. Each block is then passed to a 6- to 4-bit S-box. Finally, a straight permutation is performed with the 32-bit concatenated result from S-boxes to give a 32-bit output. The 32-bit output from the round function is then XOR-ed with the left-hand 32 bits of the plaintext to obtain the right-hand 32 bits of the round output. The left-hand 32 bits of the round output are directly taken from the right-hand 32 bits of the plaintext.

4.2 *Impact of Computational Power on DES*

The main weakness of the DES is its small key length, and the best-known practical attack on DES is an exhaustive search through its key space. The key space of DES consists of 2^{56} keys (about $7 * 10^{16}$ keys). At the time of adoption of DES, the computational power required to go through all the keys was unavailable that is now feasible. Even in 1977, Diffie and Hellman [10] suggested that the entire key space of DES can be searched in 10^5 s (about a day) by using a parallel machine. It was a suggestion but the first practical brute-force attack on DES was demonstrated in 1977 by DESCHALL project using thousands of computers coordinated across the Internet. This was followed by a number of successful attacks on DES making it possible to break DES in hours (Fig. 10).

The graph shows how brute-force attack on DES once considered computationally infeasible has become a matter of few hours with advancements in computational

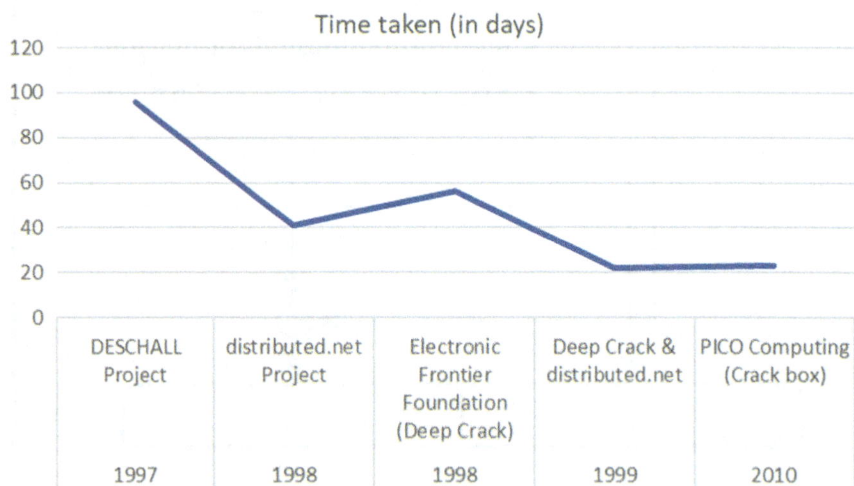

Fig. 10 Computational time of various attacks on DES

power. Even, Biham and Shamir [11] have used a ***differential cryptanalysis***[5] attack on variants of DES with 15 or fewer number of rounds that give better results than brute-force attack. They have shown that a DES reduced to 6 rounds was broken in less than 0.3 s on a personal computer using 240 cipher texts and an eight-round DES was broken in less than 2 min on a computer by analyzing 15,000 cipher texts and chosen from a pool of 50,000 candidate cipher texts. They have confirmed that the DES with 16 rounds is resistant to differential cryptanalysis attack which reveals that DES was designed to be resistant to differential cryptanalysis attack but it was not disclosed to the world.

4.3 *Tripe DES (3DES or FIPS 46–3)*

In order to remove the above-mentioned problems caused in DES due to relatively small key length, a new variant of DES called FIPS[6] 46–3 or 3DES informally was released by National Institute of Standards and Technology (NIST) in 1998 and was standardized in 1999. The original DES was permitted for legacy systems only. 3DES passes the plaintext through three iterations of DES as shown in Fig. 11.

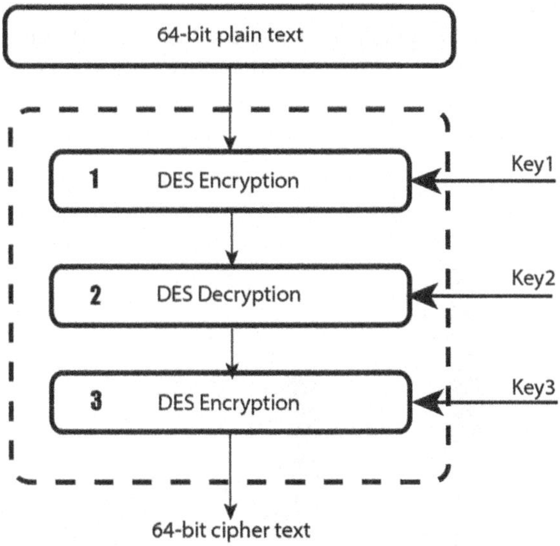

Fig. 11 Triple DES

[5]Differential cryptanalysis is a method which analyzes the effect of particular differences in plaintext pairs on the differences of the resultant cipher text pairs. These differences can be used to assign probabilities to the possible keys and to locate the most probable key.

[6]Federal Information Processing Standards (FIPS) are publicly announced standards developed by the National Institute of Standards and Technology (NIST) for use in computer systems by non-military American government agencies and government contractors.

Depending upon whether 2 or 3 independent keys are used, 3DES has two variants with different key sizes:

I. **3DES with three keys**: In this variant, 3 independent keys of 56 bits each are used giving an effective key length of 168 bits (3 * 56).
II. **3DES with two keys**: In this variant, 2 independent keys of 56 bits each are used giving an effective key length of 112 bits (2 * 56). Same key is used in the first and third iteration.

4.4 Impact of Computational Power on 3DES

The theoretical time complexity for the computational power required for the two variants of 3DES is given below:

$$\text{Variant I: } T(n) = O\left(2^{168}\right)$$
$$\text{Variant II: } T(n) = O\left(2^{112}\right)$$

The practical bounds on time complexity limits are determined by the cryptanalyst attacks. The 3DES with three keys is susceptible to a number of attacks as given in Fig. 12.

From the graph, it is evident that the only attack bound by theoretical time complexity of 2^{168} is the brute-force attack which seems computationally infeasible but not quantum safe. The man-in-the-middle attack requires computational power of 2^{112}. The other attacks on 3DES though require even less computational

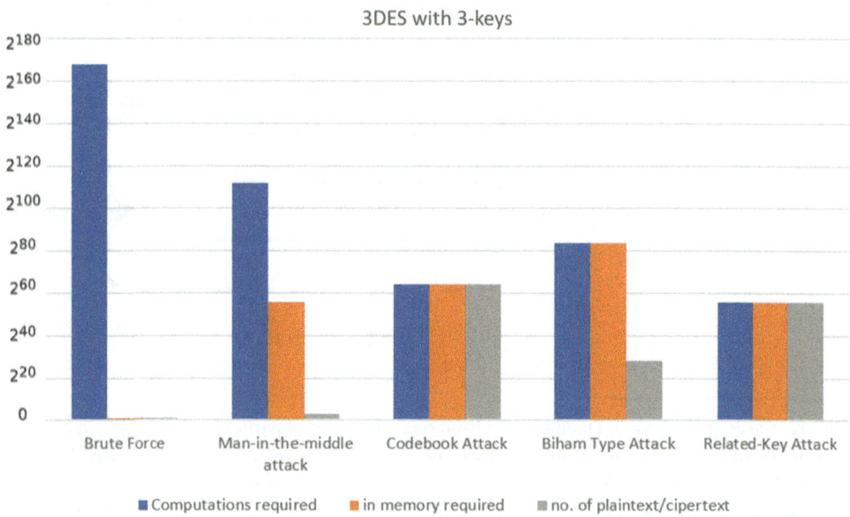

Fig. 12 Various attacks on 3DES with 3 keys. *Source* [12]

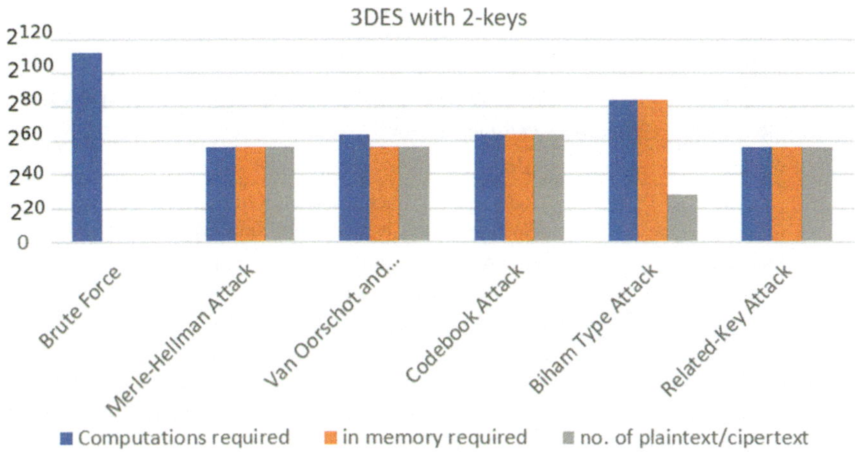

Fig. 13 Various attacks on 3DES with 2 keys [12]

power but are not feasible due to large requirement of plaintext/cipher text pairs and in-memory data. For example, the related-key attack requires a computational power of 2^{56} provided 2^{56} pairs of plaintext/cipher text pairs are available. But it is very difficult for a cryptanalyst to find these many plaintext/cipher text pairs practically. So, this variant does not provide security as expected from it and various cryptanalyst attacks limit its key size to 2^{112} only. Still 3DES with three keys is a good option for smart cards due to the limit capability of cards to encrypt under same key.

The 3DES with two keys is also susceptible to a number of attacks as given in Fig. 13.

Again, the best-known attack against 3DES with two keys is also brute-force attack requiring a computational power of 2^{112}. However, the cryptanalyst attacks again limit the practical computational complexity of this variant of 3DES. Most of these cryptanalyst attacks are unrealistic due to huge requirement of in-memory data and plaintext/cipher text pairs. The Biham-type attacks require a computational power of 2^{84} with 2^{28} plaintext/cipher text pairs. It means the attack can find a key every 2^{28} keys using 2^{84} computations and 2^{84} in-memory data. But the huge demand of memory of the order of 2^{84} is unrealistic and makes this attack practically unfeasible.

In addition to the above-mentioned attacks on 3DES, the following problems also limit its practical use:

a. ***Small block length***: The 3DES has a block length of 64 bits which is relatively small for a practical use. The smaller block length suffers same problem as in case of small key size—susceptible to brute-force or exhaustive search attack. Modern block ciphers have block length of 128 or larger as larger block size generally provides more security but with reduced speed.

Table 5 Finalist of AES competition

S. No.	Algorithm	Developers
1	Rijindal	Belgium cryptographers Vincent Rijmen and Joan Daemen
2	RC6 (Rivest Cipher 6)	RSA Inc, in USA
3	MARS	IBM
4	Twofish	Counterpane (Bruce Schneier, John Kelsey, Doug Whiting, David Wagner, Chris Hall and Niels Ferguson) in USA
5	Serpent	Ross Anderson, Eli Biham and Lars Knudsen an English/Israeli/Danish group

b. ***Reduced Speed***: The encryption and decryption speed of 3DES is relatively slower as compared to DES. It requires three block cipher operations of DES which reduce its speed by a factor of 3.

4.5 AES—Advanced Encryption Standard

The drawbacks of 3DES have led NIST to search for a new secure encryption standard. Like its predecessor, the new encryption standard called Advanced Encryption Standard (AES) was also selected through an open competition announced on January 1997 by the United States National Institute of Standards and Technology (NIST). In response, NIST received a total of 15 algorithms from some of the world's best cryptographers and cryptanalysts and the algorithms were presented at NIST conference in 1998. The algorithms were rigorously analyzed by experts of NIST, public and especially other team members. After passing through two workshops held in 1998 and 1999, the list was refined to the following five finalists in August 1999 as given in Table 5.

All the five algorithms were published, and the international community was challenged for their weakness. Finally, in August 2000, Rijindal was announced as the winner and the next encryption standard. The new data encryption standard called "AES—Advanced Data Encryption Standard (FIPS-197)" was published in November 2001 [52].

AES is a symmetric key round cipher with block size of 128 bits. It has choice of three key sizes of 128, 192 or 256 bits with 10, 12 or 14 rounds, respectively. The basic unit of operation in AES is 4×4 array of bytes, referred as "the state." The state is modified in each round using four simple, fast, invertible and byte-oriented operations. In the last round, only three of these operations are performed. Each round uses a different key derived from the original key. The AES encryption algorithm for a 128-bit key with 10 rounds proceeds as follows:

1. **State Creation**: At the beginning, the 128-bit input data is copied to the 4×4 array of bytes called **state**.

2. **Initial RoundKey Operation**: An initial round key operation is performed by way of XORing the **RoundKey** with the **state**. A **RoundKey** is a 128-bit sub-key derived from the master key and is generated by a round key generator. Like state, it is also interpreted as a 4×4 array of bytes.
3. **Transformations in $n-1$ rounds**: Now for each of the $n-1$ rounds which is 9 in this case, the following four operations are performed:

   I. **SubBytes () Transformation: SubBytes ()** is a substitution operation performed on the state where each byte of the state is replaced by another byte using an S-box. The S-box is invertible and constructed by a multiplicative inverse in the finite field GF (2^8) and an affine transformation over GF (2).

   II. **ShiftRows () transformation**: In the **ShiftRows ()** operation, a cyclic left shift operation of the bytes of the state is performed. The bytes of the first row are not shifted. The bytes of the remaining three rows, i.e., 2nd, 3rd and 4th, are shifted left by one, two and three bytes, respectively. Since the shifts are cyclic, the bytes shifted form the left side are inserted back on the right side. For example, in case of 2nd row, the fourth byte becomes the first byte.

   III. **MixColumns () transformation**: In this step, each column of the state viewed as a polynomial of degree 3 or less is treated by way of a linear transformation which is a matrix multiplication over an appropriate field. The columns are multiplied by the polynomial $c(x) = 3x^3 + x^2 + x + 2$ and then are reduced modulo $x^4 + 1$, in order to express them back as columns.

   IV. **AddRoundKey () transformation**: This is a key-dependent step in which a round key is added to the state by a simple bitwise XOR operations.

4. **Last round operations**: In the last and final round, instead of a MixColumns () transformation, an AddRoundKey () transformation is performed. So, this round has only three operations—SubBytes () transformation, ShiftRows () transformation and AddRoundKey () transformation. The first three steps being key-independent steps are susceptible to inversion by an adversary which is avoided by this key-dependent step.

The overall structure of AES-128 is given in Fig. 14.

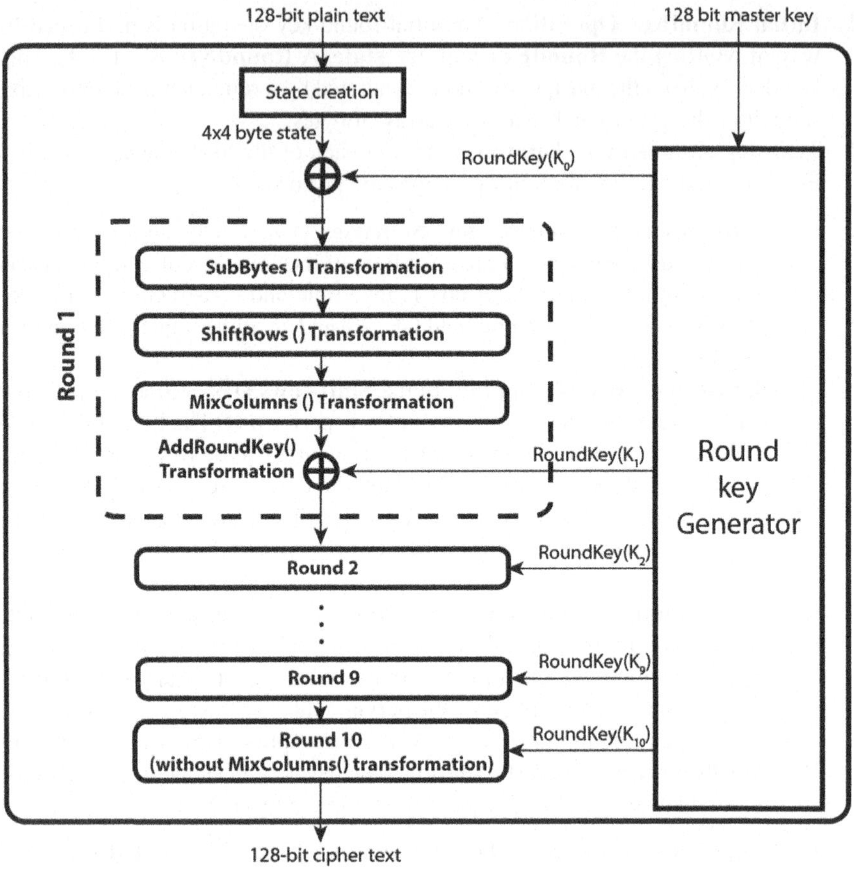

Fig. 14 Structure of AES-128

4.6 Impact of Computational Power on AES

The AES cipher has gone through rigorous and intensive scrutiny by the cryptanalyst community since its inception in 1997. It has proved to be secure against all known attacks. The AES-256 has been shown quantum safe providing a security level of 128 bits on quantum computer [3]. The best-known attack against AES is the brute-force or exhaustive search attack which requires computational power of 2^{128}, 2^{192} and 2^{256} for its three variants AES-128, AES-192 and AES-256, respectively. This much of computational power is currently infeasible. Another popular attack is the biclique attack [13] which has the computational advantage of about a factor 3–5 over brute-force attack, depending on the version. It even yields advantages of up to a factor 15 for the key recovery of the AES versions with smaller but yet secure number of rounds. The results of biclique attack are given in Fig. 15. However, even if the biclique attack reduces the computational power required by brute-force by a factor

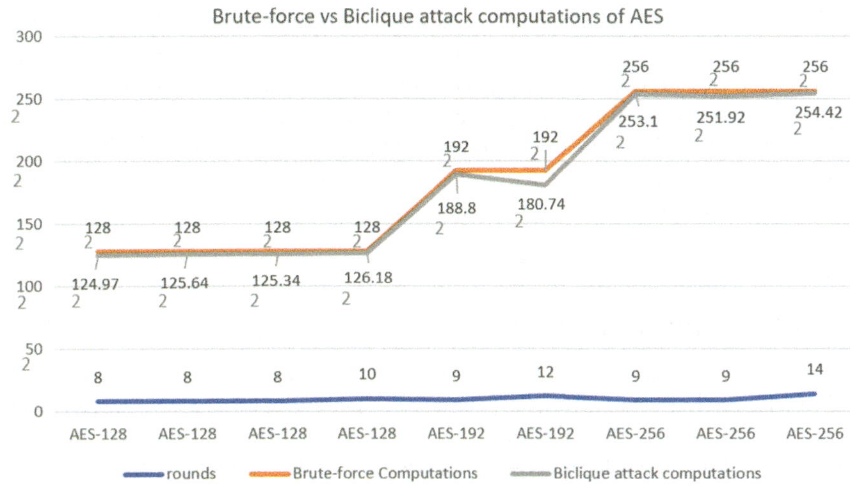

Fig. 15 Brute-force versus biclique attack computations of AES. *Data source* [13]

of 3 to 5 and even 15 in case of reduced versions but the effective computational power required is still so large that it is infeasible on the world's fastest computers. So, it does not affect the security of the AES cipher.

The various other reduced round attacks on different variants of AES are summarized in Table 6. In spite of these attacks with some significant results in terms of computations but none of these cryptanalytic attacks are practically better than brute-force attack for a key.

Table 6 Reduced round attacks on different variants of AES

AES variant	Attack	Rounds	Data	Memory	Computations	Source
AES-256	Related-key boomerang attack	14	2^{123}	2^{38}	2^{176}	[14]
	Impossible differential attack	8	$2^{89.1}$	2^{97}	$2^{229.7}$	[15]
	SQUARE attack	8	$2^{127.997}$	2^{128}	2^{204}	[16]
	Improved single key attack	8	2^{113}	2^{129}	2^{196}	[17]
	Related-key rectangle attack	10	2^{114}	?	2^{173}	[18]
AES-192	Improved single key attack	8	2^{113}	2^{129}	2^{172}	[17]
	SQUARE attack	8	$2^{127.997}$	2^{128}	2^{188}	[16]
	Related-key rectangle attack	10	2^{124}	?	2^{183}	[18]

The main security concern with the AES cipher, as put forth by Bruce Schneier and his coauthor Ferguson [53], is its simple algebraic structure which has already been detailed by Ferguson et al. [19]. They have shown that AES cipher consists of simple closed algebraic formula over the finite field with 256 elements and the whole security of AES cipher relies on solution of these algebraic formulae. Once these algebraic formulae are solved, that will be end of AES cipher. It has not happened yet but there is no guarantee that it cannot happen in the future.

4.7 RSA Public Key Cipher

RSA is a public key cryptographic system that was developed in 1977 by three MIT professors, Rivest et al. [54]. RSA came in response to Diffie and Hellman's groundbreaking work in the world of cryptography with the introduction of public cryptography in 1976 [29]. They proposed the idea of public key cryptography without an algorithm that got its first implementation in the form of RSA public cryptosystem meeting all their requirements. Unlike symmetric key cryptography, public key cryptography uses two different keys—one for encryption called the public key and other for decryption called the private key. The public key is made public so that anyone who wants to share information with the intended recipient may use it to encrypt messages. The private key is kept secret, and only the intended recipient uses it to decrypt the message. This scheme prevents the need for exchange of encryption key which is the main security bottleneck in case of symmetric key cryptography. Here are the steps involved in working of RSA public key cryptosystem:

1. **Key selection**: The first step in RSA algorithm is to select the public and private keys denoted as e and d, respectively. The following steps are used to generate these keys:

   a. Select two large prime numbers p and q.
   b. Calculate the modules n for encryption and decryption as

   $$n = p * q$$

   c. Next calculate the Euler totient function[7] for $n(\varphi(n))$ as

   $$\varphi(n) = (p - 1) * (q - 1)$$

   d. Next choose a random integer e such that $1 < e < \varphi(n)$ and e and $\varphi(n)$ are relatively prime, i.e., gcd $(e, \varphi(n)) = 1$
   e. Next calculate d, the multiplicative inverse (mod $\varphi(n)$) of e such that

[7]Euler totient function ($\varphi(n)$): The Euler totient function $\varphi(n)$ of a number n gives the count of numbers less than n that are relatively prime to n (they have no common factors). And since for a prime number p there are no factors other than p and 1, the $\varphi(p)$ is always $(p - 1)$ numbers.

$$d * e = 1 \mod \varphi(n)$$

f. With these steps, we have calculated the public key pair (*e*, *n*) and private key pair (*d*, *n*).

2. **RSA Encryption**: The following steps are used to encrypt a message *P*:

    a. Since RSA cipher is based on exponentiation in a finite (Galois[8]) field over integers (mod *p*) where *p* is a prime, the message *P* must be converted to numeric form before proceeding with encryption. This step is easy to achieve on computers.
    b. Next compute the cipher text *C* using the public key pair (*e*, *n*) as:

    $$C = P^e (\text{mod } n)$$

    C is the resultant cipher text.

3. **RSA Decryption**: Since *e* and *d* are multiplicative inverses (mod *n*), we can decrypt the message using the private key pair (*d*, *n*) as:

    $$P = C^d (\text{mod } n)$$

    where *P* is the original message retrieved from cipher text *C*.

4.8 Impact of Computational Power on RSA

The security of the RSA cipher is highly dependent on computational power. The RSA cipher is based on the mathematical assumption of prime factorization; i.e., it is easy to compute $n = p * q$ even for very large prime numbers, but it is computationally very difficult to find the prime factors of a very large composite number like *n*. So, it depends on intractability of integer factorization. Till date, there is no known efficient algorithm that can solve this problem but the increasing computational power fills this gap. Boneh and Venkatesan [55] works provide an evidence that breaking low exponent RSA (LE-RSA) may be easier than factoring. But that neither affects the security of RSA nor the intractability of integer factorization. The trail division method is the simplest but hardest way to prime factorize a composite number. It is a brute-force approach to continue dividing the number *n* by prime numbers till we find the smallest prime number *p* that divides *n*. But according to prime number theorem, there are $p/\log(p)$ number of primes up to prime number *p*. So, in case of trial division, we require to check for $p/\log(p)$ number of prime numbers to find the smallest prime divisor *p* of *n*. The computational power required to factor a composite

[8]Galois field GF (*pn*): A Galois field, denoted GF (*pn*), is a finite cyclic group with *pn* elements where for every element *x* in the set, there is an identity element for multiplication i.e., $xy = 1$ for some value *y*.

Table 7 Solved factoring challenges of RSA

RSA challenge	Digits	Date of factorization	Computing time	Method used	Source
RSA-120	120	July 9, 1993	825 MIPS[a] years	Quadratic sieve method	[58]
RSA-130	130	April 10, 1996	1000 MIPS years	Number field sieve (NFS)	[59]
RSA-140	140	February 2, 1999	2000 MIPS years	Number field sieve (NFS)	[60]
RSA-155	155	August 22, 1999	8400 MIPS years	Number field sieve (NFS)	[61]
RSA-160	160	April 1, 2003	?	Number field sieve (NFS)	[62]
RSA-200	2000	May 9, 2005	?	Number field sieve (NFS)	[63]
RSA-704	174	December 3, 2003	?	General number field sieve (GNFS)	[64]
RSA-640	192	November 8, 2005	?	Number field sieve (NFS)	[65]
RSA-768	232	December 12, 2009	?	Number field sieve (NFS)	[66]

[a]MIPS year: it means million instructions per second per year and is a measurement of computational steps for computers. 1 MIPS year is equal to the number of instructions executed during one year of computing at one million instructions per second

number n constructed as a product of two d-digit primes is of the order of 10^d which is infeasible even for moderate values of d [56].

The best-known prime factoring algorithm known till date is the number field sieve (NFS) algorithm [57] which has been successfully used to break some factoring challenges of RSA laboratories as given in Table 7. The time complexity of number field sieve algorithm for a composite number n is as given below:

i. General version of NFS: $T(n) = O\left(e^{1.92*\log(n)^{\frac{1}{3}}(\log\log n)^{\frac{2}{3}}}\right)$

ii. Special version of NFS: $T(n) = O\left(e^{1.52*\log(n)^{\frac{1}{3}}(\log\log n)^{\frac{2}{3}}}\right)$

The results of above table show how various variants of the RSA cipher have been broken by utilizing the computational power. The impact of quantum computing will even be more detrimental on RSA as research has shown that RSA is not quantum safe [3]. Shor's quantum algorithm [28] provides a polynomial time method to break RSA cipher. It requires $O(\log(n))$ computational steps to factorize a large coprime number n. At the same time, it requires approximately $2n$ qubits to factor a number n [67]. According to [68], a 100 MHz quantum computer with ≈10,000 qubits and 2×10^{11} quantum gates will be able to break RSA-2048 in less than an hour as compared to a futuristic theoretical classical computer of 1 P-Hz that will still require over 300 million years to break it. But such kind of quantum computer is still many years away

[69]. On the other hand, existence of quantum computers is no longer a theoretical concept. IBM quantum (https://quantum-computing.ibm.com/) is hosting a fleet of 18 quantum computers on its cloud with Rochester having a maximum qubit count of 53 and quantum volume of 8. Similarly, in October 2019 Google announced to have achieved quantum supremacy by benchmark testing a 54-qubit processor named "Sycamore" comprised of fast, high fidelity quantum logic gates [70]. Sycamore performed the target computation in 200 s that would take about 10,000 years to the world's fastest supercomputer to produce a similar output.

So, increasing computational power of classical computers is posing some threat to the security of RSA cipher but the real threat is the quantum computing era. The first kind of threat can be countered by increasing key size, but there is no solution to counter the second kind of threat which will result in end of RSA cipher.

The increasing key size has also been an issue with RSA cipher. There has been a regular increase in recommended key size of RSA cipher in order to maintain a level of security against increasing computational power and more powerful mathematical algorithms. Comparable to symmetric key ciphers, RSA and other asymmetric key ciphers require larger key size to provide the same level of security as provided by symmetric key ciphers. According to NIST recommendation for key management [71], an RSA key size of 3072 bits is required to achieve a security level of 128 bits as provided by AES-128-bit cipher. The comparable strengths of symmetric and asymmetric ciphers are given in Table 8.

Currently, as per NIST [72] the recommended key length for a secure RSA implementation is 2048 bits. This requirement of large key sizes has been main reason why RSA cipher has not been able to replace symmetric key ciphers like AES over the last 4 decades.

Table 8 Comparable strengths

Bits of security	Symmetric key algorithms	Finite field cryptography (FFC) (e.g., DSA, D-H)	Integer factorization cryptography (IFC) (e.g., RSA)	Elliptic curve cryptography (ECC)(e.g., ECDSA)
80	2TDEA	Public key = 1024 Private key = 160	$k = 1024$	$f = 160–223$
112	3TDEA	Public key = 2048 Private key = 224	$k = 2048$	$f = 224–255$
128	AES-128	Public key = 3072 Private key = 256	$k = 3072$	$f = 256–383$
192	AES-192	Public key = 7680 Private key = 384	$k = 7680$	$f = 384–511$
256	AES-256	Public key = 15,360 Private key = 512	$k = 15{,}360$	$f = 512+$

Source [71]

4.9 Elliptic Curve Cryptography (ECC)

The elliptic curve cryptography (ECC) is a public key cryptosystem introduced independently by two mathematicians Miller [73] and Koblitz [74] and was published in 1985. ECC is based on discrete logarithm problem in elliptic curve groups which is not only hard but till date no known sub-exponential time algorithm is available for it. The elliptic curve groups are a special class of groups consisting of points on an elliptic curve. An elliptic curve is a curve with $p \geq 5$ be a prime, defined by the equation E in the variables x and y of the form:

$$y^2 = x^3 + Ax + B \bmod p$$

where $A, B \in \mathbb{Z}_p$ are constants with the discriminant $4A^3 + 27B^2 \neq 0 \bmod p$ and a distinguished point ∞ called the point at infinity, which acts as an identify. An example of an elliptic curve over the real numbers satisfying the equation $y^2 = x^3 - x + 1$ is given in Fig. 16.

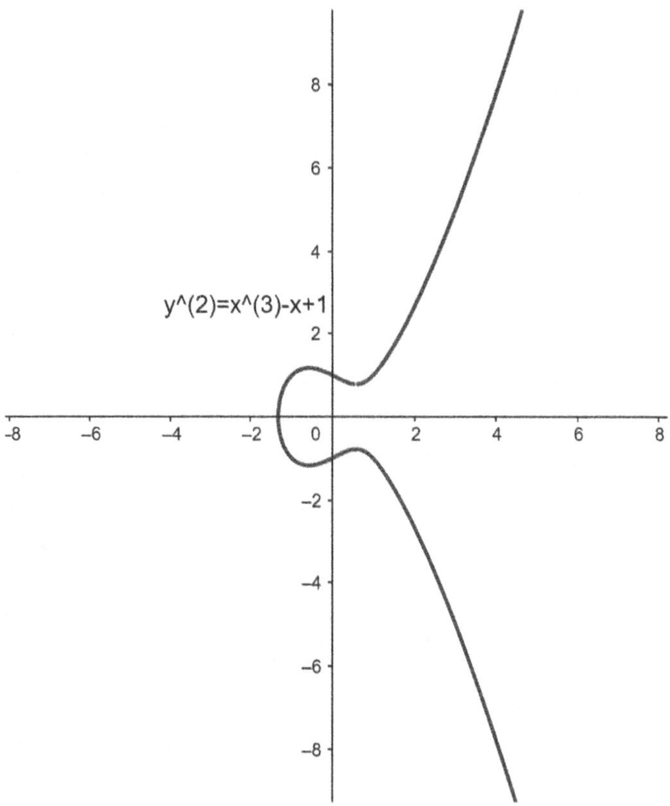

Fig. 16 Elliptic curve graph for $y^2 = x^3 - x + 1$

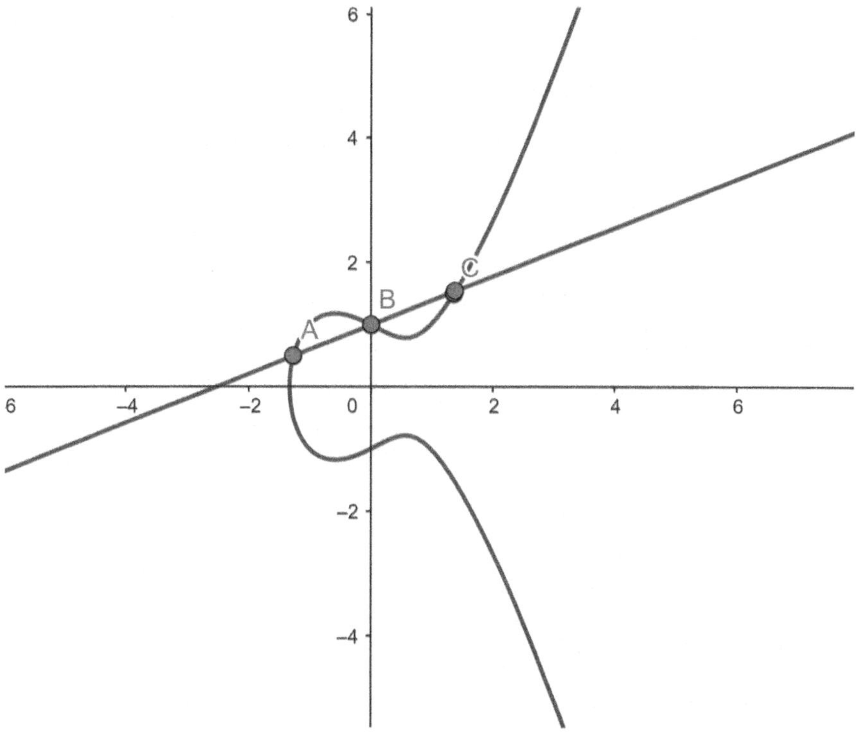

Fig. 17 Line passing through elliptic curve

There are some important and interesting properties of elliptic curves that make them best suitable for cryptography some of which are enlisted below:

i. Symmetry along x-axis, i.e., the portion of the curve below the x-axis, is a mirror image of the portion above x-axis. By choosing any point on the curve above x-axis, there is mirror point below the x-axis.
ii. Non-singular, i.e., there are no cusps or self-intersections in the curve.
iii. A non-vertical line intersects the curve in at most three points as shown in Fig. 17.
iv. A finite Abelian group is formed by the set of points of an elliptic curve defined over a finite field which is always cyclic or product of two cyclic groups [4].
v. Point Addition: The elliptic curves allow to add two points on the curve to produce result in the form of a third point on the same curve. The addition operation is performed as follows. In order to add two points P and Q in the curve, a line is drawn between these two points which according to property #iii intersects the curve at third point, say R. The reflection of the point R along the x-axis, say S, is the addition of P and Q as shown in Fig. 18.
Point addition in a special case where $Q = -P$, i.e., P and Q are vertical, produces the result of infinity ∞ (additive identity), i.e., $P + (-P) = \infty$..

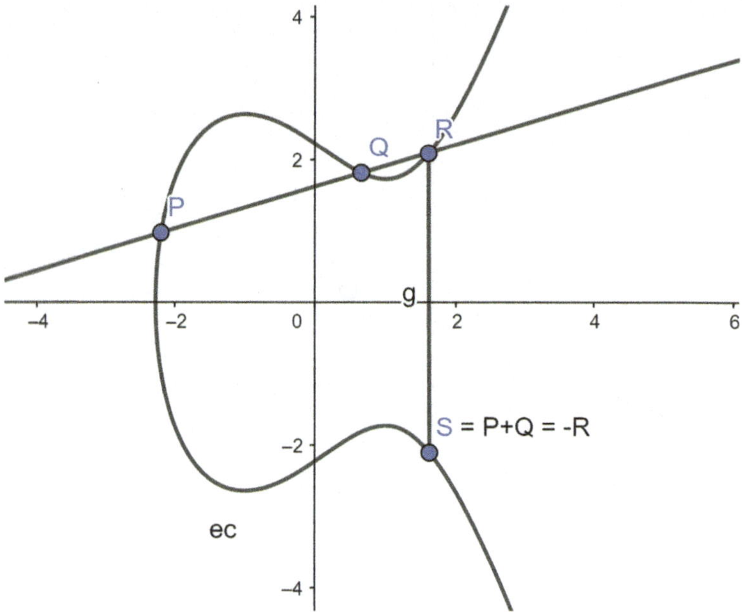

Fig. 18 Elliptic curve point addition

vi. Point doubling and multiplication: The operation of adding a point *P* on the elliptic curve to itself is called point doubling. To double a point *P* on the elliptic curve, a tangent is drawn at point *P* which intersects the curve at other point say *Q*. The reflection of the point *Q* along *x*-axis, i.e., $-Q$, gives the result as shown in Fig. 19.
Point doubling is used to perform the multiplication operation of points on elliptic curves. Especially, point doubling along with addition can be used to perform scalar multiplication, i.e., multiplying a point *P* with some scalar number d to obtain a new point d*P*. By doubling a point *P*, we get 2*P*, i.e., $P + P = 2P$. Now again if we add *P* to 2*P* by drawing a line between *P* and 2*P* which intersects the curve at third point and take its reflection across *x*-axis, we get 3*P*. Continuing this way, we can easily obtain d*P*.

vii. Elliptic curve trapdoor function: From the property #vi, it is obvious that computation of scalar multiplication on elliptic curves is easy to obtain. Given a point *P* on an elliptic curve, it is easy to compute a new point *Q* such that $Q = \mathrm{d}P$, where *d* is some scalar number. But given the result *Q* and starting point *P*, is it possible to find *d*, i.e., the number of steps taken from *P* to obtain *Q*. This problem is called elliptic curve discrete logarithm problem (ECDLP) [23] and is an intractable problem. Till date, there is no polynomial time algorithm to find *d* especially when the underlying field is very large [75]. So, ECDLP is a trapdoor function that is easy to compute in one direction ($Q = \mathrm{d}P$) but very difficult to reverse (finding *d* given *Q* and *P*).

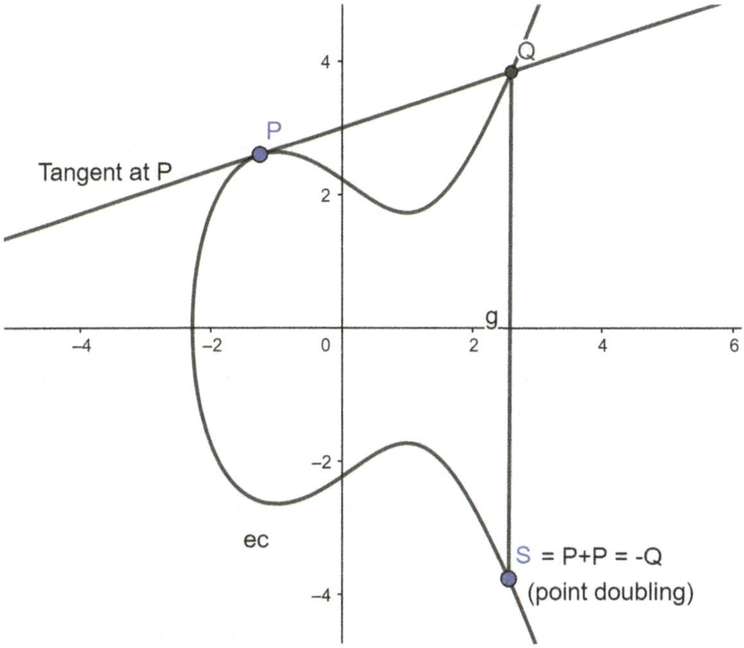

Fig. 19 Elliptic curve point doubling

The properties enlisted above especially the hardness of ECDLP are the reason for using elliptic curves in cryptographic algorithms with a high degree of security utilizing less computational and memory resources. As compared to RSA, ECC considerably uses a smaller key size to provide the same level of security as provided by RSA (Table 9). For example, to achieve a security of 128 bits, the recommended key size of ECC is 256–383 bits as compared to the 3072-bit key required by RSA cipher. This ratio of required key sizes considerably increases with increasing levels of security and improvements in computing power as shown in Table 9.

Table 9 ECC to RSA key size ratio

Bits of security	Integer factorization cryptography (IFC) (e.g., RSA)	Elliptic curve cryptography (ECC) (e.g., ECDSA)	ECC to RSA key size ration
80	$k = 1024$	$f = 160–223$	≈1:6
112	$k = 2048$	$f = 224–255$	≈1:9
128	$k = 3072$	$f = 256–383$	≈1:12
192	$k = 7680$	$f = 384–511$	≈1:20
256	$k = 15,360$	$f = 512+$	≈1:30

The capability to provide higher security with smaller key sizes utilizing least number of CPU, storage and bandwidth resources makes ECC best suitable for less powerful devices like mobile phones, chip cards, electronic commerce and wireless sensor networks. Due to these reasons, elliptic curve cryptography algorithms like the elliptic curve Diffie–Hellman (ECDH) key exchange algorithm and elliptic curve digital signature algorithm (ECDSA) for digital signature have been included by NIST in its Suite B set of recommended algorithms.

A. **Elliptic curve Diffie–Hellman (ECDH) key exchange algorithm**: The steps involved in elliptic curve Diffie–Hellman (ECDH) key exchange algorithm are given below:

   i. **Selection of Elliptic Curve and Public Domain Parameters**: Public domain parameters are agreed upon elements that define the elliptic curve. These are known to the communicating parties for implementation of ECDH key exchange protocol and may even be known to the eavesdropper. The first step is to choose an elliptic curve E defined over the finite field F_p with domain parameters (p, A, B, G, n, h) where:
   p is a prime number called the field parameter,
   A and B are the constants of the elliptic curve $E(F_p)$,
   $G \in E(Fp)$ is the generator of cyclic subgroup called the base point or generator,
   n is the order of G which is the smallest positive number such that $nG = \infty$,
   $h = \frac{|E(Fp)|}{n}$ is the cofactor and must be ≤ 4 and preferably $h = 1$ is considered most secure.

   ii. Sender Key Generation:
      a. Sender picks a random number K_s as the private key where $1 \leq K_s \geq n - 1$.
      b. Sender then calculates the public key $P_s = K_s G$.

   iii. Receiver Key Generation:
      a. Receiver picks a random number K_r as the private key where $1 \leq K_r \geq n - 1$.
      b. Receiver then calculates the public key $P_r = K_r G$.

   iv. **Exchange of Public Keys**: The sender and receiver then exchange the public keys P_s and P_r over the public channel. Even if the eavesdropper is able to intercept the public keys, he/she will not be able to recover the private keys K_s and K_r due to elliptic curve discrete logarithm problem (ECDLP).

   v. Calculation of Secret Key:
      a. The sender calculates the secret key (K) from the private key K_s and public key of receiver (P_r) as: $K = K_s P_r$.
      b. Similarly, receiver calculates the secret key (K) from the private key K_r and public key of sender (P_s) as:

$$K = K_r P_s.$$

The secret key generated by sender and the receiver is equal as shown below:
Sender: $K = K_s P_r = K_s\ (K_r G)$.
Receiver: $K = K_r P_s = K_r\ (K_s G)$.
Since points on an elliptic curve form an Abelian group, we have

$$K_s(K_r G) = K_r(K_s G) \quad \{\text{commutative property})$$

So, using elliptic curve Diffie–Hellman (ECDH) key exchange protocol, a common secret key (K) has been exchanged between the sender and receiver.

B. **Elliptic Curve Encryption Scheme**: The steps involved in encryption and decryption of a message from sender to the receiver are given below:

i. **Encryption**: Suppose m is the plaintext message and P_r is the public key of the intended receiver, then elliptic curve encryption scheme uses the following steps:
   a. The sender represents the plaintext message m as a point M on the elliptic curve.
   b. The sender selects a random positive integer k and calculates the cipher text points C_1 and C_2 as:
      $C_1 = kG$, where G is the generator.
      $C_2 = M + kP_r$.
   c. The sender sends the cipher text points (C_1, C_2) to the intended recipient.

ii **Decryption**: The receiver uses the following steps to decrypt the cipher text points (C_1, C_2) using his/her private key d:
   a. Compute:

$$m_1 = dC_1 = d(kG) = k(dG) = kP_r$$

   b. Compute the plaintext message point M from C_2 and m_1 as:

$$M = C2 - m_1 = M + kP_r - k\,\mathrm{Pr}$$

   c. Extract the original plaintext message m from M.

4.10 *Impact of Computational Power on Elliptic Curve Cryptography (ECC)*

Like the RSA cipher, the security of the ECC is also highly dependent on advancements in computational power. The security of ECC lies in the mathematical assumption of intractability of elliptic curve discrete logarithm problem (ECDLP) [76]. A number of approaches and algorithms have been used to attack ECDLP as summarized in Table 10.

Table 10 Attacks on ECDLP

S. No.	Algorithm/method	Complexity for ECDLP of order n	Source
01	Brute-force or Naïve exhaustive search	$O(n)$	
02	Baby-step giant-step (BSGS) algorithm	$O(\sqrt{n})$	[77, 78]
03	Textbook baby-step giant-step (BSGS) algorithm	Worst case: $O(2\sqrt{n})$ Average case: $O(\frac{3}{2}(\sqrt{n}))$	[79]
04	Pollard's variant of baby-step giant-step algorithm	Average case: $O(\frac{4}{3}(\sqrt{n}))$	[79, 80]
05	Two grumpy giants and a baby algorithm (variant of BSGS algorithm)	$O\left(\sqrt{\frac{n}{2}}\right)$	[79, 81]
06	Pollard's rho algorithm	$O\left(\sqrt{\frac{\pi n}{2}}\right)$(requires negligible storage)	[25]
07	Parallel Pollards rho algorithm	$O\left(\frac{\sqrt{\pi n}}{2M}\right)$ When distributed on M parallel processors	[82]
08	Pohlig–Hellman algorithm	Worst case: $O(\sqrt{n})$	[83]

Till date, there is no sub-exponential or polynomial time algorithm to solve ECDLP and the fastest known algorithm to solve ECDLP is the parallel Pollard's rho algorithm [25]. Pollard's rho algorithm is an efficient algorithm with very small storage and memory requirements. Since the algorithm can easily be parallelized on multiple processors, it can take advantage of technological enhancements in parallel computing like GPUs, clustering of standard CPUs and field-programmable gate array (FPGA)-based computing. According to Van Oorschot and Wiener [82], the speedup achieved by a parallel Pollard's rho algorithm to solve ECDLP on M parallel processors is by a factor of Chavan et al. [75] have implemented this algorithm on 256 octa-core computers having PFLOPs capacity, and the implementation results for ECDLP problems based on prime fields up to 85 bits are given in Table 11.

The results show that the computational time increases by a factor of 12 when the prime field size changes from 80 to 85. A series of challenges were published

Table 11 Implementation results of Parallel Pollard's rho algorithm to solve ECDLP

Prime field size (bits)	Total processes	Time (in sec)
70	4 * 8 = 32	4330.040
75	16 * 8 = 128	7225.267
80	16 * 8 = 128	15,880.519
85	16 * 8 = 128	190,189.061

Source [75]

Table 12 Simulation results

$\log_2(p)$ bits	#Qubits	#Toffoli gates	Toffoli depth	Simulation time (s)
110	1014	$9.44 * 10^9$	$8.66 * 10^9$	273
160	1466	$2.97 * 10^{10}$	$2.73 * 10^{10}$	711
192	1754	$5.30 * 10^{10}$	$4.86 * 10^{10}$	1149
224	2042	$8.43 * 10^{10}$	$7.73 * 10^{10}$	1881
256	2330	$1.26 * 10^{11}$	$1.16 * 10^{11}$	3848
384	3484	$4.52 * 10^{11}$	$4.15 * 10^{11}$	17,003
521	4719	$1.14 * 10^{12}$	$1.05 * 10^{12}$	42,888

Source [89]

by Certicom [84] in 1997 for the cryptographic community's understanding and appreciation of the difficulty of the ECDLP, and an update was issued in 2009 [85]. It took 549 days computational power to solve ECCp-109 challenge on a cluster of 10,000 machines [86] and 6 months of computational power to solve a 112-bit ECDLP problem using 200 PlayStation 3 [87]. So, extremely large computational power is required to solve ECDLP problems based on large underlying finite fields but there is no guarantee that such computational power will not be available in the near future. Instead, quantum computers have already made their physical existence as discussed previously. Like RSA cipher, ECC also cannot compete in the quantum computing era and research has shown that ECC is not quantum safe [3]. Research shows that even a smaller quantum computer with about 1000 qubits will be able to break a 160-bit elliptic curve cryptographic key using Shor's efficient quantum algorithm [88]. They even argued that it is easier to compute elliptic curve discrete logarithms than factoring RSA moduli at comparable classical security levels. A precise estimate of quantum resources for Shor's algorithm to compute discrete logarithms on elliptic curves over prime fields is given by Roetteler, M., et al. [89]. Their simulated results are based on quantum computing software tool suite LIQU$i|\rangle$[9] [90] that can simulate roughly 32 qubits in 32 GB RAM. As per their results to fully implement Shor's algorithm to solve an ECDLP for an elliptic curve defined over a field of prime order with n bits, overall number of Toffoli gates required is $(448 \log_2(n) + 4090)\, n^3$. The simulation results of their work for ECDLP in $E(F_p)$ on HP ProLiant DL580 Gen8 machine consisting of 4 Intel Xeon processors @ 2.20 GHz and 3 TB of memory are given in Table 12.

So, as per these results, NIST standardized curves P-224, P-256, P-384 and P-521 require 2042, 2330, 3484 and 4719 logical qubits, respectively. The deviation from these simulated logical qubit results will only be revealed when actual quantum computers with these strengths will arrive and the race to achieve quantum

[9]LIQUi|> LIQUi|> means Language-Integrated Quantum Operations and is a modular simulator and software architecture with F# as the host language, designed to control quantum hardware. It has embedded, domain-specific language designed for programming quantum algorithms and enables easy programming, compilation and simulation of quantum algorithms and circuits, and is independent of a specific quantum architecture.

supremacy has already started [70, 91, 92]. Neven's law [93] states that quantum computers are improving at a "doubly exponential rate" indicating that the impact of quantum computing on cryptography will be even worse. Their capability to solve compute-intensive problems in seconds which would otherwise take years of classical computing resources would result in failure of most widely used asymmetric cryptosystem.

5 Conclusion

In this Chapter, we have discussed and analyzed the impact of computational power on cryptography. The impact of computational power on many famous classical and modern cryptosystems has been discussed and analyzed in this Chapter. The impact of quantum computing on cryptography has also been discussed in this Chapter. The world of cryptography and computational power seems independent of one another but is practically highly dependent on one another. Advancements in one field have a profound impact on another field. The ever-increasing computational power made different technologies available at the door step of every individual resulting in high dependency on these technologies in every sphere of life. Due to advancements in computational technologies like GPUs and CUDA, desktop and laptop users are also enjoying the supercomputing technologies. Because of these fast advancements in technology, currently there are about 4.57 billion active Internet users and about 60% of the world population is already online. Almost entire population of the world lives within the reach of mobile network. Security is the key factor among all the factors on which adoption of a new technology depends. It is the world of cryptography that has enabled wide spread adaptation of these new technologies by way of providing high degree of security. Thus, a high degree of bond connects the world of cryptography with the world of computational power. This is one side of the coin. The advancements in computational technologies and the increasing computational power have highly impacted the cryptographic world both positively and negatively. By utilizing the enormous computational power, robust and secure cryptosystems can be developed. Blockchain technology is one such example where extraordinary security is achieved by utilizing the computational power of miners. On the other hand, the increasing computational power has invalidated many modern symmetric and asymmetric ciphers. The brute-force attack on Data Encryption Standard (DES) which was once considered computationally infeasible has become a matter of few hours. The theoretical time complexities of many reduced round variants of AES cipher have also been affected by way of various cryptanalyst attacks. Many variants of RSA cipher have been broken by attacking its underlying mathematical assumption of intractability of prime factorization. It has been possible only because of enormous computational power. To maintain a level of security against increasing computational power, the recommended key size of RSA cipher is regularly increased. Currently, NIST recommends a key size of 2048 bits for implementation of RSA cipher. Elliptic curve cryptography (ECC) has also been affected by the increasing computational power. The implementation of parallel Pollard's rho algorithm to solve elliptic curve discrete logarithm problem (ECDLP) on parallel processor has been utilized to break many small key size variants of ECC cipher. The recent advancements in quantum computing have put even greater threat to many famous and widely used cryptosystems. Research shows that RSA, ECDSA, ECDH and DSA are no longer safe under quantum era. AES cipher requires larger key sizes to maintain security in quantum era. It has given birth to the field of quantum safe cryptography—a new and challenging field of cryptography.

So, to conclude growing computational power is always desirable but not in case of cryptographic world where a number of most widely used cryptographic schemes rely on computational and mathematical assumptions which are facing threat due to growing computational power.

References

1. Kessler GC (1998) An overview of cryptography. In: The handbook on local area networks. Auerbach
2. Katz J, Lindell Y (2014) Introduction to modern cryptography. CRC Press
3. Mavroeidis V et al (2018) The impact of quantum computing on present cryptography. arXiv preprint arXiv:1804.00200
4. Dooley JF (2018) History of cryptography and cryptanalysis. Springer
5. Czarnul P (2018) Parallel programming for modern high performance computing systems. CRC Press
6. Ceruzzi PE (2003) A history of modern computing. MIT Press
7. Kahn D (1996) The codebreakers: the comprehensive history of secret communication from ancient times to the internet. Simon and Schuster
8. Bauer FL (2002) Decrypted secrets: methods and maxims of cryptology. Springer Science & Business Media
9. Bauer CP (2013) Secret history: the story of cryptology. CRC Press
10. Diffie W, Hellman M (1977) Exhaustive cryptanalysis of the NBS data encryption standard. IEEE Computer
11. Biham E, Shamir A (1991) Differential cryptanalysis of DES-like cryptosystems. J Cryptol 4(1):3–72
12. Nachef V, Patarin J, Volte E (2017) Feistel ciphers. Springer
13. Bogdanov A, Khovratovich D, Rechberger C (2011) Biclique cryptanalysis of the full AES. In: International conference on the theory and application of cryptology and information security. Springer
14. Biryukov A, Khovratovich D (2009) Related-key cryptanalysis of the full AES-192 and AES-256. Springer, Berlin
15. Lu J et al (2008) New impossible differential attacks on AES. Springer, Berlin
16. Ferguson N et al (2001) Improved cryptanalysis of Rijndael. Springer, Berlin
17. Dunkelman O, Keller N, Shamir A (2010) Improved single-key attacks on 8-round AES-192 and AES-256. In: Advances in cryptology—ASIACRYPT 2010. Springer, Berlin
18. Kim J, Hong S, Preneel B (2007) Related-key rectangle attacks on reduced AES-192 and AES-256, vol 4593, pp 225–241
19. Ferguson N, Schroeppel R, Whiting D (2001) A simple algebraic representation of Rijndael. In: Selected areas in cryptography. Springer, Berlin
20. Salah I, Abdullah D, Oqeili S (2006) Mathematical attacks on RSA cryptosystem. J Comput Sci 2(8):665–671

21. Hinek MJ (2009) Cryptanalysis of RSA and its variants. CRC Press
22. Washington LC (2008) Elliptic curves: number theory and cryptography. CRC Press
23. Hankerson D, Menezes AJ, Vanstone S (2006) Guide to elliptic curve cryptography. Springer Science & Business Media
24. Mahto D, Yadav DK (2017) RSA and ECC: a comparative analysis. Int J Appl Eng Res 12(19):9053–9061
25. Pollard JM (1978) Monte Carlo methods for index computation. Math Comput 32(143):918–924
26. Hosoyamada A, Sasaki Y (2018) Cryptanalysis against symmetric-key schemes with online classical queries and offline quantum computations. In: Cryptographers' track at the RSA conference. Springer
27. Chen L et al (2016) Report on post-quantum cryptography, vol 12. US Department of Commerce, National Institute of Standards and Technology
28. Shor PW (1999) Polynomial-time algorithms for prime factorization and discrete logarithms on a quantum computer. SIAM Rev 41(2):303–332
29. Diffie W, Hellman M (1976) New directions in cryptography. IEEE Trans Inf Theory 22(6):644–654
30. Kirsch Z, Chow M (2015) Quantum computing: the risk to existing encryption methods. Retrieved from: https://www.cs.tufts.edu/comp/116/archive/fall2015/zkirsch.pdf
31. Hathaway L (2003) National policy on the use of the advanced encryption standard (AES) to protect national security systems and national security information, vol 23. National Security Agency
32. Dong J et al (2017) Utilizing the double-precision floating-point computing power of GPUs for RSA acceleration. Secur Commun Netw 2017:1–15
33. Cheng W et al (2017) High-performance symmetric cryptography server with GPU acceleration. In: International conference on information and communications security. Springer
34. Zhang X et al (2019) A high-performance parallel computation hardware architecture in ASIC of SHA-256 hash. In: 2019 21st international conference on advanced communication technology (ICACT). IEEE
35. Nakamoto S, Bitcoin A (2008) A peer-to-peer electronic cash system. Bitcoin. https://bitcoin.org/bitcoin.pdf
36. Ali A et al (2019) Blockchain and the future of the internet: a comprehensive review. arXiv preprint arXiv:1904.00733
37. Zheng Z et al (2018) Blockchain challenges and opportunities: a survey. Int J Web Grid Serv 14(4):352–375
38. Scherer M (2017) Performance and scalability of blockchain networks and smart contracts
39. Bonneau J et al (2015) SoK: research perspectives and challenges for Bitcoin and cryptocurrencies. In: 2015 IEEE symposium on security and privacy
40. De Vries A (2018) Bitcoin's growing energy problem. Joule 2(5):801–805
41. Hazari SS, Mahmoud QH (2019) A parallel proof of work to improve transaction speed and scalability in blockchain systems. In: 2019 IEEE 9th annual computing and communication workshop and conference (CCWC). IEEE
42. Kasiski FW, Die Geheimschriften und die Dechiffrir-Kunst (Cryptograms and the art of deciphering): mit besonderer Berücksichtigung der deutschen und der französischen Sprache. 1863: ES Mittler und sohn
43. Friedman WF (1987) The index of coincidence and its applications in cryptanalysis. Aegean Park Press CA
44. Matthews RA (1993) The use of genetic algorithms in cryptanalysis. Cryptologia 17(2):187–201
45. Toemeh R, Arumugam S (2007) Breaking transposition cipher with genetic algorithm. Elektroni Elektrotech 79(7):75–78
46. Song J et al (2008) Cryptanalysis of transposition cipher using simulated annealing genetic algorithm. In: International symposium on intelligence computation and applications. Springer

47. Heydari M, Shabgahi GL, Heydari MM (2013) Cryptanalysis of transposition ciphers with long key lengths using an improved genetic algorithm. World Appl Sci J 21(8):1194–1199
48. Heydari M, Senejani MN (2014) Automated cryptanalysis of transposition ciphers using cuckoo search algorithm. Int J Comput Sci Mob Comput 3(1):140–149
49. Giddy J, Safavi-Naini R (1994) Automated cryptanalysis of transposition ciphers. Comput J 37(5):429–436
50. Russell MD, Clark JA, Stepney S (2003) Making the most of two heuristics: breaking transposition ciphers with ants. In: The 2003 congress on evolutionary computation, 2003. CEC'03. IEEE
51. Hameed SM, Hmood DN (2010) Particles swarm optimization for the cryptanalysis of transposition cipher. Al-Nahrain J Sci 13(4):211–215
52. Standard N-F (2001) Announcing the advanced encryption standard (AES). Fed Inf Process Stand Publ 197(1–51):3–3
53. Ferguson N, Schneier B (2003) Practical cryptography. Wiley
54. Rivest RL, Shamir A, Adleman L (1978) A method for obtaining digital signatures and public-key cryptosystems. Commun ACM 21(2):120–126
55. Boneh D, Venkatesan R (1998) Breaking RSA may not be equivalent to factoring. In: International conference on the theory and applications of cryptographic techniques. Springer
56. Lenstra AK (2004) Key length. Contribution to the handbook of information security
57. Lenstra AK et al (1990) The number field sieve. In: Proceedings of the twenty-second annual ACM symposium on theory of computing (1990)
58. Denny T et al (1996) On the factorization of RSA-120
59. Cowie J et al (1996) A world wide number field sieve factoring record: on to 512 bits, pp 382–394
60. Cavallar S et al (1999) Factorization of RSA-140 using the number field sieve. In: International conference on the theory and application of cryptology and information security. Springer
61. Cavallar S et al (2000) Factorization of a 512-bit RSA modulus, pp 1–18
62. Bahr F et al (2003) RSA-160, e-mail announcement
63. Bahr F et al (2005) Factorization of RSA-200. Public announcement, 9 May 2005
64. Bai S et al (2012) Factorisation of RSA-704 with CADO-NFS
65. Weisstein E (2005) RSA-640 factored. Math World Headline News
66. Kleinjung T et al (2010) Factorization of a 768-bit RSA modulus
67. Gidney C (2017) Factoring with $n + 2$ clean qubits and $n - 1$ dirty qubits. arXiv preprint arXiv: 1706.07884
68. Lanzagorta, M., Uhlmann J (2009) Quantum computer science. Morgan & Claypool Publishers
69. Cho A (2018) DOE pushes for useful quantum computing. American Association for the Advancement of Science
70. Arute F et al (2019) Quantum supremacy using a programmable superconducting processor. Nature 574(7779):505–510
71. Barker E et al (2006) Recommendation for key management: part 1: general. National Institute of Standards and Technology, Technology Administration
72. Barker E et al (2009) Recommendation for key management part 3: application-specific key management guidance. NIST Spec Publ 800:57
73. Miller VS (1986) Use of elliptic curves in cryptography. Springer, Berlin
74. Koblitz N (1987) Elliptic curve cryptosystems. Math Comput 48(177):203–209
75. Chavan K, Gupta I, Kulkarni D (2016) A review on solving ECDLP over large finite field using parallel Pollard's rho (ρ) method. IOSR J Comput Eng 18(2):1–11
76. Blake IF, Seroussi G, Smart NP (2005) Advances in elliptic curve cryptography, vol 317. Cambridge University Press
77. Shanks D (1973) Five number-theoretic algorithms. In: Proceedings of the second manitoba conference on numerical mathematics. Winnipeg
78. Nechaev VI (1994) Complexity of a determinate algorithm for the discrete logarithm. Math Notes 55(2):165–172

79. Galbraith SD, Wang P, Zhang F (2017) Computing elliptic curve discrete logarithms with improved baby-step giant-step algorithm. Adv Math Commun 11(3):453
80. Pollard JM (2000) Kangaroos, monopoly and discrete logarithms. J Cryptol 13(4):437–447
81. Bernstein D, Lange T (2013) Two grumpy giants and a baby. Open Book Ser 1(1):87–111
82. Van Oorschot PC, Wiener MJ (1999) Parallel collision search with cryptanalytic applications. J Cryptol 12(1):1–28
83. Pohlig S, Hellman M (1978) An improved algorithm for computing logarithms over GF (p) and its cryptographic significance (Corresp.). IEEE Trans Inf Theory 24(1):106–110
84. Certicom E (1997) Challenge, Nov 1997
85. Certicom E, Challenge, 10 Nov 2009 Update
86. Monico C (2002) The ECCp-109 challenge. Technical report. University of Notre Dame
87. Kaihara ME et al (2012) Solving a 112-bit prime Elliptic curve discrete logarithm problem on game consoles using sloppy reduction. Joppe W. Bos.
88. Proos J, Zalka C (2003) Shor's discrete logarithm quantum algorithm for elliptic curves. arXiv preprint quant-ph/0301141
89. Roetteler M et al (2017) Quantum resource estimates for computing elliptic curve discrete logarithms. In: International conference on the theory and application of cryptology and information security. Springer
90. Wecker D, Svore KM, LIQUi|>: a software design architecture and domain-specific language for quantum computing. arXiv preprint arXiv:1402.4467
91. Villalonga B et al (2020) Establishing the quantum supremacy frontier with a 281 pflop/s simulation. Quant Sci Technol 5(3):034003
92. National Academies of Sciences, Engineering, and Medicine (2019) Quantum computing: progress and prospects. National Academies Press
93. Hartnett K (2019) A new law to describe quantum computing's rise? Retrieved from: https://www.quantamagazine.org/does-nevens-law-describe-quantum-computings-rise-20190618

Chapter 4
Privacy Preserving Anti-forensic Techniques

Mohamad Ahtisham Wani

1 Introduction

In this age of digital technology, huge amount of data is generated every second by digital devices and user applications. Most of this data is real-time, pertains to user activity, is stored across multiple remote servers for easy accessibility and contains sensitive information that defines the confidential aspects of an individual or an organization, otherwise deemed classified by its owner. Confidentiality and prevention of unauthorized access to such data is imperative as it may result in manipulation of individual behaviour; loss of trade secrets, intellectual as well as physical property; extortion and targeted espionage [1, 2]. This has lead to major privacy concerns among individuals and corporate organizations as users are increasingly loosing control over authority of information they generate [3]. This can primarily be attributed to the recent boom in information technology (IT) [4] and emergence of tech companies that provide services to its users without any tangible or apparent costs. These include services like social networking platforms, online shopping catalogues, cloud storage and computing, forums, blogs, social securities, bio-metric systems and so on. To use these services users are required to generate a *username* and *password* combination by going through a detailed registration process, and involves submitting of personal details and disclosure of other private information. As a result of this, enormous amount of user data is accumulated on remote servers that is accessible via the internet and hence, vulnerable to breach. Although, sophisticated and state-of-art security measures are in place [5], breaches are inevitable and their possibility can not be ignored. Besides, most of the intruders are professional criminals and hackers motivated by financial fraud and targeted espionage. On the other hand, online access and sharing of data leave data remnants [6] on multiple devices that may not be secured, are difficult to track and may lead to serious consequences in

M. A. Wani (✉)
University of Kashmir, Srinagar, India
e-mail: ahtishamwani@gmail.com

K. J. Giri et al. (eds.), *Multimedia Security*, Algorithms for Intelligent Systems,
https://doi.org/10.1007/978-981-15-8711-5_5

wrong hands. This raises major privacy concerns as user data is transmitted over the internet and scattered over multiple remote servers and digital devices without user's knowledge [7] and authorization.

Privacy ensures that data or information must only be accessed by authorized users [8]. This includes data that is stored on any digital device or server, data that is deleted and data that is transmitted over the network. Thus privacy is a multi-dimensional domain that is not restricted to data that exists on our personal digital devices but also encompasses data that is being transmitted over a network and stored on remote servers. Recovery of deleted data also comes under the realm of privacy. Deleted data is the sole property of its user and must be protected even if its disclosure does not leads to serious security breaches and theft of intellectual or confidential information.

Privacy is not only threatened by criminals and professional hackers but also by organizations be it government or private responsible for its security. IT industries like social networking platforms, email servers and so on that provide free services to its users often sold user data to third-party organizations, government and intelligence agencies [9, 10]. Sophisticated data mining software are used to mine user data and depict user behaviour at an individual level [11]. The mined data is then used for targeted advertisement of products and services, shape public opinion or even completely sold to companies, intelligence agencies and various IT cells for millions with impunity [9]. This is because government agencies interested in intelligence, mass surveillance often collaborate and acquire services of IT industries to track suspects, monitor crime-scenes and influence public opinion to achieve their objectives [12].

On the other hand, forensic investigations can lead to serious privacy breaches if not handled properly and meticulously. Digital devices are vulnerable to breach [13] and forensic tools are tailored to identify and recover data from these devices during digital investigation. However, most of the acquired evidence is usually deleted or modified data [14] and contains personal or confidential information which may or may not be relevant to the crime-scene [15]. Although, forensic tools can extract every file that ever existed on a digital device, they are inherently incompetent to differentiate between relevant and irrelevant data with respect to the crime-scene [16]. Besides, these tools are also able to correlate evidence extracted from a digital device which may result in a dreadful invasion of one's privacy [17]. This puts digital forensics in direct conflict with privacy which defines personal data as private and sensitive information that must only be acquired after due authorization by its owner [16].

Fortunately, anti-forensics which is mainly used to counter digital forensics [18] and has recently been recognized as a legitimate field of study promises great value when it comes to data security and privacy. Anti-forensics aims to hide, conceal or alter data in such a way that its recovery is almost impossible except by its owner [19–21]. Anti-forensic techniques are so much evolved and proficient in concealing data that they have taken sophisticated digital forensic tools head-on [22] and are expected to end the golden age of digital forensics [23], making their potential for ensuring data security and privacy quite clear. Furthermore, Anti-forensics is not only restricted to secure data storage but also include techniques for secure transmission and deletion be it on smartphones or enterprise servers [24–26]. Surprisingly, anti-forensics has never been pursued in the context of privacy and no work exists that presents anti-forensics from a privacy perspective. This chapter aims to create awareness about the privacy aspect of anti-forensics by identifying various anti-forensic techniques available [27–29] and discussing their potential for data security and privacy.

The rest of this article is organized as follows. Section 2 presents the work related to privacy and anti-forensics. Section 3 discusses various anti-forensic techniques and measures available by explaining their working and usage mechanism, and their potential for data security and privacy. Section 4 discusses the role of anti-forensics in addressing privacy issues at various levels, their ease of usage and target audience. Finally Sect. 5 discusses the conclusions drawn.

2 Related Work

2.1 Privacy

Data privacy is a major concern in modern technological world [30] and many regulations have been formulated for its definition and protection [8, 15, 16, 31]. According to Westin [8], privacy is the right of an individual to control both the circumstances and the amount of information that can be shared. Adopted by United Nations in *Universal Declaration of Human Rights* [31], privacy is now considered the basic human right in most countries. However, Nieto et al. [17] argued that privacy regulations and laws are not perfect and can be subjected to varied interpretations. In [16], Nance and Ryan argued that digital forensics are in direct conflict with modern laws and regulations governing criminal procedure and evidence acquisition. They identified various legal issues associated with current practices of forensic investigation, particularly those associated with data security and privacy. In another work [15], Huang et al. summarized various aspects of acquiring forensic evidence in a crime-scene and identified strategies that may violate privacy laws leading to privacy breaches.

On the other hand, Buchi et al. [32] stressed on the importance of self awareness among users for privacy issues and possible protection measures. They argued that in order to reduce the risk of data loss to privacy breaches, users need to be technically skilled to use modern privacy protection measures.

All these studies point to the fact that privacy of data is an active and concerning issue that is increasingly becoming more challenging with each passing day. Fortunately, numerous techniques and tools that ensure data privacy and security at various levels are available [33–35]. In [33], Srinivasan and Wu proposed *SafeCode*, a powerful data and privacy protection mechanism for mobile platform. They argued that *SafeCode* guarantees privacy of data even if the device is stolen by using remote wiping procedures and can also prove to be helpful in its recovery. In [34], Laurent and Levallois identified `TOR` and `Anonymity.Online` as the two best software solution to maintain online anonymity whereas in [35], Douglas et al. provided an overview of various steganographic techniques including digital watermarking that can be used for protecting bio-metric data.

2.2 *Anti-forensics*

Anti-forensics is an emerging science that has recently been recognized as a legitimate field of study [36]. Currently there is no universally accepted and standard definition for anti-forensics as it continues to be a relatively unexplored field [37]. Without a standardized definition, several definitions and classifications exist that explain and categorize anti-forensics and its methods from varied view-points, each having merits of its own [19, 38, 39]. Dahbur and Mohammad [19] classified and categorized anti-forensics using the most general criteria based on - attacked target, employed tactics, degree of creativity and risk, and functionality. From the privacy perspective, Kessler [38] classified anti-forensics by identifying various techniques that ensure protection of one's privacy. In a similar work [39], Caloyannides argued that users have a right to ensure confidentiality of their data and explained various anti-forensic techniques that can prevent an attacker from getting access to victim's data.

Anti-forensic techniques promise great value when it comes to security and privacy. Unfortunately, anti-forensics largely remains an unexplored field when it comes to privacy. The only anti-forensic techniques that has received widespread attention from privacy experts and researchers are encryption and steganography [40–44]. In [40], Rawat and Kumar proposed a stackable enterprise-class cryptographic file system (ECFS) that uses public-key cryptography for user authentication and file sharing. To increase performance, the file system uses XTS mode of AES algorithm and kernel-keyring service. In a similar study [41], Verma et al. argued that modern digital devices need novel security solutions owing to their data capacity and networking capabilities. They proposed the use of compression and encryption techniques for efficient data security on digital devices particularly smartphones without loss of performance. In [42], Artz stressed upon the role of steganography in modern privacy concerned society. He argued that steganography does not replace encryption but enhances its capability to ensure privacy. In [43], Mohamed demonstrated the use of steganography for protecting data within image or pdf files and discussed the use of steganographic techniques for creating covert communication channels over the network. In [44], Albano et al. proposed a new anti-forensic technique for android devices that ensures secure data deletion in order to protect user's privacy. They argued that the proposed technique is capable of modifying or deleting digital evidence using simple tools without making any file system changes or using any cryptographic primitives that may otherwise raise considerable suspicion during forensic analysis.

3 Privacy Preserving Anti-forensic Techniques

Privacy is the ability of an individual or an organization to conceal private or personal information and exercise exclusive control over its accessibility. This makes privacy very challenging as it is not limited to data on storage devices but also encompasses online and deleted information. In this section, we will discuss various anti-forensic techniques that have the capacity to ensure privacy at all these levels.

3.1 Data Wiping: Secured Deletion

Conventional data deletion operations simply work by invalidating the file entry of the deleted file in the file system [45, 46]. The actual data and meta-data pertaining to the deleted file still persists on the digital device until overwritten by some other file [47] and can be recovered easily without much effort [48, 49]. As such normal deletion operations do not guarantee data security and may lead to breach of privacy. Data wiping or secured deletion as it is often called is the most basic and primitive privacy-preservation technique that not only facilitates data deletion but guarantees its privacy [50]. It involves removing or destroying all traces of data and its associated meta-data by using multiple random over-writes followed by a final over-write of 'Zeroes' making their recovery almost impossible [23]. The aim is to either destroy the data or derange its integrity beyond recovery [37]. Sophisticated recovery techniques like *Magnetic Force Microscopy* (MFM) that claim data recovery from two data over-writes fail miserably when it comes to secured-deletion. Besides a study [49] has demonstrated that a single wipe procedure is enough to make any reasonable data recovery impossible [51, 52].

Data wiping tools that ensure secured data deletion [53] are numerous and easily available. Tools like `wipe` come packaged with most of the operating systems be it on windows or linux due to increased security risks associated with leftover data and demand from users. The `wipe` can be configured to do multiple random over-writes and also provides numerous sanitization options. Furthermore, wiping techniques and tools have evolved to a point where it is possible to wipe remote data via a network [33]. `Mobile Guardian System` is one such tool that enables users to wipe data via a network, enabling them to sanitize their devices and prevent identity theft on digital devices particularly smartphones in case they are stolen or lost [54].

3.2 Boot Sector

Digital devices use file systems for data storage and management. These file systems are not only responsible for managing user data but also maintain files associated with the operating system. As such every file system reserves the first sector in each

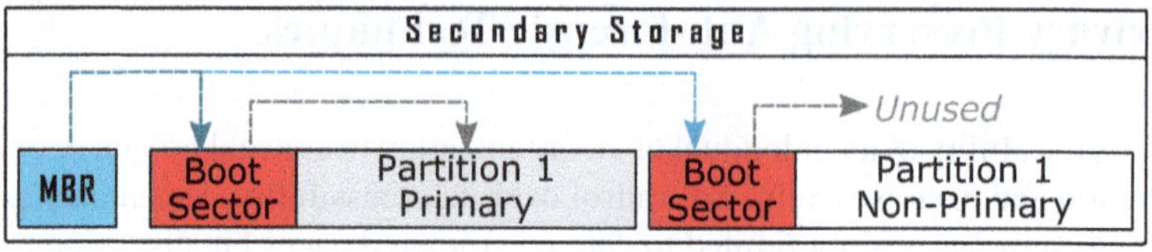

Fig. 1 Boot sector in a typical secondary storage device

partition for boot sector [55, 56] regardless of sector size supported or partitioning scheme used (MBR or GPT) as shown in Fig. 1. Boot sector contains the machine or boot code used to load the operating system in the RAM. However, except for primary partitions (that contain the operating system), this space remains unused and as such can be used to hide sensitive information [56, 57].

Data concealed within the boot sector does not cause any data overwrites or affect the normal functioning of the file system [58]. Hence, data concealed within these location remain undetected. Besides, data recovery and forensic tools assume that file systems have an exclusive access to these locations. As a result, most recovery and forensic tools overlook these locations during analysis, making them ideal for securing sensitive and confidential information.

Different file systems support different boot sector sizes for better interoperability and firmware support. The FAT file systems supports a boot sector of 512 bytes [59, 60] whereas NTFS reserves another 15 sectors [61, 62] for additional features and firmware upgrades. The B-tree file system which is one of the latest file systems offered by linux reserves a total of 64 KB of memory for boot sector [23]. As boot sector locations are reserved in every partition, this may account for a very large space on a multi-partitioned storage device particularly those using GPT partitioning schemes. Furthermore, its fixed location within the file system and availability of tools like `dd` that facilitate seamless data storage within these locations makes boot sector one of the most preferable storage space available in terms of size data security and privacy.

3.3 Reserved Locations

Reserved locations is another file system feature that promise great value when it comes to data hiding. File systems are expected to last over a period of time and evolve with the changing requirements of their users. As a result file systems receive multiple updates over their lifetime. To accommodate these updates and guarantee long-term usability of the file systems, file system designers either add complete data structures that are reserved for future use or leave extra spaces within the existing data structures for future upgrades [63]. We call these spaces as *Reserved Locations*.

Reserved locations are part of the on-disk layout of the file system and do not contain any data or meta-data defining the file systems. They are under the exclusive

```
struct btrfs_super_block {
...
       _le64 reserved[30];                      8 * 30 = 240 bytes
       ...
       struct btrfs_root_backup super_roots[4];
};

                        struct btrfs_root_backup {
                        ...
                               _le64 unused_64[4];  8*4=32 bytes
                               ...
                               u8 unused_8[10];    1*10=10 bytes
                        };
```

Fig. 2 Reserved space within the superblock of B-tree file system

control of the file system and can not be written to or read by normal file system operations, files or utilities. As such data hidden within these locations remains safe and secluded until the file system receives an update which is usually unlikely as file systems are expected to last years without any major updates or upgrades. As a result, these locations mostly remain unused while being perfectly obscured within the file system.

Reserved locations exist in every file system in numerous data structures be it Superblocks, MAC-DTS (Modification, Access and Creation Date Timestamps) structures [64], file descriptors and so on [21]. Figure 2 shows the reserved space that exists within the Superblock data structure of B-tree file system. In addition, MAC-DTS and file descriptors which are associated with every file on the file system be it user data or meta-data, account for huge amounts of reserved storage space as well and their distributed nature also makes them more secure from outside intrusion allowing for greater security and obscurity. Moreover, tools like `DataMuleFs` (linux) and `Cygwin-dd` (windows) that can store and read data within these locations are freely available and fully customizable. This makes reserved locations an ideal choice for storing personal or sensitive data while ensuring its secrecy and privacy.

3.4 Slack Space

Slack space is the disk space that cannot be used by the file system due to difference in the granularity of physical size (sector size) and block size (logical size) [57, 65]. A block is a contiguous collection of sectors defined by the file system and the smallest amount of disk space that can be allocated to a file. If a file or an entire file system does not fit perfectly into its allocated block or blocks due to difference in physical and logical sizes, the remaining left-over space remains unused and cannot be utilized by the file system. We call this space as *Slack Space* [66, 67].

Slack space is associated with every file system and exists on every storage device. Based on their location and accessibility, slack space is categorized into two types – *File Slack* and *Volume Slack*.

3.4.1 File Slack

File slack [68] is a form of internal fragmentation that exists within the file system due to different physical and logical file sizes as shown in Fig. 3. Since a file can be allocated multiple blocks and each block is made up of multiple contiguous sectors, slack space is further divided into *RAM Slack* and *Drive Slack* [56]. RAM slack exists in the last partially occupied sector and spans from the end of file to the end of that sector. On the other hand, Drive slack refers to all the sectors within the last allocated block that are completely unused or have not been overwritten by the file.

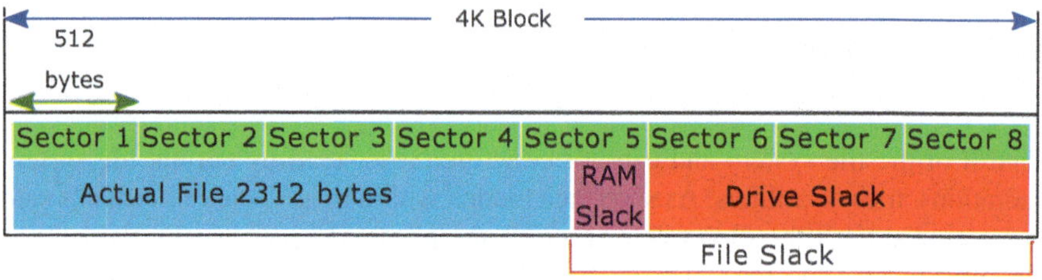

Fig. 3 Slack space generated by a 2312 byte file

3.4.2 Volume Slack

Volume slack [69] is a form of external fragmentation that spans from the end of file system to the end of partition where the file system resides. If the size of the partition is not multiple of block size, some of the sectors in the partition cannot be allocated to a block due to less number of available sectors as shown in Fig. 4. These un-allocated sectors which can not be allocated to a block and exist outside the file system are referred to as *Volume Slack*.

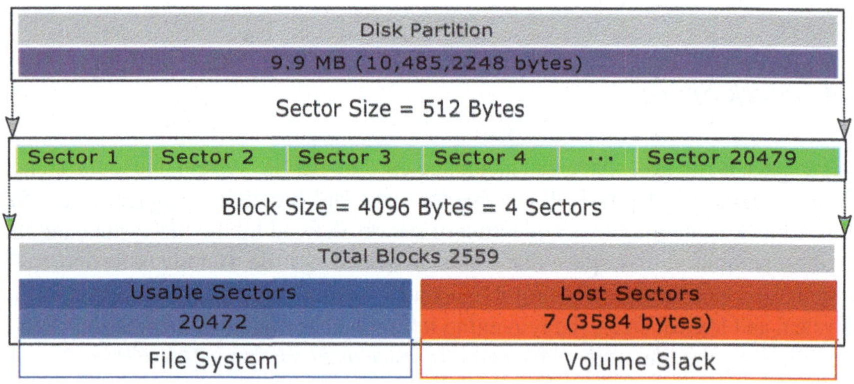

Fig. 4 Volume slack space within a disk partition

Slack space is an inherent feature of all file system and offers vast volumes of storage both in allocated and un-allocated space [70, 71]. File slack which exists in every file remains scattered all over the file system. On the other hand, volume slack is an un-allocatable space and hence, its existence is not reported by the file system. As such, identification and extraction of data from the slack space is both time consuming and computation intensive, requiring special tools and novel techniques. Besides the fact that an average computer system contains 200,000 computer and system generated files [72] and almost half of the space in the last allocated block of every file is lost to slack space, the potential of slack space for privacy and security is simply huge.

3.5 HPA and DCO

Modern hard disk drives contain reserved locations in the form of *Host-Protected Area* (HPA) and *Device Configuration Overlay* (DCO) [73]. While HPA is used to store configuration file and drive utilities, DCO is used by computer manufacturers to define the number of accessible sectors across multiple hard disk drives (HDD) so as to exhibit a uniform address space [67]. These locations are reserved by configuring the drive's controller and hence data concealed within these locations cannot be accessed even by the BIOS system [74].

The HPA was first introduced in *ATA-4 standard* [75]. Its implementation is optional and determined by the manufacturer. The locations are not deleted or altered in any way, they are just made inaccessible by the drive's controller. On the other hand, DCO was introduced in *ATA-6 standard* [75] and it allows computer manufacturers to purchase HDDs from different vendors, and then configure all the HDDs to have the same number of sectors, thereby exhibiting a uniform address space. DCO and HPA can also co-exist on the same HDD provided the DCO is configured first followed by HPA [74]. Besides, these locations particularly HPA can even be configured to span entire hard disk drives. Although, not very practical, such a possibility can be exploited by professionals to created HPA/DCO areas that are large enough to hide sensitive data without raising any suspicion.

Although, HPA and DCO are completely inaccessible [76], software capable of modifying these locations for data hiding purposes are easily available. Tools like `hdparm` (linux) and `ATATool` (windows) are capable of concealing data within the HPA or DCO, but requires great expertise and technical know-how. Furthermore, establishing the existence of these locations within the HDDs can also prove to be a challenge, making them all the more preferable for hiding data.

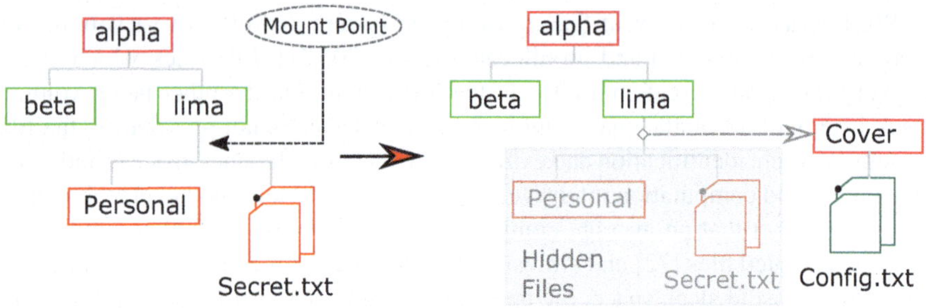

Fig. 5 File system directory structure before and after mounting

3.6 *Directory Mounting: Hiding in Plain Sight*

Probably the easiest and quite effective method of hiding data is mounting a file system on a non-empty directory [25, 75]. After successful mounting, no further access to the underlying data is possible as the mount point causes a diversion to the newly mounted directory, thereby hiding the actual contents of the mount-point in the process as shown in Fig. 5. This anti-forensic technique exploits the fact that professional hackers and forensic experts direct most of their efforts towards potential anti-forensic measures during an intrusion or investigation. As a result, anti-forensic techniques that are obviously easy and simple are less likely to be investigated [23]. Furthermore, mounting operations are supported by all operating systems on all digital device. This makes directory mounting a preferable choice of novice users in terms of ease of use and support across all platforms.

3.7 *Encryption: Second Most Anti-forensic Technique*

Encryption is considered as the second true anti-forensic technique next to wiping that defeats digital forensics [77, 78] and data recovery [79, 80]. It works by encoding data using a *key* into *cipher text* that can only be decoded by a unique *pass-phrase*. Without the necessary password, it is almost impossible to decode the data, thereby preventing unauthorized access. Encryption techniques like `bit-locker` on windows and `dm_crypt` on linux that support full disk-encryption [81] are easily available and mostly come packaged with the operating system. These techniques use *pre-boot authentication*, which requires users to enter their credentials at boot time [82]. Without the actual encryption key and password, any access to the operating system and hence the data is impossible. As a result, data access is limited and available only to authorized users. On the other hand, file systems that do not support encryption can be encrypted using a stackable file systems like `encryptfs` [83, 84]. These file systems add an encryption layer over the underlying file systems, encrypting everything including the file system [85, 86].

Encryption is challenging from the legal point of view as well [79]. Since data is encrypted using an encryption key, it is necessary to decrypt it using the same key so as to ensure that the recovered data is not compromised and its provenance can be proved during legal proceedings [80]. This requires use of actual decryption key that was used to encrypt the data, allowing an individual or organization to have an un-conditional leverage over their investigators or forensic experts citing security and privacy concerns.

Encryption not only ensures secure data storage but also data transmission [41, 87]. Encryption techniques like *Secure Socket Layer* (SSL) provide secure communication channels and prevent hackers from accessing information during transmission, thereby ensuring online privacy. Furthermore, improved *end-to-end encryption* (E2EE) techniques [88] that use asymmetric encryption protocols involve both public and private keys for encryption and decryption, making it difficult even for the server containing the transmitted information to gain access to its contents.

3.8 Steganography: A Perfect Camouflage

Steganography is a technique that involves hiding data or information including documents, images, videos and so on inside another file called a *cover or carrier file* [38, 89]. The data is hidden so precisely that it does not corrupt the contents of cover file nor alter its outward appearance [90]. This makes steganography based data hiding one of the most favoured techniques for security and privacy experts [35, 43]. To successfully reveal the data hidden using steganography, one must be aware of two key pieces of information: (1) the use of steganography and (2) the list of cover files, which is almost impossible to determine. This implies that, except for the owner or authorized user/s, it is almost impossible to establish the use of steganography on a particular file, let alone access the list of cover files that may exist on a digital device or a remote server. Contrary to encryption which hides data from unauthorized access, steganography conceals both the data and the fact that such data actually exists [91]. In addition, encryption leaves digital footprints that may attract adversaries and raise suspicion, thereby increasing the probability of attack [92].

Digital watermarking is a steganographic technique that is primarily used to ascertain ownership or copy-write by hiding the data within the *carrier file* [93]. However, digital watermarking can also be used to hide data without running the risk of privacy breach. Techniques like *Discrete Wavelet Transformation* (DWT) [94, 95], *Discrete Cosine Transform* (DCT) [96] and *Block Based DCT* [97] ensure robust data hiding without the possibility of detection during intrusion, analysis or external attacks. A channel wise watermarking scheme based on DWT [98] provides efficient data security against image processing attacks or software. Besides, the use of pseudorandom address vector (PAV) [99] and chaotic encryption techniques add an additional layer of security [100, 101], making the hidden watermark impossible to detect [102].

Furthermore, digital watermarking also promises value in secure data transmission over computer networks particularly in case of health-care facilities [103–105].

Steganography not only makes unauthorized data access difficult but challenging. Currently, there is no tool available be it forensic tools or third-party software, that can reliably detect or even predict the use of steganography on a computer system or a digital device [20]. On the contrary, steganographic tools like `OpenStego` that ensure secure data hiding are openly available. Being purely open-source, these tools are continuously under development and use sophisticated steganographic techniques available. Furthermore, most of these tools encrypt the data file before hiding, taking data concealment to a whole new level. As a result, steganography based data hiding techniques have witnessed widespread acceptance among the masses and are increasingly being used for data security and privacy [42, 106].

3.9 Onion Routing: Anonymous Communication

Onion routing is an anti-forensic technique that is used for anonymous communication and transmission of data over a network [107, 108]. It works by encapsulating data into multiple layers of encryption just like the layers of an onion so as to ensure secure and anonymous data communication [109]. The client/sender encapsulates data using multiple encryption keys depending upon the number of intermediary nodes between the client and server [110]. These layers of encryption are then peeled off one-by-one by each intermediary node until it reaches the server or its destination [111] as shown in Fig. 6. This multi-level encryption ensures that the digital footprint of the sender and the receiver in the form of IP and MAC (Media Access Control) address are not exposed, which may otherwise compromise their identity [112]. Onion routing does not only support anonymity, but also ensures that the data being transmitted is accessible only to the authorized users ie., the sender and receiver.

Onion routing systems like *TOR network* [113] are freely available and contain hundreds of nodes, processing data from thousands of unknown users. These systems also have an associated client applications (`TOR Browser` of TOR networks) [114] that are aware of the network topology and hence the number of encryptions required. These client applications allow users to connect with the onion routing systems and

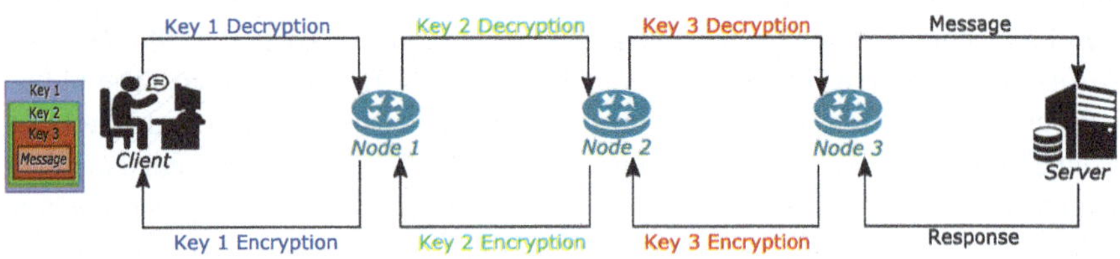

Fig. 6 Onion routing with three intermediate nodes

initiate anonymous data transmission [114]. Using `TOR` is as easy as downloading it from the Tor project website[1] and download the latest `TOR Browser` as per your operating system [115]. There is no need to configure network proxies or change any system settings, the installation takes care of everything automatically [115].

Furthermore, as only the client has access to all the encryption keys that are used to encapsulate the data before sending it over the network, the intermediary nodes are unaware of the number of encryption layers and hence, the data being transmitted. This makes it impossible for intruders or hackers to access the encapsulated data keeping both the transmission and identities of the communicating parties anonymous [110]. The effectiveness of onion routing in ensuring data security and anonymity can be realized from the fact that these systems are not only used by general public but also by government agencies be it military or intelligence [34].

4 Discussion

Table 1 provides the summary of various privacy preserving anti-forensic techniques and measures presented in this chapter. As already discussed in Sect. 1, privacy is not only limited to data that exists on a storage device, it also encompasses data that is deleted by the user or sent over a communication network. Anti-forensics has the capacity to address this multi-dimensional nature of privacy. The "Data Privacy" field in Table 1 defines how anti-forensics address current privacy issues at various levels - Destruction, Storage and Transmission. Anti-forensic techniques or measures that ensure absolute data destruction once the deletion operation is initiated are categorized as "*Secure Deletion*". Those techniques that help in data concealment be it on local digital devices or enterprise servers are categorized as "*Secure Storage*", whereas techniques that ensure secure data transmission over a communication network are categorized as "*Secure Communication*". It is clear from Table 1 that anti-forensics provides numerous techniques and measures, and promises great value in addressing the multi-dimensional nature of modern day privacy.

Anti-forensics is an emerging field and most anti-forensic measures can be executed using tools that are freely and openly available. These tools remove all the technical complexities associated with various anti-forensic measures and specialize in making the user experience easier. Unlike reserved locations, slack space and HPA/DCO, most anti-forensic techniques are "*Easy*" to use and can be executed by typical end users for privacy preservation without any technical know-how. These include data wiping; hiding data in boot sector and directory mounting; secure data storage and transmission using encryption, steganography and onion routing as illustrated in Table 1 under "Usage Mechanism".

On the contrary, using reserved locations for data concealment is relatively "*Difficult*" as the available space and its distribution varies from one file system to another. Although, tools capable of hiding data within these locations mostly come packaged

[1] www.torproject.org.

Table 1 Summary of privacy preserving anti-forensic (AF) techniques/measures

S. No.	AF measure	Description	Data privacy	Usage mechanism
1.	Data wiping	Destruction of data beyond recovery	Secure deletion	Easy
2.	Boot sector	Reserved sector for boot code that remains unused in non-primary partitions	Secure storage	Easy
3.	Reserved locations	Empty data structures within the file system used to accommodate updates received over time	Secure storage	Difficult
4.	Slack space	Unused space at the end of file (allocated) or file system (un-allocated)	Secure storage	Extremely difficult
5.	HPA / DCO	In-accessible HDD areas used to store configuration files and define disk parameters	Secure storage	Extremely difficult
6.	Directory mounting	Mounting one directory over another in order to hide its contents	Secure storage	Easy
7.	Encryption	Encoding data using a key into cipher text to prevent un-authorized access	Secure storage/Secure communication	Easy
8.	Steganography	Conceal data within a cover file without affecting its outward appearance	Secure storage/Secure communication	Easy
9.	Onion routing	Encapsulating data into multiple encryption layers before its transmission over the network	Secure communication	Easy

with the operating system, successful data hiding within these locations require users to have in-depth knowledge of file system. As such, these techniques can only be used by users with technical skills like scientists, IT administrators, researchers and so on.

On the other hand, using slack space and HPA/DCO for privacy preservation is "*Extremely Difficult*" for two reasons: (1) The highly distributed nature of slack space and in-accessibility of HPA/DCO, and (2) Non-availability of user-friendly software. This implies that, these anti-forensic measures offer best security when it comes to privacy of data. However, such privacy can only be achieved by highly technical and skilled hands. As such these locations demand attention from privacy experts and researchers so that their privacy benefits can be exploited.

From this, we infer that anti-forensics promises great value when it comes to privacy preservation. These techniques can arm users without any technical skills to take exclusive control of their data and its privacy be it on storage or over a network. Not only that, these techniques can help users in accessing online content that is blocked, prevent identification theft and online surveillance through encrypted and anonymous communication channels. This is particularly true for individuals and organizations involved in finance, journalism and military establishments. Moreover, anti-forensics is still in its infancy and is receiving increasingly more interest as it continues to take the sophisticated forensic measures head-on. However, its full potential for privacy and data security is yet to be explored by privacy experts and researchers. As such, privacy experts and researchers alike need to take a holistic view over the application and potential of anti-forensics in digital privacy, so that strides can be made in research community for its support.

5 Conclusion

Unauthorized data access is real and privacy of information is under threat. Privacy protection tools are available but struggling to gain mass acceptance as users are reluctant to trust third-party organizations with their data. Anti-forensic techniques that are used to counter forensic efforts promise great value when it comes to data security and privacy at user end. These include data wiping and shredding; hiding data in boot sector, reserved-locations, slack spaces, HPA/DCO, using directory mounting, encryption and steganography; anonymous data transmission using onion routing and tunnelling. Interestingly, tools used for these techniques are either part of the operating system or freely available and are known for their capacity to prevent unauthorized data access. Unfortunately, the privacy potential of anti-forensics has largely remained un-explored due to lack of awareness and support from privacy community. This chapter aims to create awareness among general users, scientists, privacy experts and researchers about the positive role of anti-forensics in digital privacy so that strides can be made for its support.

References

1. Memon N, Wong PW (1998) Protecting digital media content. Commun ACM 41(7):35–43
2. Hurrah NN, Parah SA, Loan NA, Sheikh JA, Elhoseny M, Muhammad K (2019) Dual watermarking framework for privacy protection and content authentication of multimedia. Future Gener Comput Syst 94:654–673
3. Geiger MEA (2006) Scrubbing stubborn data: an evaluation of counter-forensic privacy tools. IEEE Secur Privacy 4(5):16–25 https://doi.org/10.1109/MSP.2006.132
4. Parah SA, Loan NA, Shah AA, Sheikh JA, Bhat G (2018) A new secure and robust watermarking technique based on logistic map and modification of DC coefficient. Nonlinear Dyn 93(4):1933–1951
5. Hsu F-H, Wu M-H, Ou S-C, Wang S-J (2016) Data concealments with high privacy in new technology file system. J Supercomput 72(1):120–140
6. Bhat WA, Jalal ME, Khan SS, Shah FF, Wani MA (2019) Forensic analysis of sync.com and flipdrive cloud applications on android platform. Forensic Sci int 302:109845. https://doi.org/10.1016/j.forsciint.2019.06.003
7. Abdullah L, Quintero L (2019) Sealed computation: a mechanism to support privacy-aware trustworthy cloud service. Inf Comput Secur. https://doi.org/10.1109/SP.2015.45
8. Westin AF (1968) Privacy and freedom. Washington Lee Law Rev 25(1):166
9. Steel E, Fowler G (2010) Facebook in privacy breach. Wall Street J 18(1)
10. Lilley S, Grodzinsky ES, Gumbus A (2012) Revealing the commercialized and compliant facebook user. J inf Commun Ethics Soc
11. Akaichi J (2013) Social networks 'facebook' statutes updates mining for sentiment classification. In: 2013 International conference on social computing. IEEE, pp 886–891
12. Fuchs C (2011) An alternative view of privacy on facebook. Information 2(1):140–165
13. Wani MA, Bhat WA (2018) Dataset for forensic analysis of b-tree file system. Data Brief 18: 2013–2018. https://doi.org/10.1016/j.dib.2018.04.100
14. Bhat WA, Wani MA (2018) Forensic analysis of b-tree file system (BTRFS). Digit Invest. https://doi.org/10.1016/j.diin.2018.09.001
15. Huang J, Ling Z, Xiang T, Wang J, Fu X (2012) When digital forensic research meets laws. In: 2012 32nd International conference on distributed computing systems workshops. IEEE, pp 542–551
16. Nance K, Ryan DJ (2011) Legal aspects of digital forensics: a research agenda. In: 2011 44th Hawaii international conference on system sciences. IEEE, pp 1–6
17. Nieto A, Rios R, Lopez J (2019) Privacy-aware digital forensics. Secur Privacy Big Data Cloud Comput Appl 28:157
18. Gül M, Kugu E (2017) A survey on anti-forensics techniques. In: 2017 International artificial intelligence and data processing symposium (IDAP). IEEE, pp 1–6
19. Dahbur K, Mohammad B (2011) The anti-forensics challenge. In: Proceedings of the 2011 international conference on intelligent semantic web-services and applications. ACM, p 14
20. Pajek P, Pimenidis E (2009) Computer anti-forensics methods and their impact on computer forensic investigation. In: International conference on global security, safety, and sustainability. Springer, pp 145–155
21. Göbel I, Baier H (2018) Anti-forensic capacity and detection rating of hidden data in the ext4 filesystem. In: IFIP international conference on digital forensics. Springer, pp 87–110
22. Dezfoli FN, Dehghantanha A, Mahmoud R, Sani NFBM, Daryabar F (2013) Digital forensic trends and future. Int J Cyber-Secur Digit Forensics 2(2):48–77
23. Wani MA, Bhat WA, Dehghantanha A (2018) An analysis of anti-forensic capabilities of b-tree file system (BTRFS). Aust J Forensic Sci. https://doi.org/10.1080/00450618.2018.1533038
24. Rekhis S, Boudriga N (2012) A hierarchical visibility theory for formal digital investigation of anti-forensic attacks. Comput Secur 31(8):967–982
25. Wani MA, AlZahrani A, Bhat WA (2020) File system anti-forensics—types, techniques and tools. Comput Fraud Secur 2020(3), 14–19. https://doi.org/10.1016/S1361-3723(20)30030-0

26. Rekhis S, Boudriga N (2012) A system for formal digital forensic investigation aware of anti-forensic attacks. IEEE Trans Inf Forensics Secur 7(2):635–650
27. Conlan K, Baggili I, Breitinger F (2016) Anti-forensics: furthering digital forensic science through a new extended, granular taxonomy. Digit Invest 18:S66–S75
28. Botas Á, Rodríguez RJ, Väisänen T, Zdzichowski P (2015) Counterfeiting and defending the digital forensic process. In: 2015 IEEE international conference on computer and information technology; ubiquitous computing and communications; dependable, autonomic and secure computing; pervasive intelligence and computing (CIT/IUCC/DASC/PICOM). IEEE, pp 1966–1971
29. Zdzichowski P, Sadlon M, Väisänen TU, Munoz AB, Filipczak K (2015) Anti-forensic study. NATO CCDCOE (NATO Cooperative Cyber Defence Centre of Excellence), www.ccdcoe.org
30. Kokolakis S (2017) Privacy attitudes and privacy behaviour: a review of current research on the privacy paradox phenomenon. Comput Secur 64:122–134
31. Assembly UG (1948) Universal declaration of human rights, vol 302, no 2. UN General Assembly
32. Büchi M, Just N, Latzer M (2017) Caring is not enough: the importance of internet skills for online privacy protection. Inf Commun Soc 20(8):1261–1278
33. Srinivasan A, Wu J (2012) Safecode–safeguarding security and privacy of user data on stolen IoS devices. In: Cyberspace safety and security. Springer, pp 11–20
34. Laurent M, Levallois-Barth C (2015) Privacy management and protection of personal data. In: Laurent M, Bouzefrane S (eds) Digital identity management. Elsevier, pp 137–205. https://doi.org/10.1016/B978-1-78548-004-1.50004-3
35. Douglas M, Bailey K, Leeney M, Curran K (2018) An overview of steganography techniques applied to the protection of biometric data. Multimedia Tools Appl 77(13):17333–17373
36. Hassan NA, Hijazi R (2016) Data hiding techniques in Windows OS: a practical approach to investigation and defense. Syngress
37. Harris R (2006) Arriving at an anti-forensics consensus: examining how to define and control the anti-forensics problem. Digit invest 3:44–49
38. Kessler GC (2007) Anti-forensics and the digital investigator. In: Australian digital forensics conference, p 1
39. Caloyannides MA (2009) Forensics is so "yesterday". IEEE Secur Privacy 7(2) (2009)
40. Rawat U, Kumar S (2012) ECFS: An enterprise-class cryptographic file system for linux. Int J Inf Secur Privacy (IJISP) 6(2):53–63
41. Verma S, Kapoor V, Maheshwari R (2019) An enhanced cryptographic system for fast and efficient data transmission. In: International conference on advanced computing networking and informatics. Springer, pp 287–297. https://doi.org/10.1007/978-981-13-2673-8_31
42. Artz D (2001) Digital steganography: hiding data within data. IEEE Internet Comput 5(3):75–80
43. Mohamed MA (2018) Steganography for better data integrity and privacy. Ph.D. dissertation, Utica College
44. Albano P, Castiglione A, Cattaneo G, De Santis A (2011) A novel anti-forensics technique for the Android OS. In: 2011 International conference on broadband and wireless computing, communication and applications (BWCCA). IEEE, pp 380–385
45. Bhat WA, Quadri SMK (2012) After-deletion data recovery: myths and solutions. Comput Fraud Secur 4:17–20
46. Bhat WA (2015) Achieving efficient purging in transparent per-file secure wiping extensions. In: Handbook of research on security considerations in cloud computing. IGI Global, pp 345–357
47. Reardon J, Basin D, Capkun S (2013) SOK: Secure data deletion. In: IEEE symposium on security and privacy. IEEE, pp 301–315
48. Jain A, Chhabra GS (2014) Anti-forensics techniques: an analytical review. In: 2014 Seventh international conference on contemporary computing (IC3). IEEE, pp 412–418

49. Bhat W, Quadri S (2012) RESTFS: Secure data deletion using reliable & efficient stackable file system. In: 2012 IEEE 10th international symposium on applied machine intelligence and informatics (SAMI). IEEE, pp 457–462. https://doi.org/10.1109/SAMI.2012.6209010
50. Reardon J, Ritzdorf H, Basin D, Capkun S (2013) Secure data deletion from persistent media. In:Proceedings of the 2013 ACM SIGSAC conference on computer & communications security, pp 271–284
51. Sporea I, Aziz B, McIntyre Z (2012) On the availability of anti-forensic tools for smartphones. Int J Secur 6(4):58–64
52. Forte D, Power R (2007) A tour through the realm of anti-forensics. Comput Fraud Secur 6:18–20
53. Blunden B (2009) Anti-forensics: the rootkit connection. In: Black Hat USA 2009 conference proceedings. Citeseer, p 10
54. Han S-R, Noah M (2017) Android mobile guardian system security architecture for handset and data security. Bonfring Int J Ind Eng Manage Sci 7(1):25–28
55. Alazab M, Venkatraman S, Watters P (2009) Digital forensic techniques for static analysis of NTFS images. In: Proceedings of ICIT 2009, fourth international conference on information technology, IEEE Xplore
56. Huebner E, Bem D, Wee CK (2006) Data hiding in the NTFS file system. Digit Invest 3(4):211–226
57. Berghel H, Hoelzer D, Sthultz M (2008) Data hiding tactics for windows and unix file systems. Adv. Comput. 74:1–17
58. Piper S, Davis M, Manes G, Shenoi S (2005) Detecting hidden data in ext2/ext3 file systems. In: IFIP international conference on digital forensics. Springer, pp 245–256. https://doi.org/10.1007/0-387-31163-7_20
59. Lin X (2018) Examining fat file system. In: Introductory computer forensics. Springer, pp 115–144
60. Sultan S, Salman A (2017) Towards digital forensics pedagogical framework. In: Proceedings of the 2017 international conference on education and e-learning, pp 13–21
61. Lin X (2018) Examining NTFS file system. In: Introductory computer forensics. Springer, pp 163–197
62. Karresand M, Warnqvist A, Lindahl D, Axelsson S, Dyrkolbotn GO (2019) Creating a map of user data in NTFS to improve file carving. In: IFIP international conference on digital forensics. Springer, pp 133–158
63. Piper S, Davis M, Shenoi S (2010) Countering hostile forensic. Adv Digit Forensics II 222:79
64. Göbel T, Baier H (2018) Anti-forensics in ext4: on secrecy and usability of timestamp-based data hiding. Digit Invest 24:S111–S120
65. Pieterse H, Olivier M (2012) Data hiding techniques for database environments. In: IFIP international conference on digital forensics. Springer, pp 289–301
66. Quadri S, Bhat WA (2011) A brief summary of file system forensic techniques. In: Proceedings of 5th national conference, INDIACom, pp 499–502
67. Berghel H (2007) Hiding data, forensics, and anti-forensics. Commun ACM 50(4):15–20
68. Mulazzani M, Neuner S, Kieseberg P, Huber M, Schrittwieser S, Weippl E (2013) Quantifying windows file slack size and stability. In: IFIP international conference on digital forensics. Springer, pp 183–193
69. Lin X (2018) Data hiding and detection. In: Introductory computer forensics. Springer, pp 271–301
70. Carrier B (2005) File system forensic analysis. Addison-Wesley Professional
71. Göbel I, Baier H (2018) fishy-a framework for implementing file system-based data hiding techniques. In: International conference on digital forensics and cyber crime. Springer, pp 23–42
72. Srinivasan A, Nazaraj ST, Stavrou A (2013) Hideinside—A novel randomized & encrypted antiforensic information hiding. In: 2013 International conference on computing, networking and communications (ICNC). IEEE, pp 626–631. http://doi.ieeecomputersociety.org/10.1109/ICCNC.2013.6504159

73. Leng J, Li T (2018) Research on computer system information hiding anti-forensic technology. In: 8th International conference on social network, communication and education (SNCE 2018). Atlantis Press
74. Hassan NA (2019) Essential technical concepts. In: Digital forensics basics. Springer, pp 35–67
75. Krenhuber A, Niederschick A (2007) Forensic and anti-forensic on modern computer systems. Johannes Kepler Universität Linz, p 11
76. Gruhn M (2017) Forensic limbo: towards subverting hard disk firmware bootkits. Digit Invest 23:138–150
77. Al Fahdi M, Clarke NL, Furnell SM (2013) Challenges to digital forensics: a survey of researchers & practitioners attitudes and opinions. In: Information security for South Africa, 2013. IEEE, pp 1–8
78. Narayanan S (2015) Emerging challenges in digital forensics. Technical report
79. Casey E (2011) Digital evidence and computer crime: forensic science, computers, and the internet. Academic Press
80. Philipp A, Cowen D, Davis C (2009) Hacking exposed computer forensics. McGraw-Hill, Inc
81. Casey E, Stellatos GJ (2008) The impact of full disk encryption on digital forensics. ACM SIGOPS Oper Syst Rev 42(3):93–98
82. Garfinkel S (2007) Anti-forensics: techniques, detection and countermeasures. In: 2nd International conference on i-warfare and security, vol 20087, pp 77–84
83. Halcrow MA (2005) ECRYPTFS: an enterprise-class encrypted file system for linux. In: Proceedings of the 2005 linux symposium, vol 1, pp 201–218
84. Leibenger D, Fortmann J, Sorge C (2016) ENCFS goes multi-user: adding access control to an encrypted file system. In: 2016 IEEE conference on communications and network security (CNS). IEEE, pp 525–533
85. Xiao C, Zhang L, Liu W, Cheng L, Li P, Pan Y, Bergmann N (2018) NV-ECRYPTFS: accelerating enterprise-level cryptographic file system with non-volatile memory. IEEE Trans Comput 68(9):1338–1352
86. O'hare MS, Orsini RL, Davenport RS (2019) Systems and methods for a cryptographic file system layer. US Patent 10,402,582
87. Sartin B (2006) Anti-forensics-distorting the evidence. Comput Fraud Secur 2006(5):4–6
88. Chase M, Deshpande A, Ghosh E, Malaise H (2019) Seemless: secure end-to-end encrypted messaging with less trust. In: Proceedings of the 2019 ACM SIGSAC conference on computer and communications security, pp 1639–1656. https://doi.org/10.1145/3319535.3363202
89. Parah SA, Sheikh JA, Bhat G (2014) A secure and efficient spatial domain data hiding technique based on pixel adjustment. Am. J. Eng. Technol. Res. 14(2):33
90. Rekhis S, Boudriga N (2010) Formal digital investigation of anti-forensic attacks. In: Fifth IEEE international workshop on Systematic approaches to digital forensic engineering (SADFE). IEEE, pp 33–44
91. Hassan NA, Hijazi R (2017) Data hiding techniques in Windows OS. Elsevier
92. Parah S, Sheikh J, Hafiz A, Bhat G (2015) A secure and robust information hiding technique for covert communication. Int J Electron 102(8):1253–1266
93. Parah SA, Sheikh JA, Bhat G (2014) Fragility evaluation of intermediate significant bit embedding (ISBE) based digital image watermarking scheme for content authentication. In: 2014 International conference on advances in electronics computers and communications. IEEE, pp 1–6
94. Giri KJ, Peer MA, Nagabhushan P (2015) A robust color image watermarking scheme using discrete wavelet transformation. IJ Image Graph Signal Process 1:47–52
95. Giri KJ, Bashir R (2018) A block based watermarking approach for color images using discrete wavelet transformation. Int J Inf Technol 10(2):139–146
96. Parah SA, Sheikh JA, Loan NA, Ahad F, Bhat GM (2018) Utilizing neighborhood coefficient correlation: a new image watermarking technique robust to singular and hybrid attacks. Multidimens Syst Signal Process 29(3):1095–1117

97. Parah SA, Sheikh JA, Loan NA, Bhat GM (2016) Robust and blind watermarking technique in dct domain using inter-block coefficient differencing. Digit Signal Process 53:11–24
98. Giri KJ, Peer MA, Nagabhushan P (2016) A channel wise color image watermarking scheme based on discrete wavelet transformation. In: 2014 International conference on computing for sustainable global development (INDIACom). IEEE, 758–762
99. Parah SA, Sheikh JA, Assad UI, Bhat GM (2017) Realisation and robustness evaluation of a blind spatial domain watermarking technique. Int J Electron 104(4):659–672
100. Loan NA, Hurrah NN, Parah SA, Lee JW, Sheikh JA, Bhat GM (2018) Secure and robust digital image watermarking using coefficient differencing and chaotic encryption. IEEE Access 6:19876–19897
101. Parah SA, Sheikh JA, Dey N, Bhat GM (2017) Realization of a new robust and secure watermarking technique using dc coefficient modification in pixel domain and chaotic encryption. J Glob Inf Manage (JGIM) 25(4):80–102
102. Parah SA, Sheikh JA, Hafiz AM, Bhat GM (2014) Data hiding in scrambled images: a new double layer security data hiding technique. Comput Electr Eng 40(1):70–82
103. Giri KJ, Bashir R, Bhat JI (2019) A discrete wavelet based watermarking scheme for authentication of medical images. Int J E-Health Med Commun (IJEHMC) 10(4):30–38
104. Gull S, Loan NA, Parah SA, Sheikh JA, Bhat G (2018) An efficient watermarking technique for tamper detection and localization of medical images. J Ambient Intell Hum Comput 1–10
105. Parah SA, Sheikh JA, Ahad F, Loan NA, Bhat GM (2017) Information hiding in medical images: a robust medical image watermarking system for e-healthcare. Multimedia Tools Appl 76(8):10599–10633
106. Tao J, Li S, Zhang X, Wang Z (2018) Towards robust image steganography. IEEE Trans Circ Syst Video Technol 29(2):594–600. https://ieeexplore.ieee.org/document/8533349/
107. Reed MG, Syverson PF, Goldschlag DM (1998) Anonymous connections and onion routing. IEEE J Sel Areas Commun 16(4):482–494
108. Syverson P, Tsudik G, Reed M, Landwehr C (2001) Towards an analysis of onion routing security. In: Designing privacy enhancing technologies. Springer, pp 96–114
109. Ando M, Lysyanskaya A, Upfal E (2017) Practical and provably secure onion routing. arXiv:1706.05367
110. Chauhan M, Singh AK et al Survey of onion routing approaches: advantages, limitations and future scopes. In: International conference on computer networks, big data and IoT. Springer, pp 686–697
111. Cremonini M, Braghin C, Ardagna CA Privacy on the internet. In: Vacca JR (ed) Computer and information security handbook, 2nd edn. Morgan Kaufmann, Boston, pp 739 – 753. http://www.sciencedirect.com/science/article/pii/B9780123943972000428
112. Wong T, Cui H, Shen Y, Lin W, Yu T (2018) Anonymous network communication based on SDN. In: 2018 4th International conference on universal village (UV). IEEE, pp 1–5
113. Shavers B, Bair J (2016) The Tor browser. In: Shavers B, Bair J (eds) Hiding behind the keyboard. Syngress, Boston, pp 11–34
114. Macrina A (2015) The Tor browser and intellectual freedom in the digital age. Ref User Serv Quart 54(4):17
115. Shipley TG, Bowker A (2014) Working unseen on the internet. In: Shipley IG, Bowker A (eds)Investigating internet crimes. Syngress, Boston, pp 211 – 231. https://doi.org/10.1016/B978-0-12-407817-8.00009-6

Part II
Information Embedding for Secure Data Communication and Authentication

Chapter 5
Information Embedding Using DNA Sequences for Covert Communication

Shah Haris Nabi, Parsa Sarosh, Shabir A. Parah, and G. Mohiuddin Bhat

1 Introduction

In the domain of DNA steganography [1, 2], the DNA molecule can be considered as a four-symbol sequence that is capable of hiding secret information. As we know, the DNA sequence directs the biological functions of an organism

it is, therefore, essential that the data hiding process is reversible [3, 4]. Reversibility ensures that the original DNA sequence is completely retrieved at the receiver. Generally, a reference DNA sequence is used for reversible data embedding. The secret data to be hidden is initially converted into a string of binary numbers. With the help of a reference DNA sequence, the Stego-DNA sequence is generated and transmitted [5]. In this chapter, we illustrate two secure and computationally feasible methods of data hiding based on DNA sequences. Logical operations like XOR combined with DNA substitution or insertion method improve the security of the cryptosystem manifolds. Two methods of data hiding have been explained. In the first method, secret data is transmitted in the form of a Stego-DNA sequence generated with the help of a substitution table. In the second method, the Stego-DNA sequence is generated with the help of a private key 'k' used both in embedding as well as extraction procedure. The analysis of the proposed techniques reveals high security with less computation. As mentioned earlier, there are many advantages of using DNA cryptography [6, 7] for secure data transmission which includes high capacity, randomness, and simplicity. The reversibility of the DNA data hiding method is highly desirable but has not been studied extensively. These methods can

S. H. Nabi · P. Sarosh · S. A. Parah (✉)
Post Graduate Department of Electronics and Instrumentation Technology, University of Kashmir, Srinagar, India
e-mail: shabireltr@gmail.com

G. Mohiuddin Bhat
Department of Electronics and Communication Engineering, Institute of Technology, Zakoora, India

K. J. Giri et al. (eds.), *Multimedia Security*, Algorithms for Intelligent Systems,
https://doi.org/10.1007/978-981-15-8711-5_6

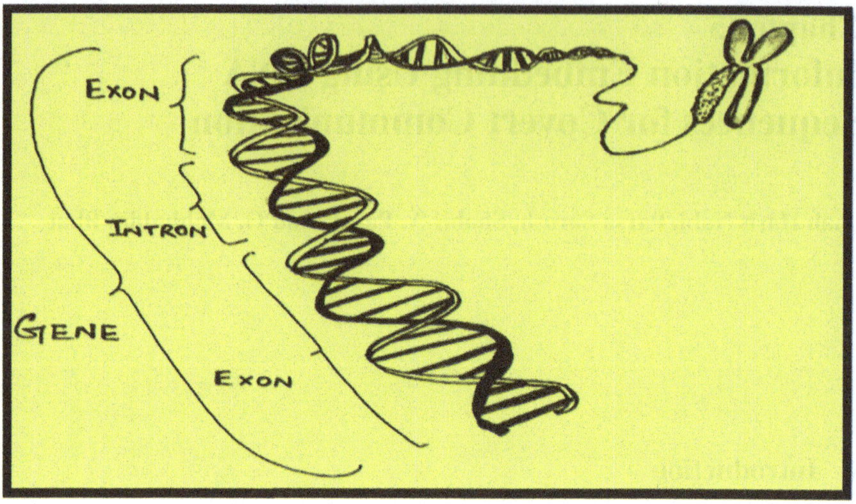

Fig. 1 Structure coding region of DNA

be used for secure data storage, transmission, and encryption with the help of DNA sequences

2 Introduction to DNA

The human genome has 23 paired chromosomes that contain the primary information and direct functions of an organism. DNA consists of two long strands of molecules entwined in the form of a double helix which has been discovered by Watson and Crick [8]. There are two copies of each chromosome which leads to about 6 billion nucleotide bases in our DNA. This is shown below in Fig. 1.

These base molecules are of only four types namely Adenine (A), Guanine (G), Thymine (T), and Cytosine (C). The DNA molecule has a complementary structure, with Adenine, Thymine as one pair, and Guanine, Cytosine as another pair. This is called Watson–Crick complementary rule as shown in Table 1 and Fig. 2.

Table 1 Structure of DNA nucleotide bases

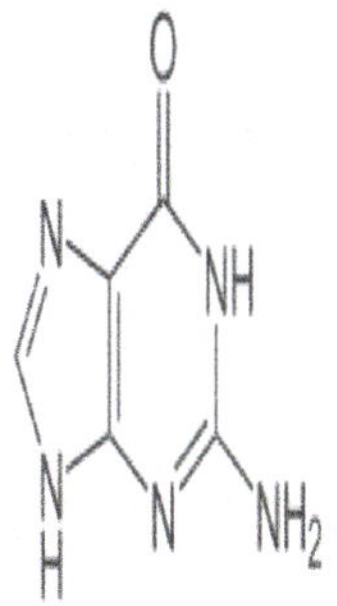	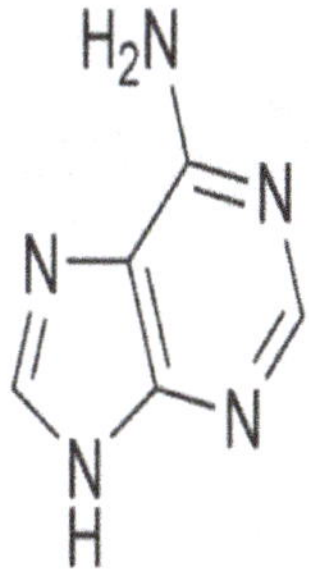	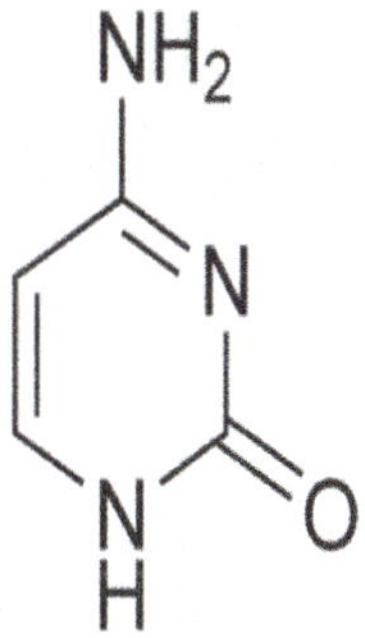	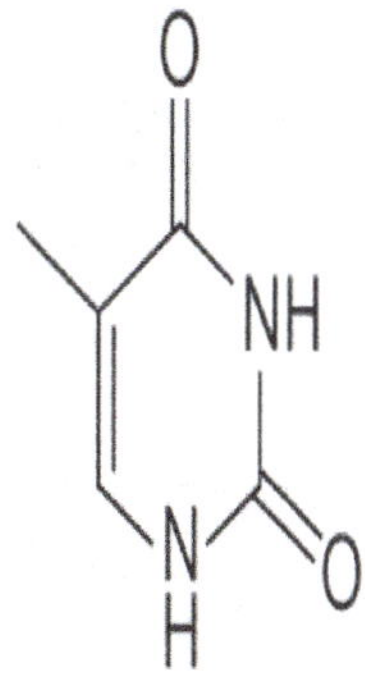
Guanine (G)	Adenine (A)	Cytosine (C)	Thymine (T)

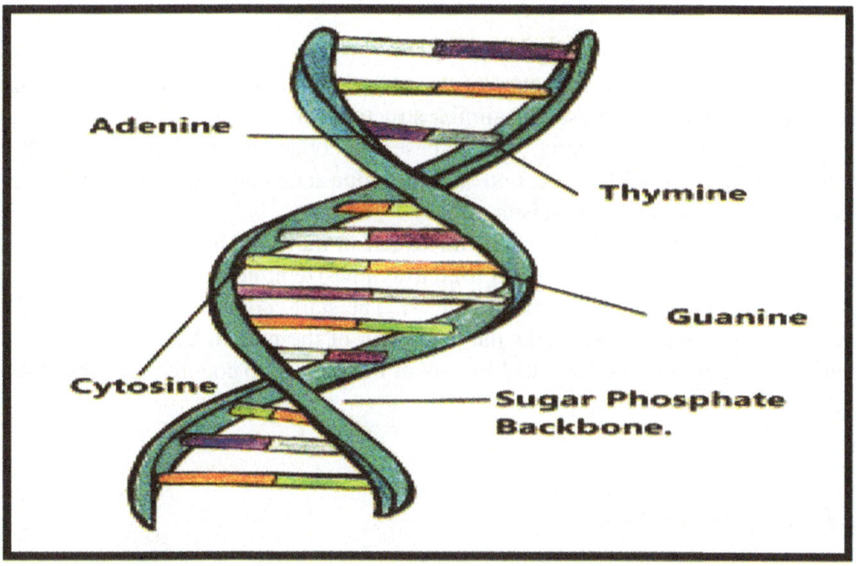

Fig. 2 DNA molecule

The DNA sequence can be broadly divided into coding and non-coding regions. The bases which code for a protein are called coding elements or exons of genes, which are collectively known as the exome. Only 1.5% of the entire DNA is made up of genes that code for proteins. The rest of the 98.5% of the genome does not code for proteins and is used for regulation of the various operations performed by an organism [9].

The RNA polymerase enzyme is used to transcribe DNA into messenger ribonucleic acid (mRNA). This enzyme splits the two strands of the DNA molecule and

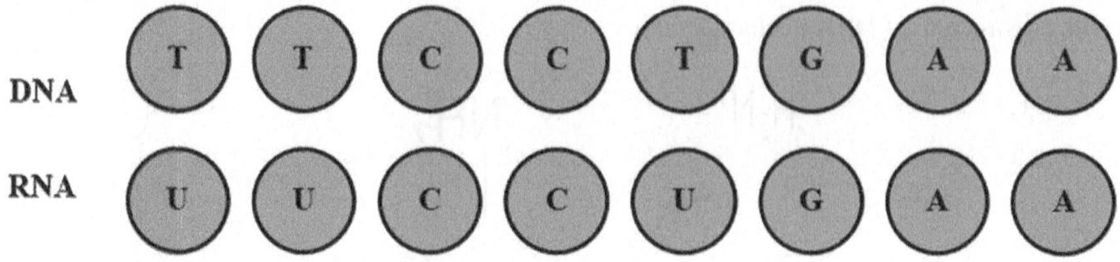

Fig. 3 DNA and corresponding RNA base sequence

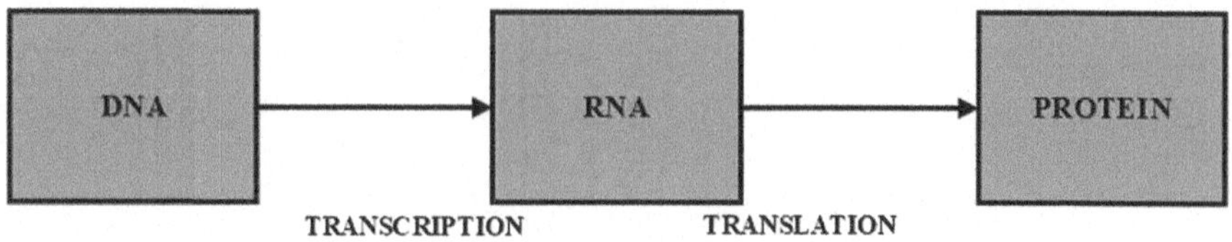

Fig. 4 Process of protein formation

reads the nucleotide base sequence of one strand. In the RNA molecule, Thymine gets replaced by another base of a similar structure called Uracil (U).

A string of DNA consisting of three base nucleotides is known as a codon that produces amino acids. The collection of these amino acids constitutes proteins, which are building blocks of an organism.

The Adenine base makes a pair with Uracil in the RNA molecule as shown in Fig. 3. The RNA molecule is further used to synthesize proteins by a process known as Translation. This is represented in Fig. 4. The start codon sequenced as 'ATG', codes for Methionine, and marks the beginning of the protein translation [9]. The end of translation can be indicated by any of the three stop codons as UAA, UGA, and UAG.

2.1 DNA Computing

The field of DNA computing came into existence when Adleman [8] described how molecules can be used to solve computational problems. The concept of DNA computing entails the use of biological molecules in place of silicon chips to perform computations. Adleman solved the directed Hamiltonian path problem using seven vertices. There are many advantages of using DNA computing like high speed, minimum storage, and power requirement. As we know that conventional computers are accurately able to perform 100 million instructions per second (MIPS), DNA computers will be many times faster with a much lower power requirement. The operations on DNA include synthesis, multiplying, annealing, and pattern extraction among others [10]. We describe a few operations performed on DNA sequences, but there are many more operations that can be performed.

i **Synthesis**: It is the process of synthesizing the small-sized single strands of DNA from a given DNA. The current laboratory techniques allow a few nucleotide base strands to be synthesized.
ii **Multiplication**: These few DNA strands can be multiplied into millions of strands using DNA polymerase. The process is known as polymerase chain reaction (PCR).
iii **Annealing**: The single-stranded DNA will itself adhere to the complementary single-strand to form a double-stranded DNA molecule.
iv **Pattern Extraction**: If we have a complex of DNA strands, we can extract a specific pattern by placing a complement of the pattern into the complex.

2.2 DNA Cryptography

Information security is of utmost importance in this digital era which identifies the need for cryptography [11]. Cryptography is a method of encoding information for secure storage and transmission [12]. The use of DNA computing in the field of cryptography has the potential to produce very strong ciphers with high security and low cracking probability. The concept of using molecular media for the security of confidential data is called DNA steganography. For data embedding, the DNA sequence undergoes a numerical transformation in which a nucleotide base is represented by a binary number [13]. The use of substitution and binary encoding can convert any cover DNA sequence into the Stego-DNA sequence [14].

As mentioned earlier, DNA-based data hiding makes use of DNA as a cover media for secure data transmission. The DNA sequence can be broadly divided into coding and non-coding regions. Secret data can be hidden in both coding and non-coding regions of the DNA sequence. If data is embedded into the coding region of the DNA sequence, then the codon degeneracy should be utilized. This helps in embedding the secret data without altering the amino acid sequence. There are many research challenges in DNA steganography compared to simple data hiding techniques like image data hiding [15–17]. The various challenges include biological function preservation of the DNA sequence, preservation of the length of DNA sequence, prevention of mutations and false start codons, and proper recovery of the DNA sequence at the receiver [14].

3 Literature Review

The field of molecular computation was invented by Adleman [8]. He solved the combinatorial problems using DNA as a computational system. DNA data hiding is a method in which the secret data is hidden into molecular media [18–20]. In such a system, the channel errors constitute the mutations of the transmitted DNA sequence. Clelland et al. [3] presented a double-steganographic method to conceal secret messages into DNA microdots. Shimanovsky et al. [21] defined two novel techniques used to hide data in DNA and RNA. The first method demonstrates how data can be hidden in non-coding DNA. In the second method, data is hidden in the coding DNA by making use of the redundancy found in codons, public-key encryption, and arithmetic coding [21]. Chen [22] presented a carbon nanotube-based message transformation and DNA-based cryptosystem. Furthermore, Guangzhao et al. [23] proposed an encryption scheme designed by using DNA digital coding with synthesis and PCR amplification. Lipton et al. [24] demonstrate how DNA computing can be used to solve the NP-complete simple search problem known as the 'SAT' problem.

Kumar and Singh [25] presented an encryption technique by making use of a one-time pad (OTP) which can secure data within DNA sequences. In their research work, the details pertaining to DNA cryptography, steganography, and encryption are presented. Tornea and Monica [26] in their research work represent a solution to the one-time pad (OTP) symmetric key generation problem using Genomic databases. Tanaka et al. [27] presents a novel method for the key distribution in a public-key system using DNA sequencing. Lai et al. [28] propose a new technique for the implementation of asymmetric encryption and signature with DNA technology. Bahig and Nassr [29] represents the encoding and decoding of data in a DNA sequence to perform the various steps of AES. The authors in [30] use the DNA digital coding and DNA fragment assembly to design a symmetric system.

Hu et al. [4] propose a reversible data hiding method based on difference expansion with an improved overflow map. Balado [5] present a study of DNA data hiding capacity under substitution mutations. They obtain the Shannon capacity of DNA data embedding wherein DNA sequences are subject to substitution mutations designed using the Kimura model. Fu et al. [31] propose a fast tamper location of batch DNA sequences based on reversible data hiding. Their proposed technique applies to large DNA databases to check whether the DNA sequences are intact during storage as well as transmission. Ma et al. [32] present a method to localize the tampered area in a DNA sequence to restore a DNA database. Chen [33] present a scheme in which nucleotide bases are converted into decimal values and difference expansion algorithm is applied for reversible data hiding. A comparative analysis of the DNA-based data hiding techniques is presented by Hamed et al. in [34]. Another DNA-based steganography technique has been presented in [35] in which the authors use both the DNA strands to hide secret data.

Prediction-error expansion (PEE)-based reversible data hiding is proposed by Hu et al. [36] wherein a prediction-error (PE) histogram is generated and secret messages are reversibly embedded by expanding and shifting the PE histogram. An improved DNA insertion algorithm is used for DNA data hiding to improve the security of the data in [37]. Lee [38] presents a multilevel histogram shifting method to hide data reversibly. Huang et al. [39] present a scheme which constructs a histogram of decimal numbers representing nucleotide bases. They embed secret data using the histogram-bin shifting method which leads to a low modification. Shiu et al. [40] presented a data hiding scheme that requires a reference DNA sequence for the detection and recovery of the transmitted data.

Lee et al. [14] presents two approaches to reversible DNA data hiding using multiple difference expansions (DE). In their proposed method, the DNA sequence is converted into decimal-coded values. Multiple watermark bits are subsequently embedded into the range of coded sequence using two methods namely DE-based multiple bits embedding (DE-MBE) and consecutive DE-MBE (CDE-MBE). Both of these techniques use comparison searching to prevent false start codon (ATG). It has been demonstrated that the CDE-MBE technique has a higher embedding capacity of 1.13 times–9.03 times than state-of-the-art methods. They can recover the cover DNA sequence even without a reference DNA and do not produce any false start codons.

Maniyath and Supriya [41] propose an image and video encryption algorithm based on DNA sequences. Their proposed work aims to significantly reduce the encryption time for images and videos. Recently, there have been many research works that try to make use of Chaos theory in combination with DNA cryptography to enhance security [42–44]. Hu et al. [45] proposed a novel image encryption method in which a pseudo-random sequence is controlled by a Chens hyper-chaos system. A cyclic operation for DNA sequences and XOR-operation is used to transpose the pixel values of the secret image and generate the cipher image. Song and Qiao [46] propose an image encryption scheme which is also based on DNA encoding and Spatiotemporal Chaos. In this method, DNA mapping, spatiotemporal chaotic system, and XOR-operation are used to generate a highly secure encrypted image. It has been demonstrated by them that the encryption algorithm is resistant to several attacks like entropy attack, differential attack, and chosen-plaintext attack among others.

Zhen et al. [47] proposed an image encryption method combining the security provided by the logistic and spatiotemporal system. Their proposed scheme combines eight DNA coding rules to efficiently permute as well as substitute the image pixel values to increase the security. Furthermore, the use of chaotic systems makes the proposed method sensitive to slight changes in the initial conditions and makes the technique resistant to many attacks. Cui et al. [48] presents a DNA information encryption scheme with high security. They propose the use of DNA digital encoding, complementary rules, and biotechnological methods to increase the confidentiality of the DNA cryptosystem [48].

Other research works related to DNA computing [10, 49, 50], encryption [51–53], DNA cryptography [54, 55], DNA watermarking [56–58], and DNA data hiding [1, 59] are also mentioned. A review of the DNA cryptography is presented in research work [60]. As we know, Cloud-based servers [61, 62] provide access to a shared pool of resources and are being used to store and process data [63, 64]. Some research works on securing data in the Cloud [65] and IoT-based servers [66] using DNA among other encryption methods are presented.

4 Proposed Method

Steganography enables confidential information to be concealed in multimedia such as an image [67–69], video, audio, or DNA sequence. DNA steganography and encryption are active areas of substantial research, as they provide high security to the data being transmitted. This is the most promising technology because of the advantages it has such as high security, high storage density, and low cracking probability. In this chapter, we illustrate two techniques for data security. The secret message transmission can be accomplished by using any of the two approaches which are explained in detail in the following subsections.

4.1 *Method 1: New Substitution Method for Data Hiding Using DNA Sequences*

In the first approach, the message is embedded into a cover DNA sequence at the sender end, and a Stego-DNA sequence is generated. Consequently, this sequence is transmitted via an unsecured channel. The block diagram of this approach is shown in Fig. 5.

The message 'M' can be a text file, audio, or video data. A DNA sequence 'D' is chosen randomly from a database that contains millions of DNA sequences. The data can be converted to binary form and substituted by corresponding nucleotide bases by a substitution lookup table. The substitution lookup table is a relational dataset where the rows represent 2-bit binary information and columns represent the corresponding nucleotide bases. In the inverse/receiver substitution lookup table, the row number represents nucleotide bases of the Stego-DNA sequence, and column

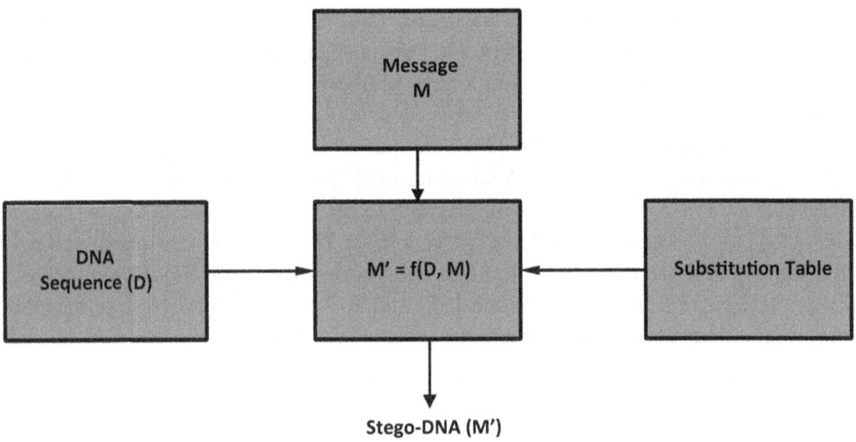

Fig. 5 Block diagram of the substitution method at the sender end

D→ M↓	A	C	T	G
00	A	C	T	G
01	C	T	G	A
10	T	G	A	C
11	G	A	C	T

Fig. 6 Transmitter substitution lookup table

number represents the nucleotide of the selected DNA sequence. Figure 6 presents the transmitter substitution table.

There are 24 possible ways to create a substitution table out of which one is selected for data manipulation. The Stego-DNA sequence generated with the help of the substitution method is transmitted through a public channel.

The Algorithm at Sender End: Conversion of message which is to be hidden in the Stego-DNA sequence

Input: Take Secret message M as 1001101010111100 and the selected DNA sequence D is ACGTACATTCCA

Step 1. Divide the message M into 2-bit length segments. Then the bitstream is as follows:
10 01 10 10 10 11 11 00
Step 2. Each segment will select one of the four possible rows of the substitution table
Step 3. The nucleotide will select one column of the substitution table. Those nucleotides of D which are not substituted are ignored (M, D) → D' (Stego-DNA)
(10, A) → T
(01, C) → T
(10, G) → C
(10, T) → A
D' = TTCATAGT
Step 4. Send the Stego-DNA sequence D' to the receiver

Output: Stego-DNA sequence D'

At the receiver, the complement of the same substitution table can be utilized to decode the message 'M' from transmitted Stego-DNA. The block diagram at the receiver is shown in Fig. 7.

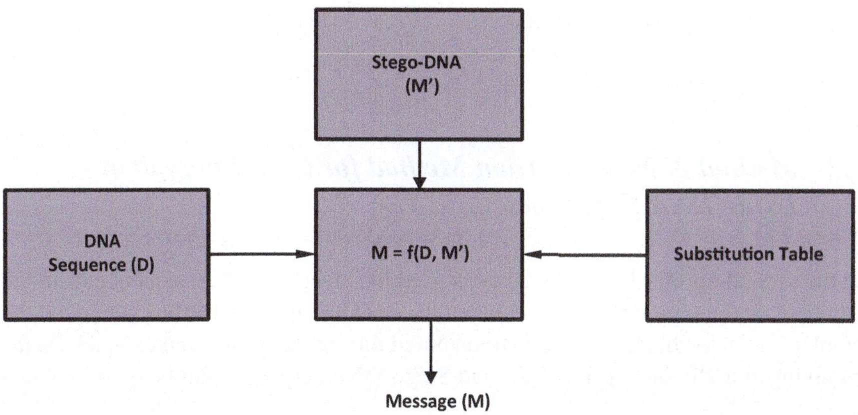

Fig. 7 Block diagram of the substitution method at the receiver end

Fig. 8 Receiver substitution lookup table

D→ D'↓	A	C	T	G
A	**00**	**11**	**10**	**01**
C	**01**	**00**	**11**	**10**
T	**10**	**01**	**00**	**11**
G	**11**	**10**	**01**	**00**

It can be seen that the extraction process is the reverse mechanism of the hiding process. When the Stego-DNA sequence is received, the message 'M' is extracted using a reference DNA sequence and the receiver substitution lookup table as shown in Fig. 8.

The algorithm applied at the receiver end is shown below.

The algorithm at Receiver End: Conversion of Stego-DNA sequence into secret message

Input: Stego-DNA sequence

Step 1. Each nucleotide of D' will select one of the four rows of the receiver substitution table.
Step 2. Similarly, the nucleotide of D will select one of the four columns of the receiver substitution table. (D', D) → M
(T, A) → 10
(T, C) → 01
(C, G) → 10
(A, T) → 10
M = 10 01 10 10 10 11 11 00
The extracted binary message M is the desired secret message.

Output: Binary message M

There is roughly 1.63×10^8 DNA sequence publicly. The probability that a hacker will make a successful guess is given as:

Probability of success = number of combinations of substitution table × 1/total number of DNA available.

$$\text{Probability of success} = (1/24) \times (1/1.63) \times 10^8$$

4.2 *Method 2: New Insertion Method for Data Encryption Using DNA Sequences*

In the second approach, a private key is used to select one DNA sequence from the database, and a secret message is hidden into it. The Stego-DNA thus generated, by employing DNA digital coding, is transmitted through an unsecured channel. At the receiving end, the same private key and Stego-DNA sequence will be used to obtain

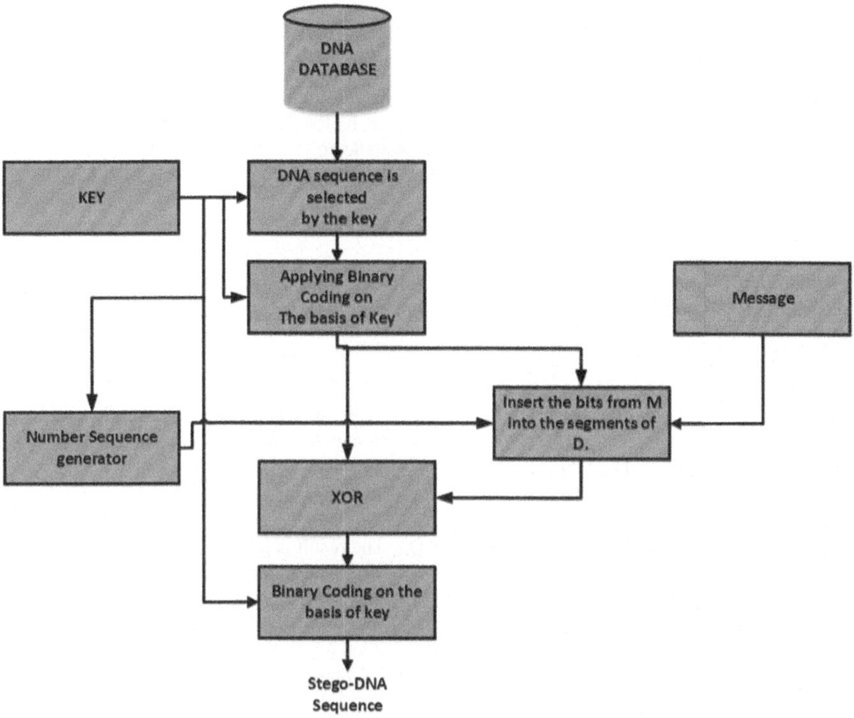

Fig. 9 Block diagram of insertion method at sender end

the secret message back in its original form. The block diagram of the method at the transmitter end is shown in Fig. 9.

The message M can be embedded into a DNA sequence, which is selected from a database using a private key. In the DNA database or DNA databank, there are different DNA sequences with varying lengths. Each DNA base can be represented by two binary digits in which A, G, C, and T are replaced by 00, 11, 10, and 01, as shown in Table 2. This coding scheme is also selected by the private key 'k'.

Process of secret data encryption is represented in the following steps:

Table 2 DNA digital coding

DNA bases	Binary value
A	00
G	11
C	10
T	01

The algorithm at Transmitter End: Data encryption

Input: Private key 'k', DNA database and secret message M = 10100110

Step 1. Divide the public key into four segments k_1, k_2, k_3, and k_4. Segment k_1 for selection of DNA sequences, k_2 for selection of binary coding rule, k_3 for the generation of random sequence L, and k_4 for the generation of random sequence p.
Step 2. Select two DNA sequences from the DNA database. Let the two selected DNA sequences are as follows:
D_A = ATCGATCG
D_B = AGCAATCC
Step 3. Code D_A and D_B into binary form with the help of the coding rule. There can be 24 possible ways to code nucleotide into a 2-bit binary code. The coding scheme is selected using k_2.
A = 00, G = 11, C = 10 and T = 01.
Step 4: Concatenate D_A and D_B to obtain D_{AB}. D_{AB} will be used in step 8.
Step 5: Generate the number sequence L_1, L_2, L_3... L_N by number seed k_3. Divide D_A and D_B into segments of lengths L_1, L_2, L_3... L_N. Split the message M into two segments M_1 and M_2.
$D_A = D_{AL1}\ D_{AL2}\ D_{AL3}\ D_{AL4}\ D_{AL5} \ldots D_{ALN}$
$D_B = D_{BL1}\ D_{BL2}\ D_{BL3}\ D_{BL4}\ D_{BL5} \ldots D_{BLN}$
M = M1 M2.
Step 6: Insert the bits of M_1, in the segments of D_A. The result is given below:
$D_{AL1}M1_1$, $D_{AL2}\ M1_2$, $D_{AL3}\ M1_3$, $D_{AL4}\ M1_4$ so on. Then, concatenate the above segments resulting in the sequence D'_A: $D_{AL1}M1_1D_{AL2}M1_2D_{AL3}M1_3D_{AL4}M1_4$.... $D_{ALN}M1_N$. Similarly, insert the bits M2, into the segments of D_B. The result is as follows $D_{BL1}M2_1$, $D_{BL2}M2_2$, $D_{BL3}\ M2_3$, $D_{BL4}M2_4$, and so on.
Then, concatenate the above segments to generate the sequence
D'_B:$D_{BL1}M2_1D_{BL2}M2_2D_{BL3}M2_3D_{BL4}M2_4$.... $D_{BLN}M2_N$.
Step 7: Generate the number sequences p_1, p_2, p_3.....p_n by number seed k_4. Then, concatenate D'_A and D'_B as follows take p_1 bits from D_A then p_2 bits from D_B then again p_3 bits from D_A, and so on.
$D'_{AB} = D'_A + D'_B$.
Step 8: Perform bitwise XOR between D'_{AB} and D_{AB}.
$(D'_{AB})' = D'_{AB}$ XOR D_{AB}
Step 9: Convert the code back into the Stego-DNA sequence using a binary coding rule from the output of XOR.
Step 10: Send the above sequence to the receiver.

Output: Stego-DNA sequence.

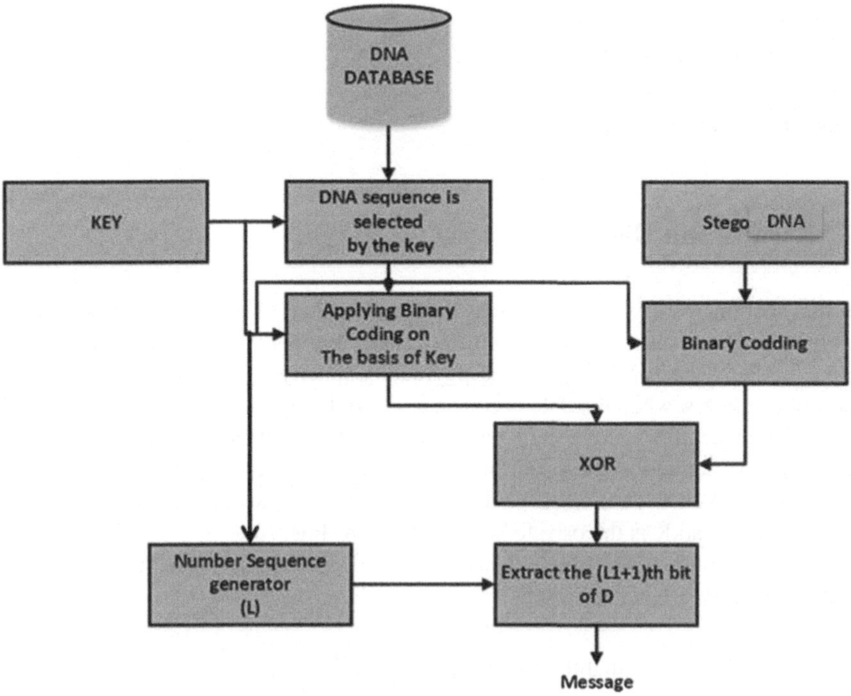

Fig. 10 Block diagram of insertion method at the receiver end

The decryption process is the reverse of the data encryption method employed at the transmitter. The receiver will extract the message M using the DNA database, private key, and the received DNA sequence $D'_{AB.}$ The block diagram of the insertion method at the receiver end has been shown below in Fig. 10.

The algorithm at Receiver End: Data decryption/extraction

Input: Private key 'k' and the DNA database

Step 1. Divide the public key into four segments $k_{1,}$ $k_{2,}$ $k_{3,}$ and $k_{4.}$ Segment k_1 is used for selection of DNA sequences, k_2 is used for selection of binary coding rule, k_3 is used for the generation of random sequence L, and k_4 is used for the generation of random sequence p
Step 2. Select two DNA sequences from the DNA database. The selection is done according to k_1
Step 3. Code $D_{A,}$ $D_{B,}$ and D'_{AB} in binary form. We use k_2 to select the coding scheme
Step 4: Concatenate D_A and D_B to obtain D_{AB}
Step 5: Perform bitwise XOR between $(D'_{AB})'$ and D_{AB} to obtain D'_{AB}
$D'_{AB} = (D'_{AB})'$ XOR D_{AB}
Step 6: Generate the number sequences p_1, p_2, p_3.....p_n by number seed $k_{4.}$ Divide the D'_{AB} into the segments of length p_1, p_2, p_3.....$p_{n.}$ Concatenate odd indices to obtain D'_A and even indices to obtain D'_B
Step 7: Generate the number sequence $L_{1,}$ $L_{2,}$ L_3... L_N by number seed $k_{3.}$ Extract the $(L_1 + 1)$th bit of D'_A to get M1 and similarly extract the $(L_1 + 1)$th bit of D'_B to get M2
Step 8: Concatenate M1 and M2 to get the desired message

(continued)

(continued)

The algorithm at Receiver End: Data decryption/extraction
Input: Private key 'k' and the DNA database
Output: Secret binary message M

5 Results and Discussion

In this section, we try to demonstrate the performance of the proposed work. The evaluation of the substitution method for data hiding is done using parameters like capacity, BPN, encryption time, and payload. Capacity 'C' refers to the length of the selected DNA when the data is concealed into it. Payload is known as the length of the sequence after recovering a particular sequence of DNA and is denoted by P. BPN may be defined as the count of bits concealed in every character. If the DNA sequence is taken as D, |D| will be the total length of the sequence, M is a message of length |M|, and K is the private key of length |K| bits.

The robustness of the method can be calculated by the probability of guess. Mathematically, we can represent

$$\text{Probability of Guess} = \frac{1}{1.163 \times 10^{8}} \times \frac{1}{24}$$

The performance of the method can be shown in Table 3.

For demonstration, several experiments have been carried out using the Python programming language. Encryption time is the time taken by CPU to encrypt the message, and decryption time is the time required to reverse the encryption process. The encryption time, decryption time, and BPN of the proposed method have been calculated and are presented in Table 4.

Table 3 Performance of the method

Method type	Capacity C	Payload P	BPN
Substitution	D	0	M/D

Table 4 Results of the substitution method

DNA sequence	Size (bytes)	Message (bytes)	Encryption time	Decryption time	BPN
BN000065	325750	420912	0.26482	0.21875	2
AC153526AC153526.13	206787	363214	0.234372	0.21875	2
AC167221AC167221.9	211669	304376	0.203125	0.15625	2
AC168908AC168908.3	206875	284900	0.1875	0.1620	2
AC168897AC168897.3	202281	284900	0.1875	0.15625	2

Table 5 Performance of the method

Type of method	Capacity (C)	Payload (P)	BPN
Insertion	D + M/2	M/2	M/(D + M/2)

Table 6 Results of insertion method

DNA sequence	Size (bytes)	Message (bytes)	Encryption time	Decryption time	BPN
BN000065	325,750	284,900	3.03125	1.0135	0.87
AC153526AC153526.13	206,787	167,920	1.953125	0.296875	0.65
AC167221AC167221.9	211,669	165,773	1.84375	0.28125	0.61
AC168908AC168908.3	206,875	165,773	1.8415	0.296875	0.74
AC168897AC168897.3	202,281	165,773	1.703125	0.265625	0.83

For evaluation of the insertion method for data encryption, the same parameters like capacity, BPN, encryption time, and payload have been used as shown in Table 5. For the second method, it can be seen that the robustness of the method depends upon the length of the key used.

Mathematically,

$$\text{Probability of Guess} = \frac{1}{2^K}$$

Where K represents the length of the key (in bits).

The experimental results of the proposed method have been demonstrated in Table 6.

6 Conclusion and Future Work

In this Chapter, we devise a state-of-the-art algorithm for data hiding using DNA sequences. The reference DNA sequence is obtained from a DNA database. There are about 1.63×10^8 DNA sequences publically available. The massive amount of DNA sequences available makes it increasingly difficult for an attacker to ascertain the existence of a secret message. Two data hiding algorithms have been proposed. The first one makes use of a substitution table and a reference sequence to convert secret data into Stego-DNA. In the second method for hiding data, DNA insertion method is applied. A private key is used to select the reference sequence of DNA and to select the binary coding for the DNA sequence. It is clear that with the help of DNA sequences, we can efficiently hide secret data. Furthermore, confidential documents and images can also be hidden efficiently inside a DNA sequence. We can see that, DNA manipulations can be used to study encryption and genetic programming for securing sensitive data. The proposed method can be utilized for secure data storage, transmission, and encryption with the help of DNA sequences.

Acknowledgements The authors would like to thank the Department of Science and Technology (DST) New Delhi, Government of India, for providing financial support under the DST Inspire Fellowship Scheme.

References

1. Wang Z, Zhao X, Wang H, Cui G (2013) Information hiding based on DNA steganography. In: 4th IEEE international conference on software engineering and service science (ICSESS), pp 946–949
2. Lee SH (2014) DWT based coding DNA watermarking for DNA copyright protection. Inf Sci 273:263–286
3. Clelland CT, Risca V, Bancroft C (1999) Hiding messages in DNA microdots. J Nat 399(6736):533–534
4. Hu Y, Lee H-K, Li J (2007) DE-based reversible data hiding with improved overflow location map. IEEE Trans Circ Syst Video Technol 19(2):250–60
5. Balado F (2013) Capacity of DNA data embedding under substitution mutations. IEEE Trans Inf Theory 52(2):928–941
6. Sukumaran SC, Mohammed M (2017) DNA cryptography for secure data storage in cloud. Int J Netw Secur 20(3):447–454
7. Biswas MR, Alam KMR, Akber A, Morimoto Y (2017) A DNA Cryptographic technique based on dynamic DNA encoding and asymmetric cryptosystem, 978-1-5386-3288-8/17/IEEE
8. Adleman ML (1994) Molecular computation of solutions to combinatorial problems. Science 266:1021–1024
9. Topol E (2011) The creative destruction of medicine. A member of perseus books group. Basic Books, New York
10. Amos M (ed) (2004) Cellular computing. Series in systems biology. Oxford University Press, USA
11. Giri KJ, Bashir R (2017) Digital watermarking: a potential solution for multimedia authentication. In: Dey N, Santhi V (eds) Intelligent techniques in signal processing for multimedia security. Studies in Computational Intelligence, vol 660. Springer, Cham
12. Alassaf N, Gutub A, Parah SA et al (2019) Enhancing the speed of SIMON: a light-weight-cryptographic algorithm for IoT applications. Multimed Tools Appl 78:32633–32657. https://doi.org/10.1007/s11042-018-6801-z
13. Giri KJ, Peer MA, Nagabhushan P (2014) A robust color image watermarking scheme using discrete wavelet transformation. IJ Image Gr Signal Proces 1:47–52
14. Lee SH, Lee EJ, Hwang WJ, Kwon KR (2017) Reversible DNA data hiding using multiple difference expansions for DNA authentication and storage. Multimed Tools Appl. https://doi.org/10.1007/s11042-017-5379-1
15. Parah SA, Ahad F, Sheikh JA, Loan NA, Bhat GM (2017) Information hiding in medical images: a robust medical image watermarking system for E-healthcare. Multimed Tools Appl 76(8):10599–10633
16. Parah SA, Sheikh JA, Akhoon JA, Loan NA (2018) Electronic health record hiding in images for smart city applications: a computationally efficient and reversible information hiding technique for secure communication, Futur Gener Comput Syst https://doi.org/10.1016/j.future.2018.02.023
17. Parah SA, Sheikh JA, Ahad F, Bhat GM (2018) High capacity and secure electronic patient record (EPR) embedding in color images for iot driven healthcare systems. Internet of things and big data analytics toward next-generation intelligence, Edited book, pp 409–437

18. Verma AK, Dave M, Joshi C (2008) DNA cryptography: a novel paradigm for secure routing in MANETs. J Discret Math Sci Cryptogr 11(4):393–404
19. Gehani A, Thomas L, Reif J (2004) DNA-based cryptography. Aspects of molecular computing. Springer, Berlin, Heidelberg, pp 167–188
20. Boneh D (1996) Breaking DES using molecular computer. American Mathematical Society, pp 37–65
21. Shimanovsky B, Feng J, Potkonjak M (2002) Hiding data in DNA. Procs. of the 5th Intl: Workshop in Information Hiding, pp 373–386
22. Chen J (2003) A DNA-based biomolecular cryptography design. In: Proceedings of IEEE international symposium, vol 3, pp III-822
23. Guangzhao C, Qin L, Wang Y, Zhang X (2008) An encryption scheme using DNA technology. In: 3rd IEEE International Conference on Bio-inspired computing: theories and applications, BICTA, pp 37–42
24. Lipton JR (1995) Using DNA to solve NP-complete problems. Science 268:542–545
25. Kumar D, Singh S (2011) Secret data writing using DNA sequences. In: IEEE international conference on Emerging trends in networks and computer communications (ETNCC), pp 402–405
26. Tornea O, Monica BE (2013) Security and complexity of a DNA-based cipher. In: Roedunet international conference (RoEduNet), 11th IEEE International Conference, pp 1–5
27. Tanaka K, Okamoto A, Saito I (2005) Public-key system using DNA as a one-way function for key distribution. Biosystems 81(1):25–29
28. Lai X, Lu M, Qin L, JunSong H, XiWen F (2010) Asymmetric encryption and signature method with DNA technology. Sci China Inf Sci 53(3):506–514
29. Bahig HM, Nassr DI (2019) DNA-based AES with silent mutations. Arab J Sci Eng 44:3389–3403. https://doi.org/10.1007/s13369-018-3520-8
30. Zhang Y, Fu B, Zhang X (2012) DNA cryptography based on DNA fragment assembly, Information science and digital content technology (ICIDT), 8th IEEE International Conference, vol 1, pp 179–182
31. Fu J, Zhang W, Yu N, Ma G, Tang Q (2014) Fast tamper location of batch DNA sequences based on reversible data hiding. 7th international conference on biomedical engineering and informatics (BMEI), pp 868–872
32. Ma G, Tang Q, Zhang W, Yu N (2013) Tamper restoration on DNA sequences based on reversible data hiding. Int Conf Biomed Eng Inf (BMEI) 1–13:484–489
33. Chen T (2007) A novel biology-based reversible data hiding fusion scheme. Front Algorithm (FAW) Lect Notes Comput Sci, vol 4613, pp 84–95
34. Hamed G et al (2016) Comparative study for various DNA based steganography techniques with the essential conclusions about future research. 11th International Conference on Computer Engineering and Systems (ICCES). IEEE
35. Marwan S, Shawish A, Nagaty K (2017) Utilizing DNA strands for secured data-hiding with high capacity. Int J Interact Mob Technol 11(2)
36. Hu X, Zhang W, Li X, Yu N (2015) Minimum rate prediction and optimized histograms modification for reversible data hiding. IEEE Trans Inf Forensic Secur 10(3):653–664
37. Malathi P et al (2017) Highly improved DNA based steganography. Procedia Comput Sci 115:651–659
38. Lee SH (2018) Reversible data hiding for DNA sequence using multilevel histogram shifting. Secur Commun Netw. https://doi.org/10.1155/2018/3530969
39. Huang YH, Chang CC, Wu CY (2014) A DNA-based data hiding technique with ow modification rates. Multimed Tools Appl 70(3):1439–1451
40. Shiu HJ, Ng KL, Fang JF, Lee RCT, Huan CH (2010) Data hiding methods based upon DNA sequences. Inf Sci 180(11):2196–2208
41. Maniyath SR, Supriya M (2011) An uncompressed image encryption algorithm based on DNA sequences. CCSEA, CS & IT-CSCP, pp 258–270
42. Li X, Zhou C, Xu N (2018) A Secure and efficient image encryption algorithm based on DNA coding and spatiotemporal chaos. Int J Netw Secur 20(1):110–120

43. Enayatifar R, Abdullah AH, Isnin IF (2014) Chaos-based image encryption using a hybrid genetic algorithm and a DNA sequence. Opt Lasers Eng 56(5):83–93
44. Zhang Q, Guo L, Wei X (2010) Image encryption using DNA addition combining with chaotic maps. Math Comput Model 52(1112):2028–2035
45. Hu T, Liu Y, Gong LH, Ouyang CY (2016) An image encryption scheme combining chaos with cycle operation for DNA sequences. Nonlinear Dyn pp 1–16
46. Song C, Qiao Y (2015) A novel image encryption algorithm based on DNA encoding and spatiotemporal chaos. Entropy 17(10):6954–6968
47. Zhen P, Zhao G, Min L, Jin X (2016) Chaos-based image encryption scheme combining DNA coding and entropy. Multimed Tools Appl 75(11):6303–6319
48. Cui G, Han D, Wang Y, Wang Y, Wang Z (2014) An improved method of DNA information encryption. BIC-TA, CCIS 472:73–77
49. Sarker A, Ahmed T, Rashid SMM, Anwar S, Jaman L, Tara N, Alam MM, Babu HMH (2011) Realization of reversible logic in DNA computing, IEEE 11th international conference on bioinformatics and bioengineering (BIBE)
50. Xue B, Xiyu L (2013) Design of E-commerce development roadmap for SMEs using DNA computing technique. IEEE 17th international conference on computer supported cooperative work in design (CSCWD)
51. Amin ST, Saeb M, Salah EG (2006) A DNA-based implementation of YAEA encryption algorithm. In: Computational intelligence, pp 120–125
52. Azoug SE, Bouguezel S (2016) A non-linear pre-processing for Opto-digital image encryption using a multiple-parameter discrete fractional Fourier transform. Opt Commun 359(1):85–94
53. Chai X, Gan Z, Yang K, Chen Y, Liu X (2017) An image encryption algorithm based on the memristive hyperchaotic system, cellular automata, and DNA sequence operations. Sig Process Image Commun 52:6–19
54. Qi O, Kaplan DP, Shumao L, Libchaber A (1997) DNA solution of the maximal clique problem. Science 278(5337):446–449
55. Sabari P, Setua KS (2012) DNA cryptography. In: Electrical & Computer Engineering (ICECE), 7th IEEE International Conference, pp 551–554
56. Heider D, Barnekow A (2007) DNA-based watermarks using the DNA-crypt algorithm. BMC Bioinf 8(176)
57. Heider D, Barnekow A (2011) DNA watermarking: challenging perspectives for biotechnological applications. Curr Bioinforma 6(3):375–382
58. Giri KJ, Peer MA, Nagabhushan P (2014) A channel wise color image watermarking scheme based on discrete wavelet transformation. In: 2014 international conference on computing for sustainable global development (INDIACom), pp 758–762. https://doi.org/10.1109/indiacom.2014.6828064
59. Tang Q, Ma G, Zhang W, Yu N (2014) Reversible data hiding for DNA sequence and its applications. Int J Dig Crime Forensic Arch 6(4):1–13
60. Anwar T, Paul S, Singh S (2014) Message transmission based on DNA cryptography: review. Int J Bio-Sci Biotechnol 6215–222. https://doi.org/10.14257/ijbsbt.2014.6.5.22
61. Brindha K, Jeyanthi N (2017) Securing portable document format file using extended visual cryptography to protect cloud data storage. Int J Netw Secur 19(5):684–693
62. Patel D (2017) Accountability in cloud computing by means of a chain of trust dipen contractor. Int J Netw Secur 19(2):251–259
63. Giri KJ, Bashir RA (2018) A block-based watermarking approach for color images using discrete wavelet transformation. Int J Inf Technol 10:139–146. https://doi.org/10.1007/s41870-017-0075-y
64. Giri KJ, Bashir R, Bhat JI (2019) A discrete wavelet-based watermarking scheme for authentication of medical images. Int J E-Health Med Commun (IJEHMC) 10(4):30–38. https://doi.org/10.4018/ijehmc.2019100103
65. Vaanchig N, Xiong H, Chen W, Qin Z (2018) Achieving collaborative cloud data storage by key-escrow-free multi-authority CP-ABE scheme with dual revocation. Int J Netw Secur 20(1):95–109

66. Tai WL, Chang YF (2017) Comments on a secure authentication scheme for IoT and cloud servers. Int J Netw Secur 19(4):648–651
67. Parah SA, Ahad F, Sheikh JA, Bhat GM (2017) Hiding clinical information in medical images: a new high capacity and reversible data hiding technique. J Biomed Inform 66:214–230
68. Parah SA, Ahad F, Sheikh JA, Bhat GM (2017) A new reversible and high capacity data hiding technique for E-healthcare applications. Multimed Tools Appl 76(3):3943–3975
69. Parah SA, Sheikh JA, Akhoon JA, Loan NA, Bhat GM (2018) Information hiding in edges: a high capacity information hiding technique using hybrid edge detection. Multimed Tools Appl 77(1):185–207

Chapter 6
Information Hiding in Color Images Utilizing Multi-bit Substitution Patterns

Javaid A. Kaw, Shabir A. Parah, and G. M. Bhat

1 Introduction

The paradigm shift from conventional modes of technology to electronic service-based technology has resulted in exponential growth of multimedia industry. In modern approach, the baseline multimedia content is either in the form of text, audio or video. Exchange of this content over the network infrastructure has increased at a tremendous rate and information flow, whether private or public, usually take one of these multimedia forms. E-services are being governed by some set rules and an authorized access to the beneficiary. Pushing these set rules by some unauthorized maloperations can eventually put the valuable multimedia content at risk, especially in fields where this unauthorized access like information theft, copyrighted content duplication and distribution may result in capital losses, loss of security and life as well. A malicious attack may compromise the security of millions of individuals and cost multimedia industry billions of dollars. In light of the afore said discussion, the transit of information with both secrecy and security is a matter of utmost concern. The security of data has been conventionally dealt with cryptography; however, data hiding has flourished as an alternative in last decade. The former makes the message illegible and draw attention of the eavesdropper while as latter is an innocent looking media with no heed paid from the adversary. In order to combat unauthorised access of data, copying distribution and its transmission throughout the globe research community has paid tremendous attention for the same. Data hiding can be carried out in every type of digital file like text, audio, protocol, image and video. Digital images due to their tremendous size and bulk availability have gained a special focus. Video is a continuous repetition of still images and hence can also be put in the category of digital images. With the e-counterpart of all the conventional walks of life and their

J. A. Kaw · S. A. Parah (✉) · G. M. Bhat
Department of Electronics and Instrumentation Technology, University of Kashmir, Hazratbal, Srinagar 190006, India
e-mail: shabireltr@gmail.com

K. J. Giri et al. (eds.), *Multimedia Security*, Algorithms for Intelligent Systems,
https://doi.org/10.1007/978-981-15-8711-5_7

globalization, secure and safe data transmission has gained an utmost priority particularly in healthcare where transmitting a medical image from one corner to other along with the requite data has paved a smooth way for telemedicine and remote healthcare. Classification of data hiding techniques where a data hiding technique uses covert communication through steganography and watermarking is represented in [1]. A digital image loaded with the secret information is termed as a stego image, the hidden data is camouflaged so that it does not reflect any change in the media and no one suspects its innocence and the existence of the embedded data. Unlike watermarking, faithful data hiding techniques govern the recovery of original data from the stego image at all costs, while as in watermarking, even if the retrieved watermark may not be as clear as original, the image's ownership could still be judged. Data hiding is generally categorized in two domains, spatial domain and transform domain. In the former, the data is embedded in the pixels with minimal complexity and high data capacity [2–8], while as in the latter, the data is embedded in the coefficients after certain transformations like discrete cosine transform (DCT) or discrete wavelet transform (DWT) [9–11]. In present technology scenario and to acknowledge the demand of certain critical fields such as in the transmission of military, medical and forensic images, besides hidden data retrieval of original images has attained much importance. For such cases, reversible data hiding (RDH) is used. In RDH, a stego image is generated from a cover image in an imperceptible way and transmitted through a wired or wireless medium, and the receiving end ensures the lossless recovery of both, the embedded data and the original image. RDH is a paramount requirement for the application in healthcare, military, forensic and defense, and a lot of work has been carried out in this direction. Without the recovery of original image, the essence of these fields could hardly be achieved [12–18]. Besides RDH transmission, the lossless recovery of the data is very important and is taken care of by making the scheme unsusceptible to different intentional and natural attack. Different algorithms have been designed in different techniques to add multiple security levels to the transmitted data [19–22]. Most of the work in RDH has been carried out on grayscale images used as host. Hiding data in color images using RDH approach is relatively less visible in the literature [23, 24]. The reasons being that the deformations in a color image with certain perceptual quality parameters may be evident to human naked eye whereas a grayscale image under similar parameters may pass the subjective quality test. Since a color image is made up of red, green and blue planes and the intensity level of each color decides the visibility of an RGB image to human eye. The contribution of color planes and sensitivity of human eye to these planes has been shown in Table 1.

Table 1 Contribution of color planes and sensitivity of human eye

Description	Red (%)	Green	Blue
Contribution in a color pixel	30	59	11
Sensitivity of human eye retina cone cells	33	65	2

In this chapter, we propose a reversible data hiding technique for color images with different multi-bit substitution patterns with an application specific feasibility in terms of imperceptibility and embedding capacity. The data hiding can be done with a choice of number of bits to be embedded in a particular color plane.

Rest of the chapter is organized as: Sect. 2 introduces the related work. Section 3 describes the proposed scheme. Section 4 shows the experimental results and discussions and the chapter concludes in Sect. 5.

2 Related Work

Over the period of time, embedding data in digital images with focus on gray images has remained a dominant focus for the research community throughout the globe. Literature survey also reveals that glut of literature is available for data hiding in grayscale images compared to color images. Further reversible data hiding in color images has not been well researched compared to reversible data hiding in grayscale images. However, these schemes utilize the transform domain. The framework of embedding strategy reveals that RDH methods can be mainly categorized into different categories like: lossless compression-based methods, difference expansion (DE)-based methods, histogram modification-based methods and prediction-error expansion (PEE)-based methods. Proposed by Fridrich et al. [25] in lossless compression-based RDH scheme, the compression of a proper bit plane with minimum redundancy is presented, and it however does not support high capacity and low distortion performances due to the weak correlation within a bit plane. An improved method with image compression using mapping optimization is proposed by Qin et al. [26]. Zhang et al. [27] combined various state of art with the recursive compression to get the rate distortion bound. Tian [28] proposed the first DE technique with high capacity and low distortion RDH scheme. The scheme takes advantage by expanding the difference between the two neighboring pixels of pixel pairs; however, the embedding capacity could not exceed 0.671 bpp, and a maximum PSNR of 34.80 dB is achieved. First technique where histogram modifications are used was proposed by Ni et al. [29], in this method in an image, the zero or the minimum point of the intensity histogram was utilized, and data embedding was ensured by slightly modifying the image. The improvement in embedding performance using histogram modification was proposed by Luo et al. [30] where the embedding manipulation on the prediction-error histogram is done rather than the intensity histogram. Recently, RDH methods with prediction-error expansion (PEE), treated as an extension of DE methods and methods based on histogram modification, have gained considerable attention from RDH researcher community. It was Thodi and Rodriguez [31] who first proposed PEE-based RDH method where PEE was used for expansion of embedding instead of DE. The local correlation in PEE is being computed using a large neighborhood while as DE uses the correlation between two adjacent pixels. In recent years, the method proposed by Thodi and Rodriguez was improved in two directions: one, that many data embedding techniques were proposed like double-layered embedding, adaptive embedding, 2D difference-histogram modification, 2D pairwise prediction-error modification, multiple histogram modification and pixel-value-ordering technique incorporated two-dimensional pairwise prediction-error expansion [32–37] and the other, that some PEE methods exploited advanced prediction techniques [38, 39] to generate a steeper prediction-error histogram. Shuyuan et al. [40] ensure a good embedding capacity of 810,757 bits in a 512 × 512 host image and a PSNR of 37.14 dB, and the author has fully taken the advantage of correlation of R, G and B planes of a color image with high perceptual quality. The above published literature demonstrates a significant amount of work done in data hiding with gray images in focus. A very few methods are being seen to have dedicated to color images. The techniques, however, could be applied to color images by equally splitting the data for embedding individually, channel by channel. Recently, the strategy of channel-dependent payload partition and adaptive embedding was used in state-of-the-art color image RDH algorithm [41] for improvement in the embedding performances using inter-channel correlation with the minimization of the altered pixels for single channel embedding and assignment of different payloads for RGB channel, adaptively. The technique shows improved embedding capacity with good PSNR. However, in contrast with PSNR, the embedding capacity is not at par with the proposed scheme.

3 Proposed Scheme

Data hiding images with color images as host usually ensure a high data embedding capacity and better perceptual quality, the fact being reversible data hiding in color images is an interesting preposition, and the embedding can take place in all the three different color planes. In contrast with grayscale images, color images with their large image size with three different color planes, the applications with a demand of large payload capacity embedding in color imaged are highly encouraged, followed with property of reversibility with lossless extraction of the original image. The block diagram of the proposed scheme is shown in Fig. 1.

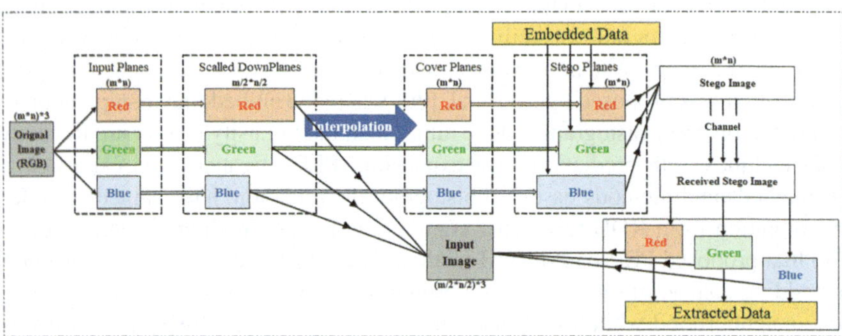

Fig. 1 Block diagram of the proposed scheme

The red, green and blue color planes are scaled down to form an input image of size $m/2 \times n/2$ which after interpolation again gets scaled up to $m \times n$ and acts as a host for the data to be embedded. The stego image is then constructed by combining three resultant stego planes of red, green and blue after involving certain multiple bits. At the receiver, the input image is obtained from the segregated planes of the received stego image, and further, the data is extracted separately from the three stego planes. The section is further divided as: 3.1 describes the process of Interpolation, multi-bit substitution pattern is described in 3.2, and the process of data embedding and data extraction are discussed in 3.3 and 3.4, respectively.

3.1 Interpolation

In a reversible hiding scheme, generation of cover image is a must. Reversibility can only be achieved by keeping the scope for the originality of the input image and is generally done by sparing the original pixels as unaltered and the embedding data in the pixels resulted during the process of interpolation, the pixels can be treated as redundant with respect to original image but act as the carrier for host data (payload) ensuring reversibility but unfortunately taking a toll on the perceptual quality of the image. The limitations of HVS allow a compromise between the payload and perceptual quality, and hence, the perceptual quality is taken care off while attaining optimal payload, security and robustness. Different interpolation techniques have been devised in different state-of-art techniques, and the technique neighbor mean interpolation (NMI) proposed by Jung and Yoo is one of them.

For the generation of cover, all the three planes are interpolated. The process of interpolation is similar in all the three cases, and thus, for the sake of comprehension we have considered interpolation of red plane with a 2×2 input image block, for illustration as shown in Fig. 2. The figure also shows the nomenclature of the pixels with a 2×2 block where it can be observed that every first pixel of the 2×2 block also termed as the seed pixel is dedicated to maintain the originality and rest of the three pixels act as host for the data to be embedded. Generation of this 2×2 cover

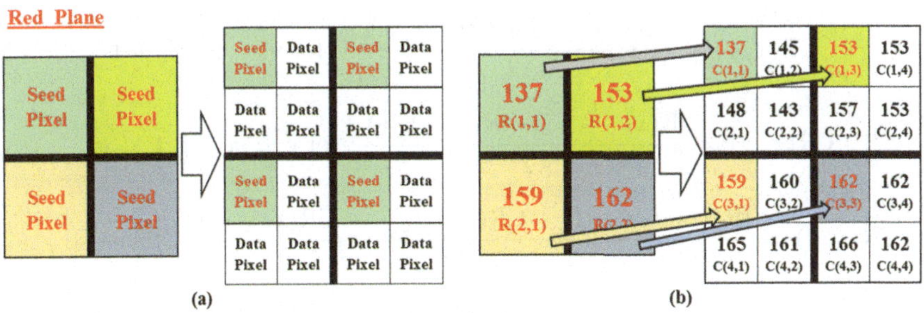

Fig. 2 Interpolation/upscaling 2×2 block of red plane

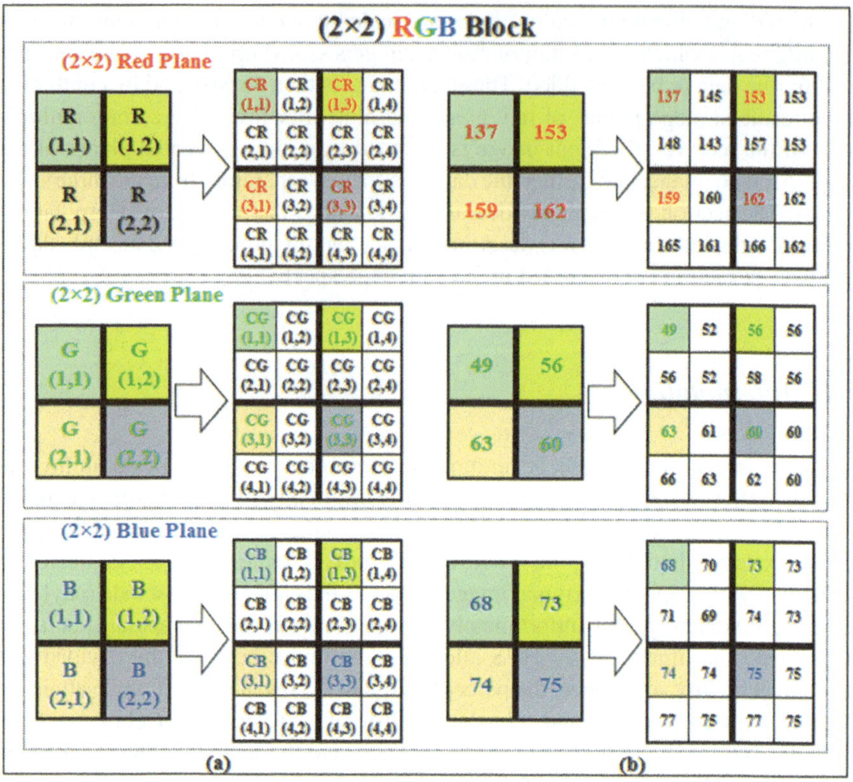

Fig. 3 Interpolation/upscaling 2 × 2 RGB image block

block is governed by simple mathematical equations as shown below by Eqs. (1) through (4). Generation of cover block for rest of the planes shall follow the same analogy and is illustrated in Fig. 3.

$$\mathrm{CR}(x, y) = \mathrm{IR}(x, y) \quad \text{for}\, x = m, y = n \tag{1}$$

$$\mathrm{CR}(x, y) = \frac{\mathrm{IR}(x, y-1) + \mathrm{IR}(x, y+1)}{2} \quad \text{for}\, x =, y = n+1 \tag{2}$$

$$\mathrm{CR}(x, y) = \frac{\mathrm{IR}(x-1, y) + \mathrm{IR}(x+1, y)}{2} \quad \text{for}\, x = m+1, y = n \tag{3}$$

$$\mathrm{CR}(x, y) = \frac{\mathrm{IR}(x-1, y-1) + \mathrm{CR}(x-1, y) + \mathrm{CR}(x, y-1)}{3} \text{Otherwise} \tag{4}$$

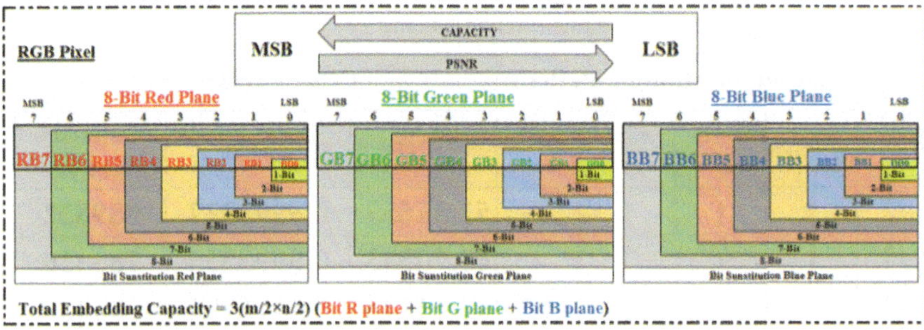

Fig. 4 Multi-bit substitution pattern

3.2 Multi-bit Substitution Pattern

The embedding capacity in a data hiding system is proportional to the bits involved in the process, greater the number of significant bits involved result in higher data embedding capacity but with reduction in image quality. Involvement of the significant bits in the process of data embedding is generally application specific, in some applications, demand for high payload capacity is observed, while as in some, the parameters like image quality and security are of prime importance. In a multi-bit substitution pattern when using LSB, the payload is minimum, while as the maximum image quality is achieved and as we move toward MSB, the payload capacity increases with a decrease in perceptual quality, the two are always in an inverse proportion, increase in one will always result in a decline toward other and vice versa. With respect to the image quality, the deformations with greater than 34 dB are measured objectively as the perceptual deformations are not visible to human visual system (HVS) and below 34 dB are subjective in nature and are generally visible to human open eye (Fig. 4).

The substitution of multi-bits is chosen with respect to desirable quality parameters of PSNR and payload. Table 2 illustrates all possible substitution patterns along with their respective parameters like PSNR, SSIM, payload, and depending on the selection of significant bits in different planes as illustrated in Fig. 4, the overall embedding capacity/payload of the cover is calculated using Eq. (5).

$$\text{Payload} = 3\left(\frac{m}{2} * \frac{n}{2}\right)(X\,\text{Bit}\,R + X\,\text{Bit}\,G + X\,\text{Bit}\,B) \tag{5}$$

The payload is explained as follows:

Example: 1 Embedding 2-data bits in red plane, 1-data bit in green plane and 3-data bits in blue plane provides an embedding capacity of 11,79,646 bits and PSNR of 42.85 dB.

where
X Bit R = Bits embedded in Red plane

Table 2 Different possible multi-bit substitution patterns

S No	Multibit Substitution Patterns				Other Parameters		
	Red bpp*3	Green bpp*3	Blue bpp*3	RGB bpp	PSNR	SSIM	Capacity (bits)
1	1	1	1	9	52.37557	0.9998327	589822
2	1	1	2	12	48.70583	0.9995533	786430
3	1	2	1		43.53648	0.9996794	786430
4	2	1	1		48.71330	0.9995989	786430
5	1	1	3	15	43.52571	0.9984393	983038
6	1	2	2		46.74373	0.9993987	983038
7	1	3	1		43.53336	0.9990683	983038
8	2	1	2		46.74294	0.9993181	983038
9	2	2	1		46.72979	0.9994445	983038
10	3	1	1		43.51466	0.9986731	983038
11	1	2	3	18	42.84867	0.9982919	1179646
12	1	3	2		42.82958	0.9987833	1179646
13	2	1	3		42.85479	0.9982200	1179646
14	2	2	2		45.39643	0.9991654	1179646
15	2	3	1		42.84694	0.9988371	1179646
16	3	1	2		42.83939	0.9983996	1179646
17	3	2	1		42.83585	0.9985165	1179646
18	1	3	3	21	40.81367	0.9976837	1376254
19	2	2	3		42.25533	0.9980662	1376254
20	2	3	2		42.24559	0.9985591	1376254
21	3	1	3		40.82601	0.997288	1376254
22	3	2	2		42.2341	0.9982424	1376254
23	3	3	1		40.80402	0.9979073	1376254
24	2	3	3	24	40.45561	0.9974570	1572862
25	3	2	3		40.44633	0.9971465	1572862
26	3	3	2		40.43062	0.9976382	1572862
27	3	3	3	27	39.19285	0.9965482	1769470

X Bit G = Bits embedded in Red plane

X Bit B = Bits embedded in Red plane

Table 2 also reveals that maximum embedding capacity of 1,769,470 bits is attained with multi-bit substitution pattern with bpp-27 while as minimum of 589,822 bits with multi-bit substitution pattern with bpp-9. The perceptual quality is highest in multi-bit substitution pattern with bpp-9 and lowest in bit substitution pattern with bpp-27; however, from the table, it is also observed that the multi-bit substitution pattern with minimum perceptual quality of 39.19285 dB does not exceed the limitations of HVS, and hence, the difference between the cover and the stego image could not be distinguished with human open eye.

3.3 Data Embedding

Being computationally efficient and easy to implement, LSB substitution for embedding data is used. The number of significant bits used for embedding in the planes shall depend on the selection of the multi-bit substitution pattern. The example given below in Fig. 5 illustrates the process of data embedding substitution pattern with bpp-18, as can be seen in Table 2 where embedding 6-bits in an RGB pixel is done with 2, 1 and 3 bits in red, green and blue planes, respectively.

In Fig. 5, a data stream (1 0 1 1 0 0 1 1 0 1 0 0 1 1 …….) has been taken for illustration. The data to be embedded may be the combination of both the secret data and fragile watermark interleaved with the data in a way so that the watermark bits are spread throughout the whole image so that malicious attack, if any, on any portion of the image is detected. The number of embedding bits is decided by the

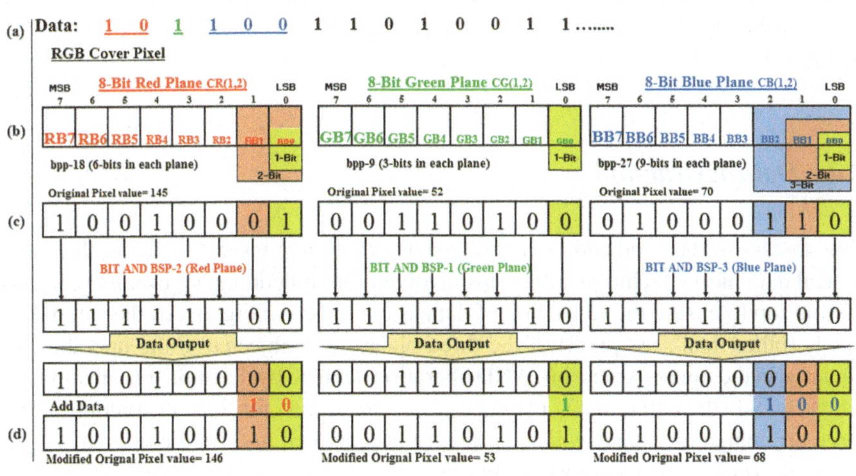

Fig. 5 Data embedding. **a** Data stream. **b** Multi-bit substitution pattern. **c** Original cover pixels. **d** Modified cover pixels

Table 3 Data embedding algorithm

Data Embedding Algorithm:	
INPUT: Original RGB Image (m×n×3), Watermark, Secret Data	
Step: 1	*Take an original input RGB image 'I' of size m×n×3.*
Step: 2	*Segregate the image into Red, Green and Blue Planes and generate a down sampled input image of size m/2×n/2×3.*
Step: 3	*Generate cover 'C' of size m × n, by interleaving columns and rows at alternate level using NMI.*
Step: 4	*Take a fragile watermark sized (N×M).*
Step: 5	*Convert the watermark to a single row vector.*
Step: 6	*Decide the BSP with the number of bits to be embedded in R, G and B planes of the cover image./*
Step: 7	*Take the secret data as a bit stream of single row vector and interleave the watermark bits in the data at regular intervals. The intervals would be decided by the selected Multi-bit Substitution Pattern.*
Step: 8	*Set the least significant bits of all the pixels of all the three planes after BSP of the cover to zeros except seed pixel.*
Step: 9	*Replace the significant bits as per BSP with the data (watermark interleaved).*
Step: 10	*Repeat Step 9 till all bits of the secret data are embedded to obtain stego-image.*
OUTPUT: Stego Image (m×n)	

multi-bit substitution pattern decided by the number of bits embedded in each plane as shown in Fig. 5b. The first six bits are embedded in the cover pixels shown in Fig. 5c resulting in modified cover pixels shown in Fig. 5d. The figure shows one pixel for illustration, and same analogy shall follow with rest of the data pixels of the cover block. The steps involved in the process of data embedding are mentioned in Table 3.

3.4 Data Extraction

The process of data embedding and data extraction are converse of each other. In a typical data hiding technique, the steps involved for embedding are usually repeated bottom top to ensure the extraction of data as explained in Fig. 6. All the stego data pixels (other than seed pixel) of the three planes are subjected to the operation as depicted in the figure. After a bit, AND operation decided by the multi-bit substitution pattern yields the output data. The seed pixels are spared for the construction of the three original color planes which are latter concatenated to form the original image 'I' of size ($m/2 \times n/2$) as shown in Fig. 6.

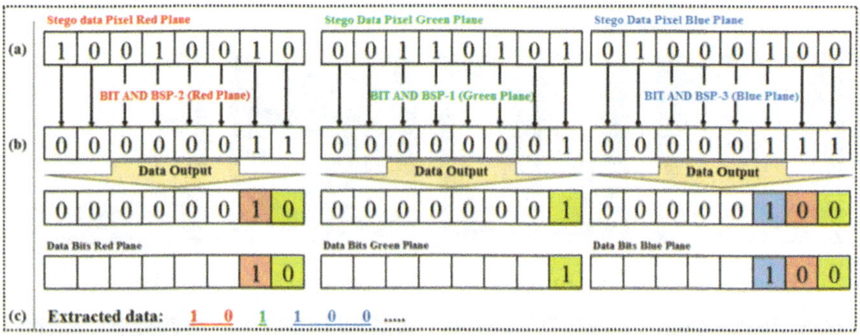

Fig. 6 Data extraction. **a** Stego data pixel. **b** After bit AND with multi-bit substitution pattern for plane. **c** Extracted data

Table 4 Data extraction algorithm

Data Extraction Algorithm:	
INPUT: Stego Image (m×n)	
Step: 1	*Receive the stego image sized (m×n).*
Step: 2	*Extract the original image 'I' of size (m/2 × n/2) from the seed pixels.*
Step: 3	*Extract the data from corresponding planes after executing the bit AND operations as decided by the Multi-bit Substitution Pattern.*
Step: 4	*Save the extracted data in a data extraction row matrix.*
Step: 5	*Extract the fragile watermark by subjecting the data to certain matrix operations in accordance with BSP.*
Step: 6	*Reshape the watermark data to the size (M×N).*
Step: 7	*Reshape the remaining data obtained in step-4 into a row matrix to yield the embedded secret.*
OUTPUT: Original RGB Image (m×n×3), Input RGB Image (m/2×n/2×3), Watermark (M×N) and Secret Data.	

The recovery of secret data and reconstruction of watermark is obtained after segregation of secret data and interleaved watermark bits. The steps involved in the extraction process are mentioned in Table 4.

4 Results and Discussions

This section presents the results with discussions. The algorithm has been tested with all the multi-bit substitution pattern; however, the results have been consolidated and compared with multi-bit substitution pattern with bpp-27. The test images with dimensions (512 × 512) used for the analysis were obtained from open source image data base are shown in Fig. 7 (Image-1 through Image-12). Besides test images,

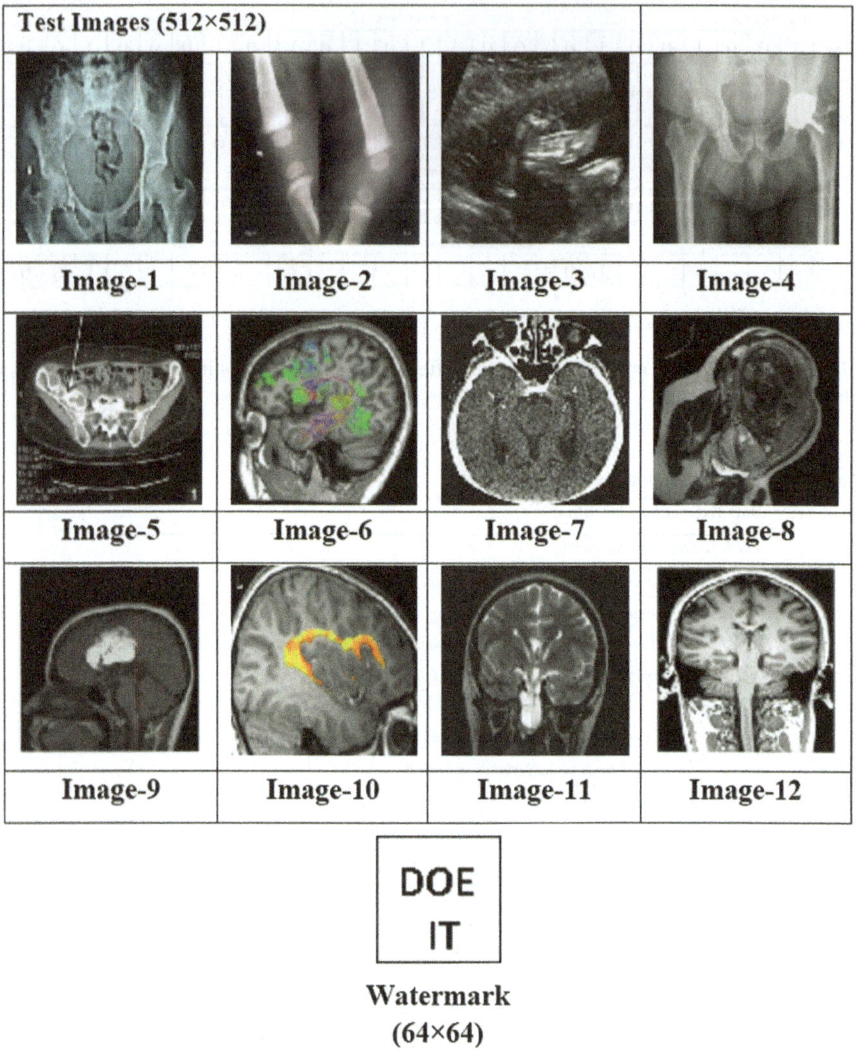

Fig. 7 Test Images (512 × 512) and Watermark (64 × 64)

the figure also shows a 64 × 64 fragile watermark. The logo and data have been concatenated in a way that the logo gets spread throughout the image so that tampering with any of the image portion gets reflected at the extraction of watermark. For objective quality analysis, the parameters like peak signal-to-noise ratio (PSNR) and structural similarity measure index (SSIM) have been used, while as for authenticity analysis, bit error rate (BER %) and normalized cross-correlation (NCC) have been put into force.

4.1 Imperceptibility Analysis

Imperceptibility deals the subjective quality of the images. With reference to the cover image, the perceptual quality of the stego image should be such that it should not draw attention of the attacker. The value of parameter PSNR gives the perceptual quality of an image, stego with respect to the cover. Higher values of PSNR mean better perceptual quality, while as the lower PSNR values witness poor visual quality of the image. Figure 8 reveals the results for multi-bit substitution pattern with bpp-9, bpp-12, bpp-15, bpp-18, bpp-21, bpp-24 and bpp-27; however, change in substitution pattern is application specific and can be decided with prerequisites for PSNR and data embedding capacity.

Table 5 illustrates the comparison of the proposed with some state-of-art techniques. From the comparison, it is evident that the proposed scheme shows improvements in both PSNR and embedding capacity. PSNR of the technique proposed by Jang and Yoo [43] is higher than proposed, but it is for relatively lesser payload.

Multibit Substitution Patterns	Other	Image-1	Image-2	Image-3	Image-4	Image-5	Image-6	Image-7	Image-8	Image-9	Image-10	Image-11	Image-12	Average Values
111 (9 bpp)	PSNR	52.376	52.395	52.386	52.394	52.373	52.395	52.394	52.373	52.4	52.384	52.399	52.416	**52.39042**
	SSIM	0.999	0.997	0.997	0.996	0.994	0.997	0.998	0.991	0.992	0.996	0.996	0.995	**0.995667**
112 (12 bpp)	PSNR	48.71	48.69	48.79	48.72	48.95	48.75	48.7	48.34	48.56	48.655	48.702	48.3781	**48.66246**
	SSIM	0.975	0.948	0.934	0.891	0.933	0.966	0.828	0.938	0.952	0.952	0.968	0.94	**0.935417**
113 (15 bpp)	PSNR	43.54	43.52	43.62	43.55	43.38	43.71	43.55	42.79	42.98	43.432	43.634	42.833	**43.37615**
	SSIM	0.975	0.948	0.934	0.891	0.933	0.966	0.828	0.938	0.952	0.952	0.968	0.94	**0.935417**
123 (18 bpp)	PSNR	42.831	42.824	42.883	42.854	42.79	42.979	42.848	42.127	42.335	42.733	42.92	42.156	**42.69**
	SSIM	0.99	0.968	0.976	0.962	0.957	0.976	0.982	0.902	0.923	0.963	0.967	0.942	**0.959**
133 (21 bpp)	PSNR	40.839	40.821	40.771	40.837	40.659	40.995	40.835	40.045	40.236	40.697	40.901	40.066	**40.64183**
	SSIM	0.986	0.957	0.966	0.949	0.937	0.968	0.976	0.869	0.895	0.95	0.956	0.922	**0.94425**
233 (24 bpp)	PSNR	40.456	40.443	40.359	40.441	40.332	40.615	40.441	39.676	39.864	40.33	40.526	39.68	**40.26358**
	SSIM	0.985	0.952	0.961	0.943	0.926	0.964	0.974	0.848	0.879	0.943	0.951	0.91	**0.936333**
333 (27 bpp)	PSNR	39.193	39.185	39.065	39.171	39	39.338	39.173	38.365	38.546	39.038	39.255	38.373	**38.97517**
	SSIM	0.979	0.934	0.946	0.923	0.892	0.949	0.965	0.801	0.837	0.923	0.936	0.882	**0.913917**

Fig. 8 Perceptual quality parameters with bpp-9, bpp-12, bpp-15, bpp-18, bpp-21, bpp-24 and bppp-27

Table 5 Comparison with some state of art

Scheme	Type	PSNR (dB)	Capacity (bpp)
Tian et al. [28]	Reversible	34.80	0.671
Shyuan et al. [40]	Irreversible	37.14	1.546
Tai et al. [42]	Reversible	37.98	0.493
Jang and Yoo [43]	Irreversible	41.46	0.766
Proposed (Multi-bit substitution pattern with bpp-27)	Reversible	38.97517 (Average)	2.25

4.2 Reversibility analysis

In order to prove reversibility, we need to recover all the three planes of the image at the receiving end. The proposed scheme recovers all the color planes of the image to its original form and is latter concatenated to form the original RGB image. Figures 9 and 10 illustrated the fact where the corresponding pixels of all the three planes (red, green and blue) on subtraction with the pixels of original color planes results in a black image, which means that the recovered RGB planes are same as that of original. The recovered planes when concatenated to form the original color image also yield a black image when subtracted with the original one as shown in Figs. 9 and 10. Moreover, the reversibility test stands true for all multi-bit substitution pattern; however, results with bpp-27 are shown, and rest can be obtained with the similar analogy.

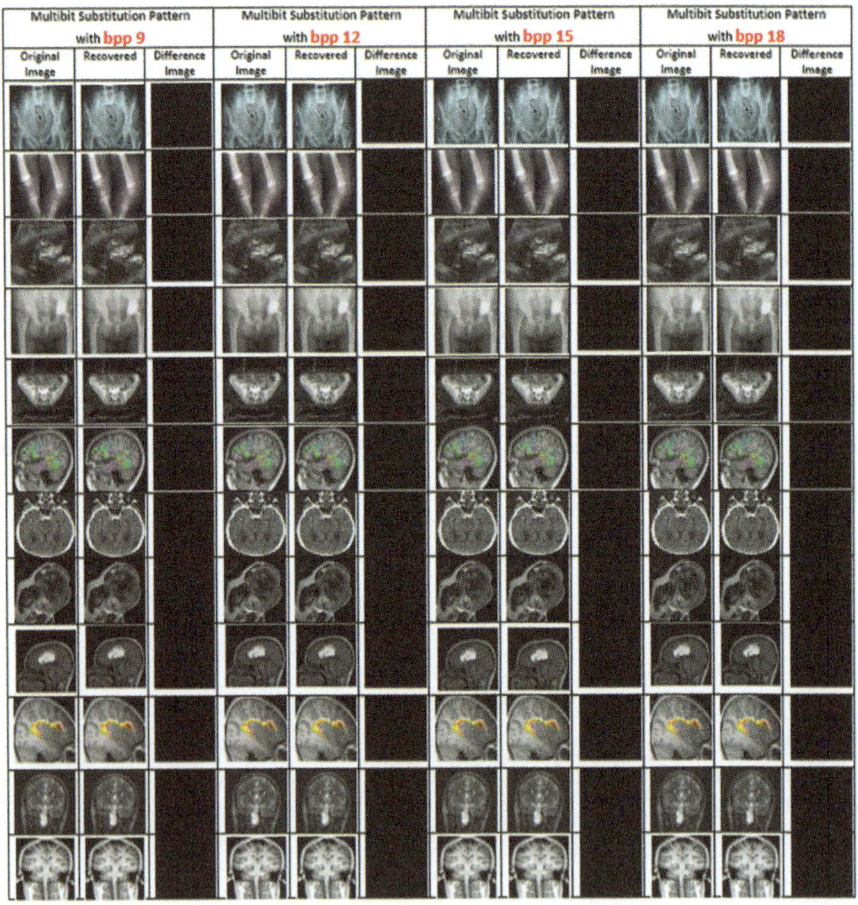

Fig. 9 Reversibility analysis with bpp-9, bpp-12, bpp-15 to bpp-18

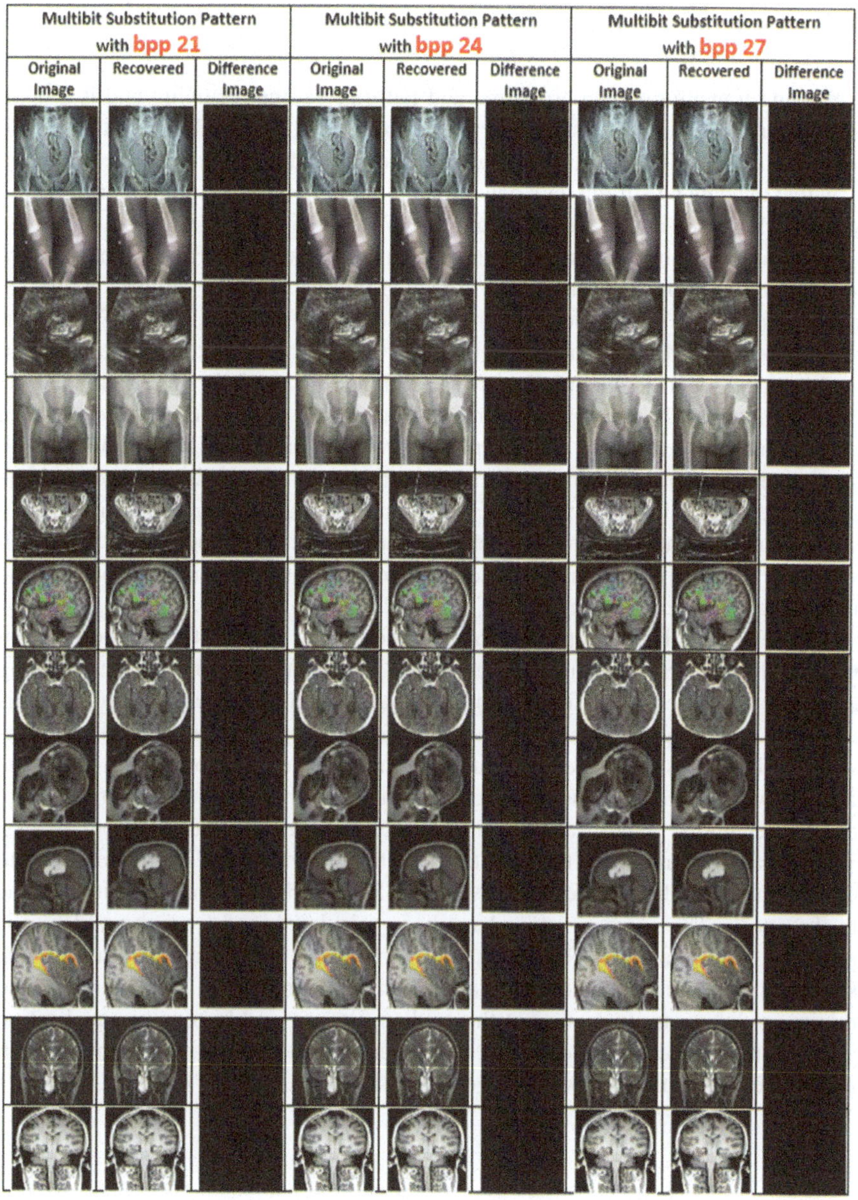

Fig. 10 Reversibility analysis with bpp-21, bpp-24 and bpp-27

4.3 Authentication and Fragility analysis

Receiving transmitted data error free and faithfully at the receiving end is a mark of efficient data hiding system. For authenticity of the embedded data, a fragile watermark/logo of size 64 × 64 is embedded in the RGB cover image in such a way that the data bits of the logo get scattered throughout all the color planes. These watermark bits are interleaved at regular intervals of the unit row data vector, and the interval is decided by the multi-bit substitution pattern selected. For example, in bpp-9, the watermark bits are interleaved at regular intervals of 143, while as in bpp-18, the watermark bits are interleaved at regular intervals of 287 and in bpp-27 at regular intervals of 575. Malicious attack, if any, may result in whole or partly deformation of the image due to erroneous scattered watermark bits, both the watermark and the data extracted at the receiving end may not resemble the one embedded with the cover. Failure in extracting the original watermark means that during transmission some intentional or un-intentional error has occurred in the media file and could be deliberate attempt by the adversary. For authenticity analysis, various geometric and signal processing attacks have been carried out with the RGB stego image, and in all such cases, a highly deteriorated watermark was extracted showing that our scheme is capable of content authentication.

Figures 11 and 12 reveal that the images when subjected to different attacks reveal the fact that watermark could not be extracted in its original form at the receiving end and can be seen in the figures. In the only case with no attach, the watermark is extracted to its original form hence making the technique fragile to all other attacks shown. The analysis has been carried out at multi-bit substitution pattern with bpp-27; however, the fragility check was carried out with other substitution patterns also, and the technique stands valid for all fragility tests carried with different attacks. Tables 6 and 7 summarize the effect on parameters with bpp-27 when exposed to certain attacks and the effect on PSNR and SSIM along with the data bits lost due to the attack. The average values have been calculated, and from the tables, it is observed that average PSNR is maximum with median filtering and minimum when subjected to geometric attack of rotation. Similarly, for SSIM, the maximum and minimum values are seen against the median filtering and histogram equalization attacks, and with respect to data loss, it is observed that data loss is minimum in salt and pepper and maximum when stego image subjected to JPEG compression.

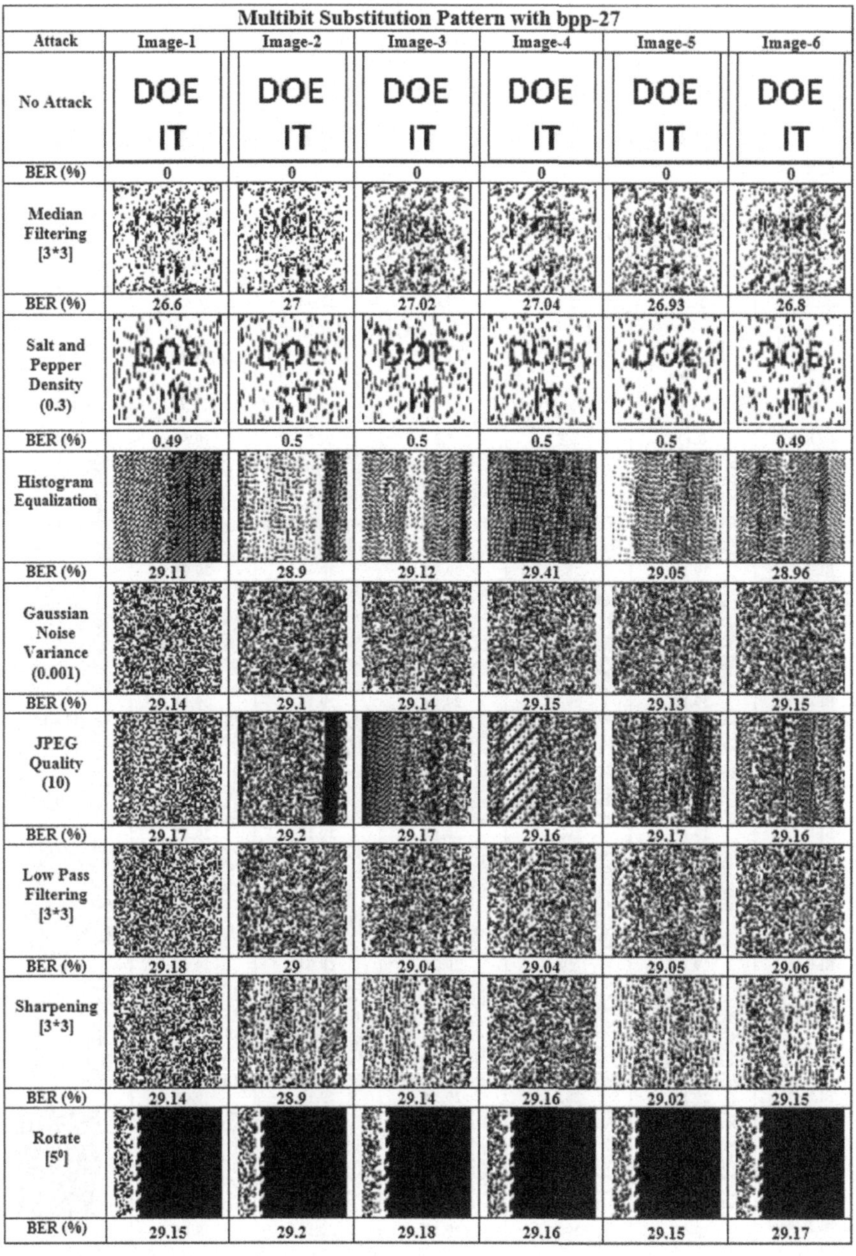

Multibit Substitution Pattern with bpp-27						
Attack	Image-1	Image-2	Image-3	Image-4	Image-5	Image-6
No Attack	DOE IT	DOE IT	DOE IT	DOE IT	DOE IT	DOE IT
BER (%)	0	0	0	0	0	0
Median Filtering [3*3]						
BER (%)	26.6	27	27.02	27.04	26.93	26.8
Salt and Pepper Density (0.3)						
BER (%)	0.49	0.5	0.5	0.5	0.5	0.49
Histogram Equalization						
BER (%)	29.11	28.9	29.12	29.41	29.05	28.96
Gaussian Noise Variance (0.001)						
BER (%)	29.14	29.1	29.14	29.15	29.13	29.15
JPEG Quality (10)						
BER (%)	29.17	29.2	29.17	29.16	29.17	29.16
Low Pass Filtering [3*3]						
BER (%)	29.18	29	29.04	29.04	29.05	29.06
Sharpening [3*3]						
BER (%)	29.14	28.9	29.14	29.16	29.02	29.15
Rotate [5^0]						
BER (%)	29.15	29.2	29.18	29.16	29.15	29.17

Fig. 11 Authentication/fragility analysis fir image-1 through image-6

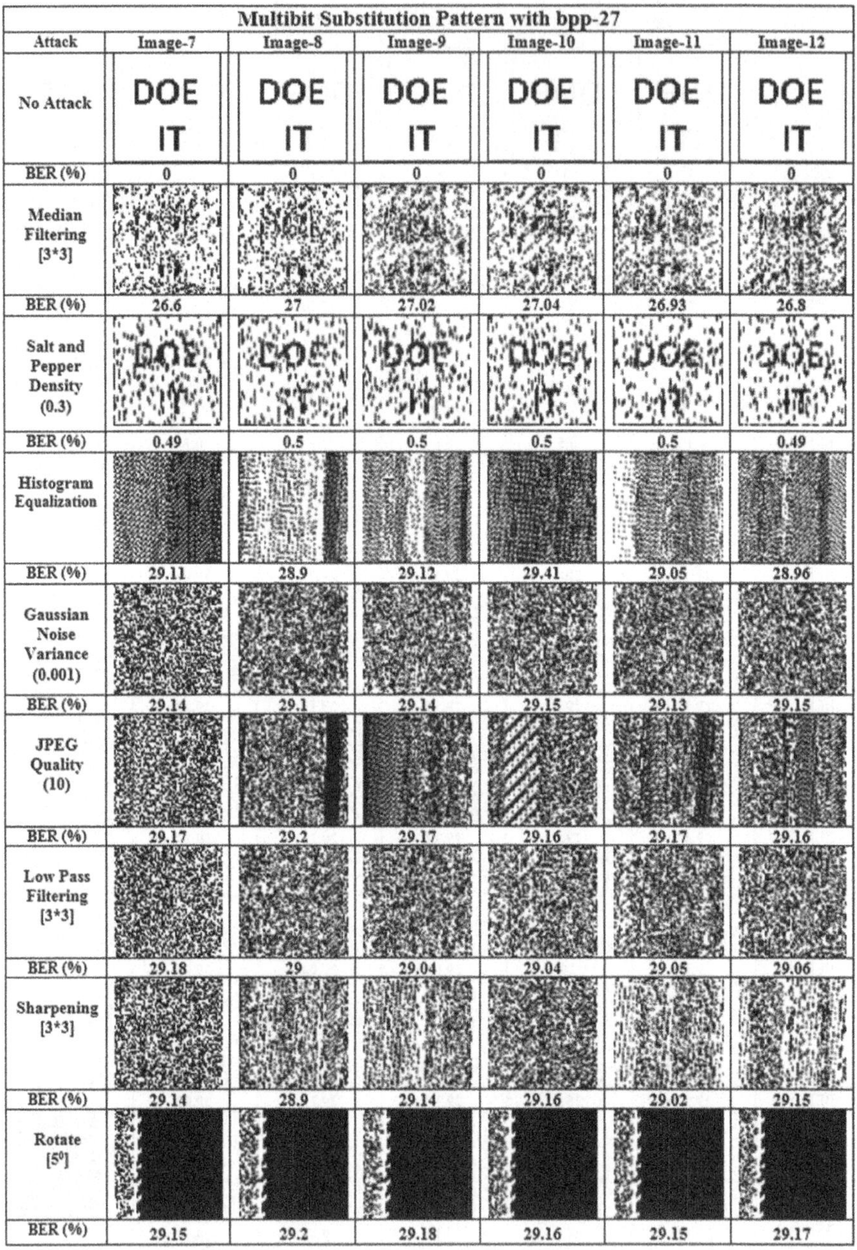

Fig. 12 Authentication/fragility analysis fir image-7 through Image-12

Table 6 Authentication analysis with different attacks

Multibit substitution pattern with bp		P27						
Attack	Authenticity" parameters	Image 1	Image-2	Image-3	Image-4	Image-5	Image-6	Image-7
No attack	PSNR	39.16031	39.2156S	39.16203	39.1505	39.10562	39.18275	39.162815
	SSIM	0.9928419	0.9647482	0.9989349	0.9489681	0.9985488	0.9917702	0.98263535
	Eroneous data bit	0	0	0	0	0	0	0
	PSNR	31.61348	39.01065	35.93016	34.56859	35.67713	32.99995	34.96749333
[3*3]	SSIM	0.9576811	0.9819381	0.9960817	0.9673782	0.9961919	0.983225	0.980416
	Eroneous data bit	469,442.0024	476,755.0024	477,050.0024	477,589.0024	475,784.0024	473,668.0024	475,048.0024
Salt and pepper density [0 J]	PSNR	25.21065	24.28642	25	24.73504	24.71919	24.81552	24.79447
	SSIM	0.8726386	0 7,039,427	09,320,254	0 669,043	09,328,809	0 8,519,802	0.827085133
	Eroneous data Bit	8851.0024	8899.0024	8666.0024	8771.0024	8951.0024	8995.0024	8855.5024
Histogram equalization	PSNR	1S.5594	10.01852	14.11285	11.14669	1,441,465	19.98517	14.70621333
	SSIM	0.7408595	0.2702864	0.2093324	0.3192283	0.360638	0.5123391	0.40211395
	Eroneous data bit	513,643.0024	510,911.0024	513,656.0024	519,523.0024	512,912.0024	510,392.0024	513,506.1691
Gaussian noise Variance (0.001)	PSNR	29.51607	29.72221	29.51678	29.50007	29.57633	29.51807	29.558255
	SSIM	0.9258456	0.7683467	0.9713925	0.7015794	0.9721684	0.9155133	0.87580765
	Eroneous data bit	514,595.0024	513,436.0024	514,546.0024	514,870.0024	514,209.0024	514,801.0024	514,409.5024
JPEG quality (10)	PSNR	28.58909	33.44065	32.2459	31.64567	32.16304	30.34637	31.40512

(continued)

Table 6 (continued)

Multibit substitution pattern with bp		p27						
Attack	Authenticity" parameters	Image 1	Image-2	Image-3	Image-4	Image-5	Image-6	Image-7
	SSIM	0.9111607	09,175,179	09,869,732	09,004,888	09,868,266	0.9576536	0.9434368
	Eroneous data bit	514,613.0024	515,074.0024	514,780.0024	515,659.0024	514,672.0024	515,026.0024	514,970.6691
Low pass filtering [3*3]	PSNR	30.49826	36.92973	32.92481	31.12392	33.05355	31.60141	32.68861333
	SSIM	0.9478685	0.9764948	0.9917485	0.9542363	0.9919174	0.9775511	0.973302767
	Eroneous data bit	514,664.0024	512,530.0024	513,316.0024	512,203.0024	512,622.0024	512,844.0024	513,029.8357
Sharpening [3*3]	PSNR	24.64797	29.87881	27.61073	26.35256	27.61822	25.60894	26.95287167
	SSIM	0.8957518	0.8586705	0.9724849	0.7986804	0.9732335	0.9256425	0.904077267
	Eroneous data bit	514,958.0024	509,745.0024	514,902.0024	514,527.0024	512,631.0024	514,429.0024	513,532.0024
Rotate [5°]	PSNR	15.52488	17.68881	14.98956	13.33677	13.71616	12.60916	14.64422333
	SSIM	0.5144418	0.6587417	0.8187483	0.5524781	0.7428905	0.5195466	0.6344745
	Eroneous data bit	514,651.0024	514,625.0024	514,899.0024	514,416.0024	514,839.0024	514,430.0024	514,643.3357

Table 7 Authentication analysis with different attacks

Multibit substitution pattern with bp		p 27					
Attack	Authenticity parameters	Image-8	Image-9	Image-10	Image-11	Image-12	Average values image 1–12
No attack	PSNR	39.16101	39.21638	39.16273	39.1512	39.10632	39.16144542
	SSIM	0.9935419	0.9654482	0.9996349	0.9496681	0.9992488	0.982165779
	Eroneous data bit	0	0	0	0	0	0
Median filter [3*3]	PSNR	31.61918	39.01135	35.93086	34.56929	35.67783	35.13174694
	SSIM	0.9583811	0.9826381	0.9967817	0.9680782	0.9968919	0.980473 5 83
	Eroneous data bit	469,442.0031	476,755.0031	477,050.0031	477,589.0031	475,784.0031	475,163.0027
Salt and pepper density [0.3]	PSNR	25–21,135	2,428,712	25.0007	24.73574	24.71989	24.7930075
	SSIM	0.8733386	0.7046427	0.9327254	0.669743	0.9335809	0.825302211
	Eroneous data bit	SS51.0031	8899.0031	8666.0031	8771.0031	8951 0031	8843.877692
Histogram equalization	PSNR	18.5601	10.01922	14 11,355	11.14739	14.41535	14.26659194
	SSIM	0.7415595	0.2709864	02,100,324	0.3199283	0,361,338	0.393220188
	Eroneous data bit	513,643.0031	510,911.0031	513,656.0031	519,523.0031	512,912.0031	513,765.6832
Gaussian noise "Variance (0.001)	PSNR	29.51677	29.72291	29.51748	29.50077	29.57703	29.56189542
	SSIM	0.9265456	0.7690467	0.9720925	0.7022794	0.9728684	0.872790513
	Eroneous data bit	514,595.0031	513,436.0031	514,546.0031	514,870.0031	514,209.0031	514,376.S777
JPEG qualitv (10)	PSNR	28.58979	33.44135	32.2466	31.64637	32.16374	31.49364083
	SSIM	0.9118607	0.9182179	0.9876732	0.9011888	0.9875266	0.942543733
	Eroneous data bit	514,613.0031	515,074.0031	514,780.0031	515,659.0031	514,672.0031	514,966.0582
Low pass filtering [3*3]	PSNR	3,049,896	36.93043	32.92551	31.12462	33.05425	32.77950528
	SSIM	0.9485685	0.9771948	0.9924485	0.9549363	0.9926174	0.973240406

(continued)

Table 7 (continued)

Multibit substitution pattern with bp		p 27					
Attack	Authenticity parameters	Image-8	Image-9	Image-10	Image-11	Image-12	Average values image 1–12
	Eroneous data bit	514,664.0031	512,530.0031	513,316.0031	512,203.0031	512,622.0031	5,130,453,221
Sharpening [3*3]	PSNR	24.64867	29.87951	27.61143	26.35326	27.61892	27.06515764
	SSIM	0.8964518	0.8593705	0.9731849	0.7993804	0.9739335	0.902571831
	Eroneous data bit	514,958.0031	509,745.0031	514,902.0031	514,527.0031	512,631.0031	5,134,572,527
Rotate [S°]	PSNR	15.52558	17.68951	14.99026	1,333,747	13.71686	14.81410361
	SSIM	0.5151418	0.6594417	0.8194483	0.5531781	0.7435905	0.644343492
	Eroneous data bit	514,651.0031	514,625.0031	514,899.0031	514,416.0031	514,839.0031	514,661.1138

5 Summary

A reversible data hiding technique for color images is being presented in this chapter. The selection of a particular bit substitution pattern depends on the application. The technique shall be highly useful in the cases where reversible data embedding in color images is required. Data embedding can take place with all or any of the color planes with respect to the contribution of color planes in the overall image and the sensitivity of human eye to these colors or both. Depending on the requirements, the provision of trade-off between perceptual quality and the embedding capacity needs to be put in place. Experimental results show that this technique has a good perceptual quality and data embedding capacity while ensuring reversibility. The proposed scheme performs better compared to state of the art. In addition, the proposed scheme is capable of detecting any tamper, caused due to various signal processing and geometric attacks.

References

1. Petitcolas FAP, Anderson RJ, Kuhn MG (1999) Information hiding—a survey. Proc IEEE 87:1062–1078
2. Chan CK, Cheng LM (2004) Hiding data in images by simple LSB substitution. Pattern Recogn 37:469–474
3. Chang CC, Hsiao JY, Chan CS (2003) Finding optimal least-significantbit substitution in image hiding by dynamic programming strategy. Pattern Recogn 36:1583–1595
4. Thien C-C, Lin J-C (2003) A simple and high-hiding capacity method for hiding digit-by-digit data in images based on modulus function. Pattern Recogn 36:2875–2881
5. Wang RZ, Lin CF, Lin JC (2001) Image hiding by optimal LSB substitution and genetic algorithm. Pattern Recogn 34:671–683
6. Parah SA, Sheikh JA, Bhat GM (2012) Data hiding in intermediate significant bit planes, a high capacity blind steganographic technique. In: International conference on emerging trends in science, engineering and ...
7. Parah SA, Sheikh JA, Bhat GM (2012) High capacity data embedding using joint intermediate significant bit (ISB) and least significant bit (LSB) technique. J Inf Eng Appl. ISSN (Paper) 2224-5782
8. Loan NA, Hurrah NN, Parah SA, Sheikh JA (2017) High capacity reversible stenographic technique based on image resizing and pixel permutation. In: Fourth international conference on image information processing, ICIIP
9. Chang CC, Chen TS, Chung LZ (2002) A steganographic method based upon JPEG and quantization table modification. Inf Sci 141:123–138
10. Iwata M, Miyake K, Shiozaki A (2004) Digital steganography utilizing features of JPEG images. IEICE Trans Fundam E87- A:929–936
11. Noda H, Spaulding J, Shirazi MN, Kawaguchi E (2002) Application of bit-plane decomposition steganography to JPEG2000 encoded images. IEEE Sig Process Lett 9:410–413
12. Parah SA, Ahad F, Sheikh JA, Loan NA, Bhat GM (2017) Pixel repetition technique: a high capacity and reversible data hiding method for e-healthcare applications. Intelligent techniques in signal processing for multimedia security, 371–398
13. Parah SA, Ahad F, Sheikh JA, Loan NA, Bhat GM (2017) A new reversible and high capacity data hiding technique for E-healthcare applications. Multimedia Tools Appl 76(3):3943–3975
14. Parah SA, Sheikh JA, Ahad F, Loan NA, Bhat GM (2017) Information hiding in medical images: a robust medical image watermarking system for E-healthcare. Multimedia Tools Appl 76(8):10599–10633

15. Parah SA, Sheikh JA, Akhoon JA, Loan NA (2018) Electronic health record hiding in images for smart city applications: a computationally efficient and reversible information hiding technique for secure communication. Fut Gener Comput Syst
16. Loan NA, Parah SA, Sheikh JA, Akhoon JA, Bhat GM (2017) Hiding electronic patient record (EPR) in medical images: a high capacity and computationally efficient technique for e-healthcare applications. J Biomed Inf 73:125–136
17. Parah SA, Ahad F, Sheikh JA, Bhat GM (2017) Hiding clinical information in medical images: a new high capacity and reversible data hiding technique. J Biomed Inform 66:214–230
18. Kaw JA, Loan NA, Parah SA, Muhammad K, Sheikh JA, Bhat GM (2019) A reversible and secure patient information hiding system for IoT driven e-health. Int J Inf Manage 45:262–275
19. Parah SA, Sheikh JA, Hafiz AM, Bhat GM (2015) A secure and robust information hiding technique for covert communication. Int J Electron 102(8):1253–1266
20. Parah SA, Bhat GM, Sheikh JA (2014) A secure and efficient data hiding technique based on pixel adjustment: American. J Eng Technol Res 14(2):38–44
21. Parah SA, Sheikh JA, Assad UI, Bhat GM (2017) Realisation and robustness evaluation of a blind spatial domain watermarking technique. Int J Electron 104(4):659–672
22. Parah SA, Sheikh JA, Loan NA, Bhat GM (2017) A robust and computationally efficient digital watermarking technique using inter block pixel differencing. Multimedia Forens Secur, 223–252
23. Parah SA, Sheikh JA, Bhat GM (2013) Data hiding in color images: a high capacity data hiding technique for covert communication. Comput Eng Intell Syst 4(13):107–115
24. Hussan M, Parah SA, Qurashi GJ (2019) Realization of a reversible data hiding framework using histogram Bin shifting for color images. In: Fifth international conference on image information processing, ICIIP
25. Fridrich J, Goljan M, Du R (2002) Lossless data embedding—new paradigm in digital watermarking. Eurasip J Appl Sig Process 2:185–196
26. Qin C, Chang C-C, Chen Y-C (2013) Efficient reversible data hiding for VQ compressed images based on index mapping mechanism. Sig Process 93:2687–2695
27. Zhang W, Hu X, Li X, Yu N (2013) Recursive histogram modification: establishing equivalency between reversible data hiding and lossless data compression. IEEE Trans Image Process 22(7):2775–2785
28. Tian J (2003) Reversible data embedding using a difference expansion. IEEE Trans Circ Syst Video Technol 13(8):890–896
29. Ni ZC, Shi YQ, Ansari N, Su W (2006) Reversible data hiding. IEEE Trans Circ Syst Video Technol 16(3):354–362
30. Luo L, Chen Z, Chen M, Zeng X, Xiong Z (2010) Reversible image watermarking using interpolation technique. IEEE Trans Inf Foren Secur 5(1):187–193
31. Thodi DM, Rodriguez JJ (2007) Expansion embedding techniques for reversible watermarking. IEEE Trans Image Process 16(3):721–730
32. Sachnev V, Kim HJ, Nam J, Suresh S, Shi YQ (2009) Reversible watermarking algorithm using sorting and prediction. IEEE Trans Circ Syst Video Technol 19(7):989–999
33. Li X, Yang B, Zeng T (2011) Efficient reversible watermarking based on adaptive prediction-error expansion and pixel selection. IEEE Trans Image Process 20(12):3524–3533
34. Li X, Zhang W, Gui X, Yang B (2013) A novel reversible data hiding scheme based on two-dimensional difference-histogram modification. IEEE Trans Inf Foren Secur 8(7):1091–1100
35. Ou B, Li X, Zhao Y, Ni R, Shi Y (2013) Pairwise prediction-error expansion for efficient reversible data hiding. IEEE Trans Image Process 22(12):5010–5021
36. Li X, Zhang W, Gui X, Yang B (2015) Efficient reversible data hiding based on multiple histograms modification. IEEE Trans Inf Forens Secur 10(9):2016–2027
37. Ou B, Li X, Wang J (2016) High-fidelity reversible data hiding based on pixel-value ordering and pairwise prediction-error expansion. J Vis Commun Image Represent 39:12–23
38. Qin C, Chang C, Huang Y, Liao L (2013) An inpainting-assisted reversible steganographic scheme using a histogram shifting mechanism. IEEE Trans Circ Syst Video Technol 23(7):1109–1118

39. Lu T-C, Tseng C-Y, Deng K-M (2014) Reversible data hiding using local edge sensing prediction methods and adaptive thresholds. Sig Process 104:152–166
40. Shen S, Huang L, Tian Q (2015) A novel data hiding for color images based on pixel value difference and modulus function. Multimedia Tools Appl 74:707–728
41. Ou B, Li X, Zhao Y, Ni R (2015) Efficient color image reversible data hiding based on channel-dependent payload partition and adaptive embedding. Sig Process 108:642–657
42. Tia WL, Yeh CM, Chang CC (2009) Reversible data hiding based on histogram modification of pixel differences. IEEE Trans Circ Syst Video Technol 19(6):906–910
43. Jung KH, Yoo KY, Data hiding method using image interpolation. Computer

Chapter 7
Data Embedding in Color Images: A Secure Data Communication Framework Based on Modular Arithmetic

Bilal Ahmad, Asra Kamili, Solihah Gull, and Shabir A. Parah

1 Introduction

Digital multimedia can now be effortlessly shared over the Internet because of advancements in the field of signal processing and networking. However, the ease of transmission and availability of editing tools pose an immense threat to the security and privacy of data being transmitted [1–5]. A range of cryptographic techniques is being used to protect the privacy of data being shared by encrypting the data [6–8]. However, with a substantial increase in the computing power of modern computers, the security of data obtained using cryptographic techniques is under threat [9, 10]. In addition to this, encrypted secret messages arouse more suspicion and result in sudden attacks from hackers because of their unrecognizable form. Hence, whenever we want to transmit a huge quantity of secret data securely, steganography methods are quite effective than cryptographic methods. Steganography is a data hiding technique wherein secret data is embedded into some publicly available cover media. In steganography, cover media is not important but only the embedded data is useful [11]. Cover media such as audio, video, image, or text act as the carrier only. The main advantage of steganography over cryptography is that the existence of secret data is made imperceptible, and no clue is given to adversary regarding secret data transmission. Researchers from all over the world have shown immense interest in developing and exploring image steganography for securely transmitting data. We can categorize image steganography techniques into spatial domain and transform domain techniques based on the domain used for embedding secret data.

B. Ahmad · A. Kamili · S. Gull · S. A. Parah (✉)
Department of Electronics and Instrumentation Technology, University of Kashmir, Hazratbal, Srinagar 190006, India
e-mail: shabireltr@gmail.com

K. J. Giri et al. (eds.), *Multimedia Security*, Algorithms for Intelligent Systems,
https://doi.org/10.1007/978-981-15-8711-5_8

Among different image steganography methods based on spatial domain, the least significant bit modification is the simplest one but has disadvantages associated with it which include low embedding capacity and less imperceptibility. Keeping these points in view, a spatial domain-based technique which provides stego images of high quality for a reasonably high payload based on modular arithmetic has been proposed in this chapter. The proposed technique makes use of color images as cover media instead of grayscale images. Color images because of their three constituent bit planes provide a good cover media for hiding secret data. We have structured the remaining part of the chapter as follows: Sect. 2 describes the literature survey, and Sect. 3 describes the proposed scheme along with algorithms used for embedding and extraction. Experimental results have been presented in Sect. 4. At last, the chapter has been concluded in Sect. 5.

2 Literature Survey

We can securely transmit a huge amount of secret data using different image steganographic schemes. Steganography can be used for secret communication quite effectively because in steganography no clue is given to adversaries regarding secret data transmission. We can hide secret data into either the spatial domain or the transform domain. Both domains have associated advantages and disadvantages. Less complexity and high payload are associated with spatial domain-based steganographic methods while transform domain-based methods are more robust and secure [12]. These image steganography methods based on either of the two domains can be assessed by subsequent key features: (1) Payload, which gives information about the amount of data that can be embedded secretly. (2) Imperceptibility, which denotes the quality of stego image, obtained after secret data embedding and (3) Security, which gives information about how securely the scheme can resist different detection attacks. Every steganographic method must accomplish all the above objectives but it is tough for one method to meet all the three requirements. Hence, there is always a trade-off between the three objectives. In literature related to steganography, different steganographic schemes use different approaches for embedding secret data. The ease of embedding and extraction makes spatial domain-based steganographic techniques more popular than those based on the transform domain. Some spatial domain-based steganographic methods include LSB substitution methods [13–19], pixel value-differencing methods [20–24], modification direction-based methods [25, 26], multiple-base notation-based methods [27, 28], pixel pair-matching-based methods [29], gray-level modification-based methods [30], pixel-prediction-based [31], histogram-based methods [32, 33], edge-based methods [34], pixel indicator-based methods [35], and color model-based methods [36]. The most conventional and common method of data hiding in the spatial domain is the least significant bit

(LSB) substitution method [13]. We simply replace the LSB's of cover media pixels by secret data bits. The method is capable of transmitting a huge amount of secret data with ease and less complexity. The scheme makes use of multiple planes for embedding depending on payload and PSNR requirements. Different variants of LSB substitution have been proposed. The number of pixels used for embedding is decided based on image features such as texture, edge, intensity, and brightness [14]. In [16], an approach based on ± 1 LSB has been proposed with enhanced imperceptibility and security. The visual quality enhancement is achieved because LSB's of only one-third pixels are replaced by secret data. But unlike other LSB-based methods, this method results in low capacity. In [17], LSB substitution method based on a word hunt puzzle has been proposed. The scheme proposed in this paper reduces the number of alterations per pixel to a greater extent than other LSB substitution methods which in turn increase the imperceptibility. The LSB-WH-based method is resistant to only one detection attack, namely chi-square. To make the scheme more secure, secret data is first encrypted using some of the LSB-based methods [18, 19]. The main shortcoming of LSB-based methods is that their embedding capacity is directly related to imperceptibility. As soon as we try to increase the payload, the stego image quality gets degraded. Hence, new methods based on different approaches other than LSB have been proposed to overcome the visual quality issue associated with LSB-based methods. One such method is pixel value differencing. In [20], a scheme based on pixel value differencing has been proposed, wherein the difference between neighbouring pixels decides the number of bits that can be embedded secretly. The scheme divides the image in a zigzag direction into non-overlapping blocks and then inter-pixel difference decides the embedding capacity depending on preselected range intervals. The range intervals are chosen based on how sensitive the human visual system (HVS) is. The difference depends on the texture of the image and the greater the difference, the greater is the capacity. The PVD-based techniques offer high imperceptibility than LSB-based techniques but the main disadvantage in this technique is that histogram analysis can beat the main purpose of steganography which is the concealment of the secret data. Problems of overflow and underflow as well are associated with PVD. To resolve the problems associated with PVD, some researchers have combined PVD with some other techniques [21, 22]. In [23], a technique has been proposed which uses modulus function in addition to PVD for embedding secret data with enhanced capacity. The scheme exploits the correlation between the three planes of color image for embedding secret data. The overflow and underflow problems associated with PVD are removed by different correction processes. In [24], a scheme utilizing the horizontal and vertical edges of the image for secret data embedding has been proposed. The scheme uses two different approaches based on adaptive PVD for embedding: one using 2×2 pixels block and other using a 3×3 block. The stego images got using this method are of high quality. In [25], a tri-way PVD-based technique has been proposed to tackle the overflow and underflow problems of PVD with higher payload and imperceptibility. The scheme computes the best possible pixel values to embed secret data in each block.

The schemes that have been discussed above have associated advantages and disadvantages. Some have high imperceptibility but low capacity while some provide huge embedding capacity but are not secure. Some others accomplish both high imperceptibility and high capacity but are computationally more complex. Keeping all these things in mind, we have proposed a high capacity steganographic framework using modular arithmetic.

3 Proposed Scheme

Research toward data hiding has become more and more popular because of the fast-growing Internet technology. The key challenges for an efficient data hiding technique are capacity and imperceptibility. Ideally, both the capacity and imperceptivity should be as high as possible but there exists a trade-off between capacity and imperceptibility, i.e., if we try to increase the capacity, the imperceptivity decreases. In this chapter, we present a data hiding technique based on modulus4 (mod4) operation for embedding data in color images, which has a high capacity as well as an optimum imperceptivity. We have embedded secret data bits into the cover image by following the pixel intensities of the cover image. Each pixel of the color image comprises three colors, i.e., Red (R), Green (G), and Blue (B). Further, 8 bits are required to represent each primary color. In the proposed technique, the concept of "mod-factor" is introduced for embedding the secret data bits into the cover image which is calculated as

$$\text{Mod - factor} = 2^n \tag{1}$$

where n is the number of bits that can be embedded into each component of a pixel. We have taken Mod-factor as 4 which implies that we can embed 2 bits of secret data into each component of a pixel, and hence, a total of 6 bits can be embedded into each pixel, resulting in embedding capacity of 6 bits per pixel (bpp).

1. **Embedding Algorithm**
   The block diagram of the proposed scheme is Fig. 1. The cover image is portioned into its three planes, i.e., Red, Green, and Blue. Secret data bits are embedded into each plane by changing the pixel values according to the embedding algorithm, and finally, we get a stego image in which our secret message is hidden. Steps of the data embedding are explained in Algorithm 1.

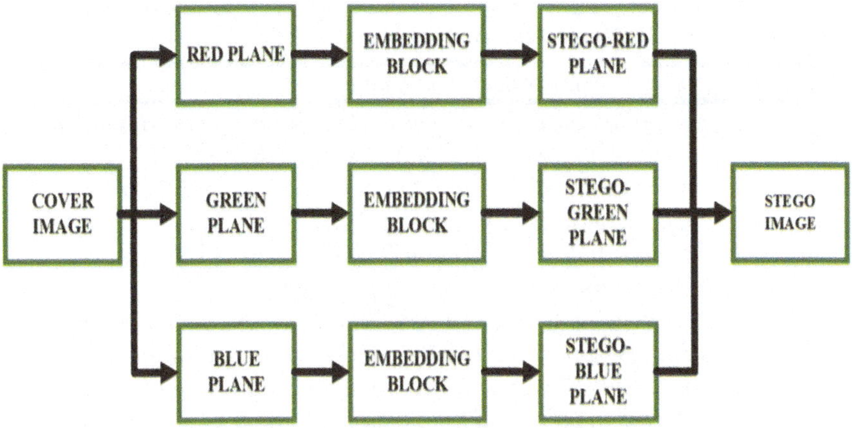

Fig. 1 Block diagram of the proposed scheme

Algorithm 1: The process used for embedding data.

INPUT: ***Cover Image (color), Secret data***

1. *The input cover image of size M×N is decomposed into its constituent planes i.e. Red, Green, and Blue plane.*
2. *Convert the secret data into base 3 values implying that each secret data digit Sd is of 2 bits.*
3. *Compute mod4 of R, G and B components of a pixel, denoted by Rm, Gm, and Bm respectively.*
4. *For embedding a secret data bit into the R component of a pixel, the difference D between the secret data digit Sd and the mod4 value of R component (Rm) is calculated as depicted in fig. 2 and fig. 3 and is given as*

$$D = S_d - Rm$$

5. *Embed the secret data bits into the G and B components of a pixel by calculating the difference between the corresponding secret data digit and the mod4 value Gm and Bm of G and B components respectively. The corresponding difference value D is added to G and B components resulting in Stego_G and Stego_B components.*
6. *Repeat steps from 3 to 6 until all the secret data digits are embedded.*
7. *Use constituent stego planes to get the stego image.*

OUTPUT: Stego Image

Embedding Example

Let $R = 40$, $G = 90$, $B = 59$ and let secret data digits be 2, 3, and 0 then

$$Rm = \text{mod}(40) = 0$$

$$Gm = \text{mod}(90) = 2$$

$$Bm = \text{mod}(59) = 3$$

$$D_1 = 2 - Rm = 2 - 0 = 2$$

$$D_2 = 3 - Gm = 3 - 2 = 1$$

$$D_3 = 0 - Bm = 0 - 3 = -3$$

Therefore,

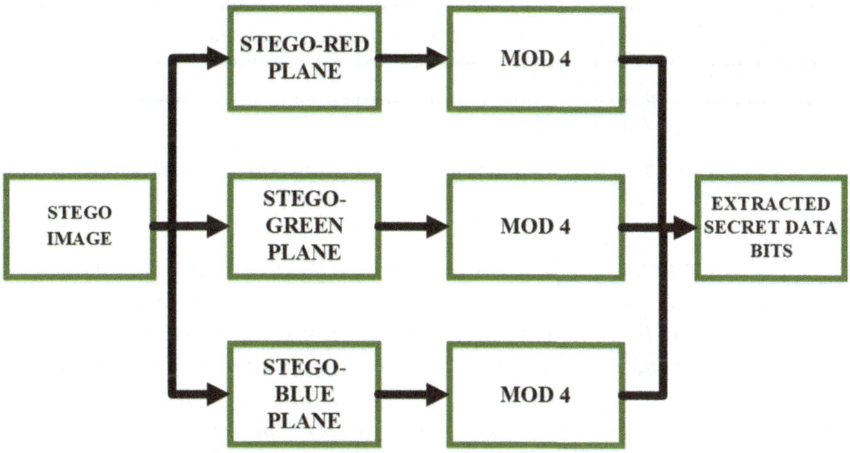

Fig. 2 Block diagram for the extraction process

$$\text{Stego\_}R = R + D_1 = 40 + 2 = 42$$

$$\text{Stego\_}G = G + D_2 = 90 + 1 = 91$$

$$\text{Stego\_}B = B + D_3 = 59 + (-3) = 56$$

2. **Extraction Algorithm**
   The block diagram of the extraction process has been shown in Fig. 2. Algorithm 2 explains how data is extracted from stego image planes.

Algorithm 2: The process used for extracting secret data

*INPUT: **Stego Image***

8. *Read the stego image.*
9. *Partition the stego image into its constituent planes i.e. Red, Green and Blue plane.*
10. *Compute mod4 of Stego_R, Stego_G and Stego_B components of a stego pixel, denoted by Stego_Rm, Stego_Gm and Stego_B respectively. Stego_Rm is the secret data digit hidden in the Stego_R component of the stego pixel. Stego_Gm is the secret data digit hidden in the Stego_G component of the stego pixel. Stego_Bm is the secret data digit hidden in the Stego_B component of the stego pixel.*
11. *Repeat step 2 until all the secret data digits are extracted.*
12. *Replace each secret data digit by its 2-bit binary equivalent to get the secret data bits.*

OUTPUT: Original Image, Secret data

Extraction Example: From the above example, we get

$$\text{Stego\_}R = 42$$

$$\text{Stego\_}G = 91$$

$$\text{Stego\_} = 56$$

then

$$\text{Stego\_}Rm = \text{mod4}(42) = 2$$

$$\text{Stego\_}Gm = \text{mod4}(91) = 3$$

$$\text{Stego\_}Bm = \text{mod4}(56) = 0$$

Therefore, we extract the secret data digits hidden in the stego pixel and these are 2,3, and 0.

4 Experimental Results

We have evaluated the proposed scheme for four standard color images namely Lena and Airplane of size 512 × 512, Baboon of size 225 × 225, and Pool of size 510 × 383. Simulations have been carried out using MATLAB software. The variety of image quality indices used to assess the effectiveness of the proposed scheme are discussed below:

1. **Peak Signal-to-Noise Ratio (PSNR)**: It is one of the most common and important image quality metrics. PSNR is used to measure the quality of the stego image obtained concerning the Cover image. If C represents the original image and S represents the stego image, then PSNR is given as

$$\text{PSNR (dB)} = 10\log_{10}\frac{255^2}{\text{MSE}} \tag{2}$$

where MSE is the mean square error and is given as:

$$\text{MSE} = \frac{1}{MN}\sum_{i=1}^{M}\sum_{j=1}^{N}[C(i,j) - S(i,j)]^2 \tag{3}$$

where $M \times N$ is the size of C and S. PSNR should be as high as possible because higher PSNR value leads to a good quality stego image.

2. **Normalized correlation (NC)**: It is also a metric used to measure the similarity between the original image and the stego image. It is defined as

$$\text{NC} = \frac{\sum_i\sum_j C(i,j)S(i,j)}{\sum_i\sum_j[C(i,j)]^2} \tag{4}$$

3. **Bit Error Rate (BER)**: It is the ratio of secret data bits with incorrect extraction to the total number of secret bits embedded. Lesser the value of BER, more accurate is the extraction of secret data bits. It is defined as

$$\text{BER}(\%) = 1/m\left[\sum_{j=1}^{n} B(j) \oplus \text{Be}(j)\right] \times 100 \tag{5}$$

where m is the total number of secret data bits that are embedded, $B(j)$ is an original jth bit (i, j) and $\text{Be}(j)$ is the jth extracted bit.

4. **Normalized Absolute Error (NAE)**: It is the arithmetic difference between the original image and the stego image. We can use NAE to estimate the quality of the stego image obtained. It is given as:

$$\mathrm{NAE} = \frac{\sum_{i=1}^{M}\sum_{j=1}^{N}|C(i,j) - S(i,j)|}{\sum_{i=1}^{M}\sum_{j=1}^{N}|C(i,j)|} \tag{6}$$

NAE should be as small as possible.

5. **Structural Content (SC)**: The similarity between two images can be measured using this quality metric. The metric measures the similarity between two images by dividing them into small regions containing low-level significant information related to the structure of the image and comparing the results obtained. The images are then considered similar if we obtain a greater number of common regions between the two images. It is defined as:

$$\mathrm{SC} = \frac{\sum_{i=1}^{M}\sum_{j=1}^{N} C(i,j)^2}{\sum_{i=1}^{M}\sum_{j=1}^{N} S(i,j)^2} \tag{7}$$

The proposed mod4 technique has been tested on different color images of different sizes for evaluating imperceptivity and payload capacity. The comparison results of the proposed scheme with other existing schemes show that our technique provides high capacity and stego images obtained are of high quality as well. The maximum number of bits that can be embedded in $M \times N$ color image is given as:

$$n = 6MN \tag{8}$$

The imperceptivity and extraction errors of the proposed scheme are evaluated for different images using different payload capacities. This is done to compare our results with certain existing techniques.

The results obtained for each image when the maximum capacity of an image is used for embedding, i.e., 6 bits per pixel is illustrated as:

1. **Lena image**: For test image Lena of size 512 × 512 × 3, the maximum amount of secret data bits that can be embedded using the proposed scheme is 1,572,864. On embedding this amount of data, a high-quality stego image of PSNR = 44.14 dB is obtained and the secret data is retrieved with BER = 0. The original image along with its corresponding planes and stego image along with its corresponding planes are shown in Fig. 3, and further, their histograms are shown from Figs. 4, 5 and 6. From the images obtained for each plane, it is clear that the changes brought in the original image because of embedding are not recognizable by the human eye.
2. **Airplane image**: The second test image used is the airplane of size 512 × 512 × 3, which means the total number of bits that can be embedded is 1,572,864. When this amount of data is embedded, a high-quality stego image of PSNR = 44.12 dB is obtained and the secret data is retrieved from the stego image without any error. The various planes of test image airplane, and their corresponding stego-images are shown in Fig. 7. Figure 8 shows the histogram of original and stego-red plane

of the airplane image, while as histograms pertaining to green and blue planes are depicted in Figs. 9 and 10.

3. **Baboon image**: A standard baboon image of size 225 × 225 × 3 is also used as one of the test images. As 6 bits per pixel are embedded, therefore, the maximum number of bits that are embedded in this image are 303,750. After embedding this much of data, a high-quality stego image of PSNR = 44.2 dB is obtained and the secret data is retrieved without any error. The original image and planes along with stego image and stego planes are shown in Fig. 11, and corresponding histograms are shown from Figs. 12, 13 and 14.
4. **Pool image**: The experimental results are also obtained for a standard pool image of size 510 × 383 × 3. The maximum capacity of this image is 1,171,980. After

Fig. 3 Original image and planes along with stego image and stego planes

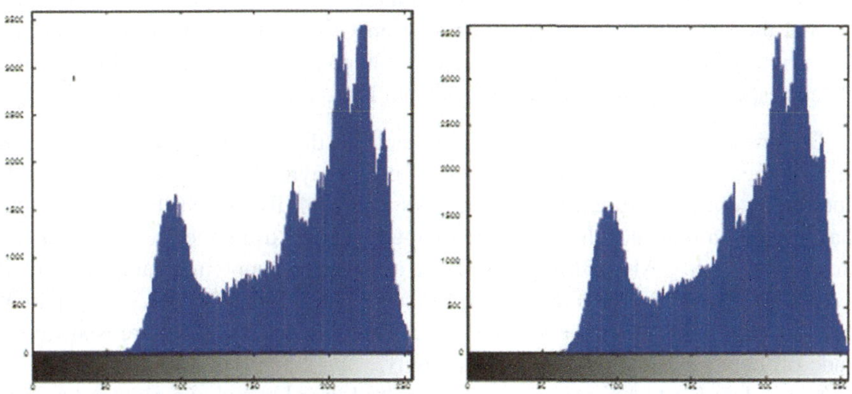

Fig. 4 Histogram of original and stego red plane

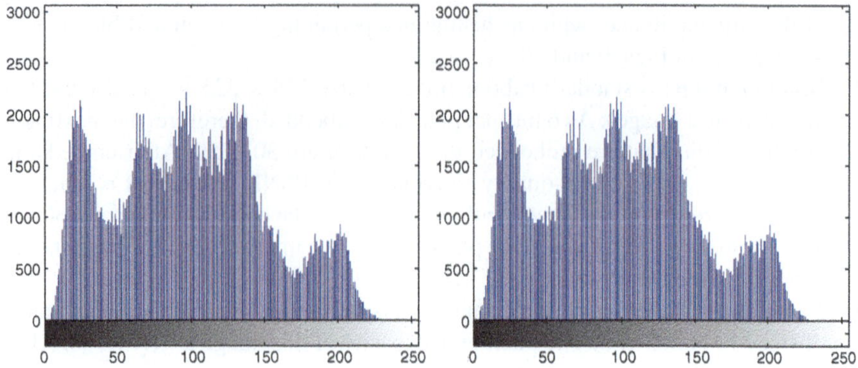

Fig. 5 Histogram of original and stego green plane

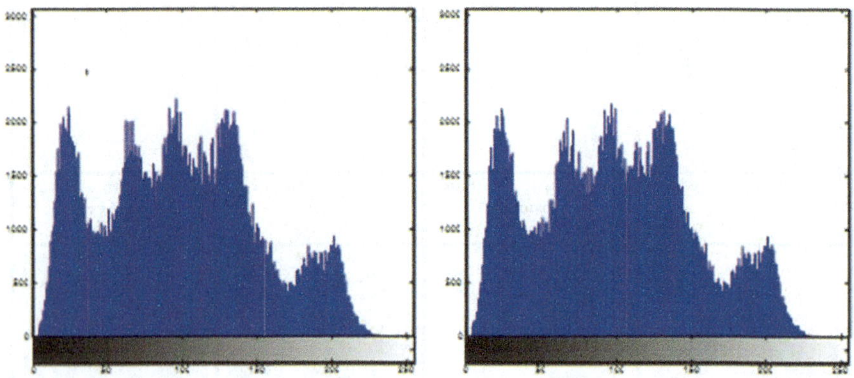

Fig. 6 Histogram of original and stego blue plane

embedding this much of data, a stego image of PSNR = 44.12 dB is obtained and the secret data is extracted without any error. The original image and its various planes along with stego image and stego planes are shown in Fig. 15, and corresponding histograms are shown from Figs. 16, 17 and 18.

The proposed technique has been evaluated for different payload capacities for each image. The results obtained against a given payload are reported in Table 1. The PSNR obtained for each case is always above 44 dB, and MSE is always less than 2.5 which indicates that the stego images obtained using the proposed scheme are of good quality. Also, BER and NAE are always zero, implying that the extracted secret data exactly matches the embedded secret data. From Table 1, it is apparent that as payload increases, the PSNR decreases, which proves that there is always a trade-off between capacity and imperceptivity.

To validate our technique, we have compared the obtained experimental results with some already existing spatial domain-based image steganographic techniques. The corresponding comparison results obtained have been presented in Table 2. The

Fig. 7 Original image and planes along with and stego image and stego planes

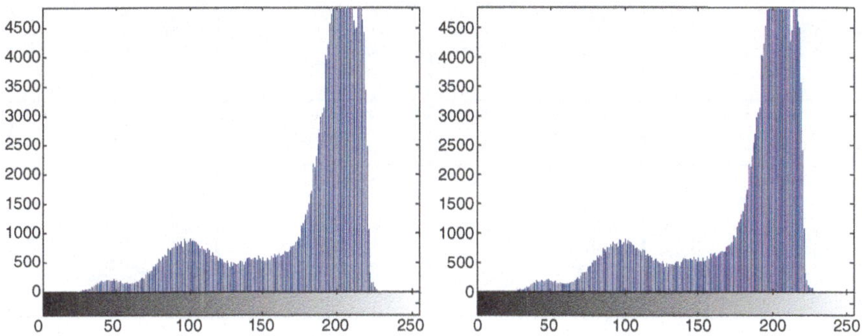

Fig. 8 Histogram of original and stego red plane

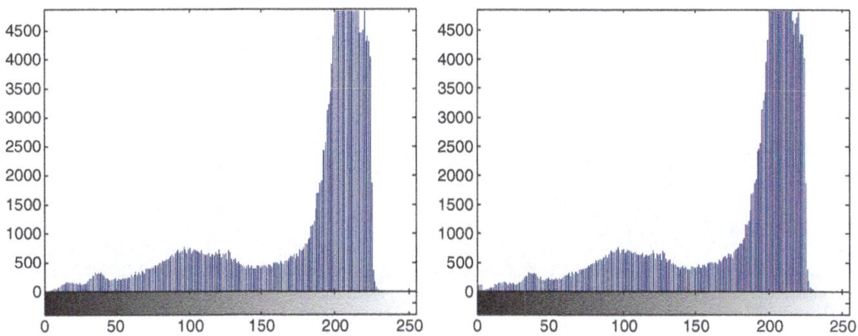

Fig. 9 Histogram of original and stego green plane

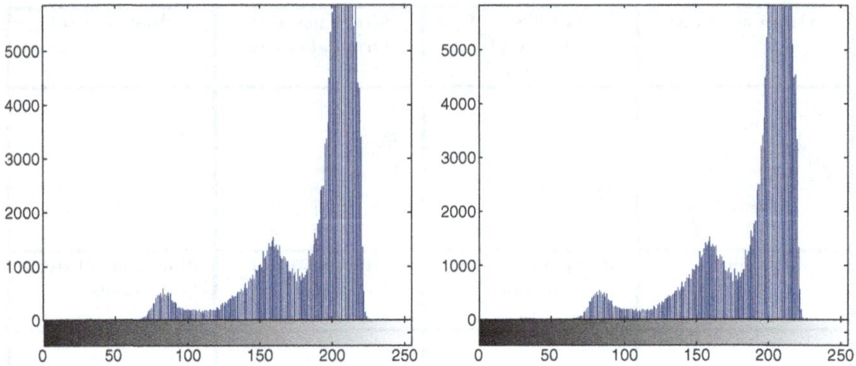

Fig. 10 Histogram of original and stego blue plane

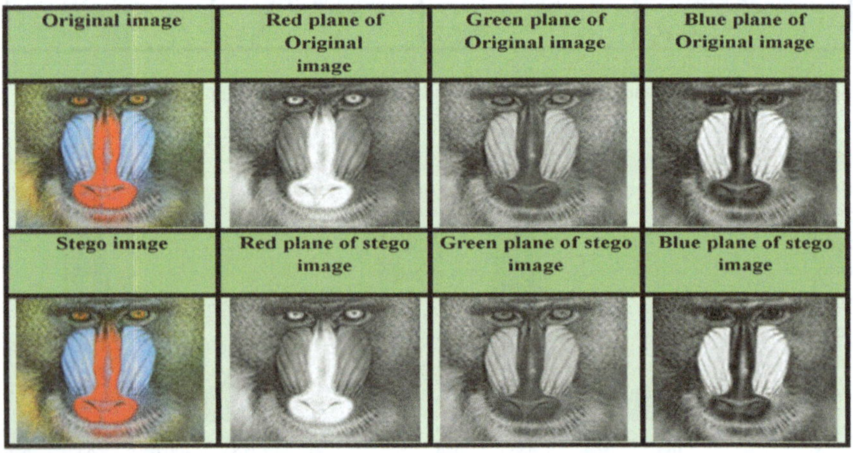

Fig. 11 Original image and planes along with and stego image and stego planes

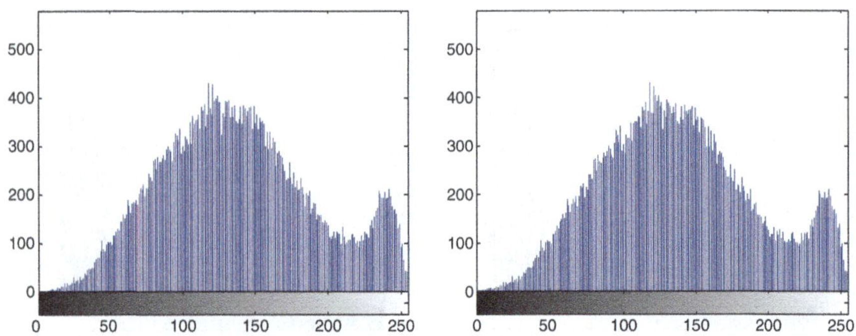

Fig. 12 Histogram of original and stego red plane

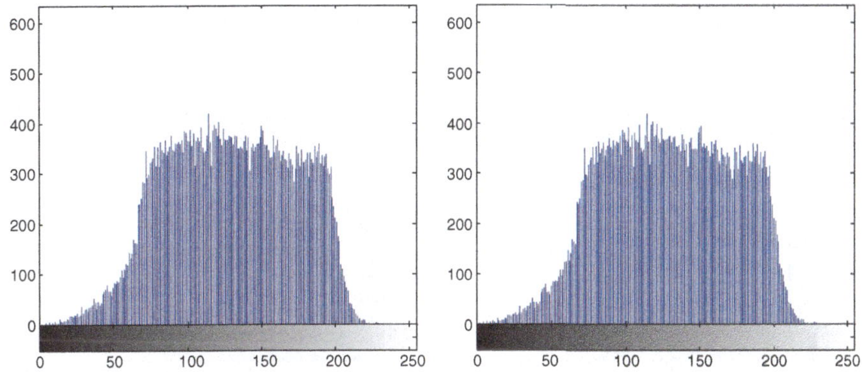

Fig. 13 Histogram of original and stego green plane

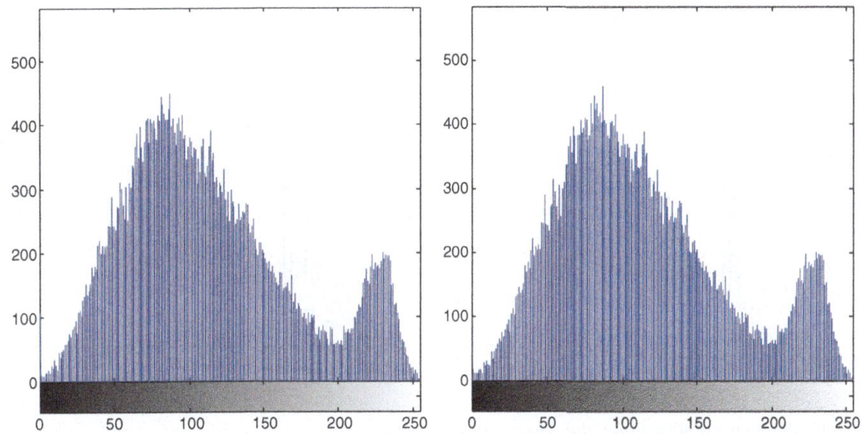

Fig. 14 Histogram of original and stego blue plane

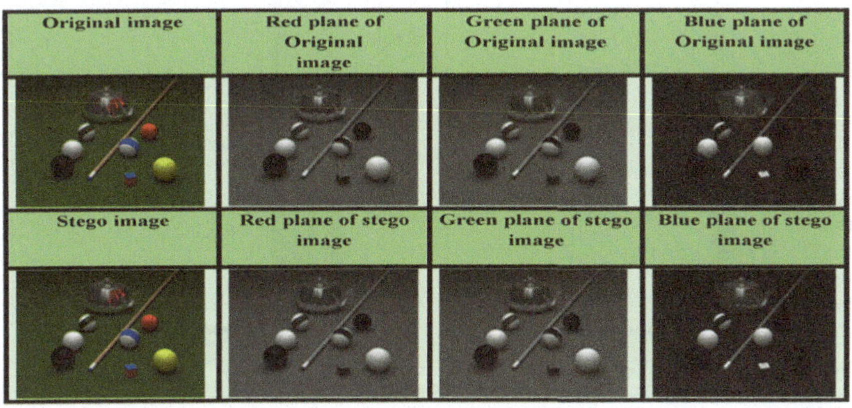

Fig. 15 Original image and planes along with and stego image and stego planes

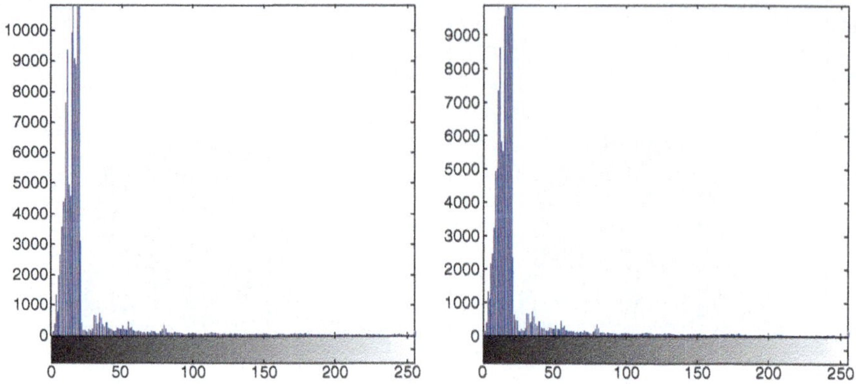

Fig. 16 Histogram of original and stego red plane

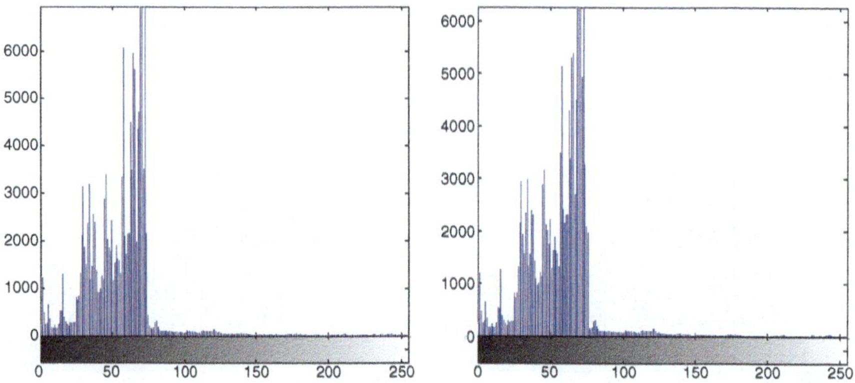

Fig. 17 Histogram of original and stego green plane

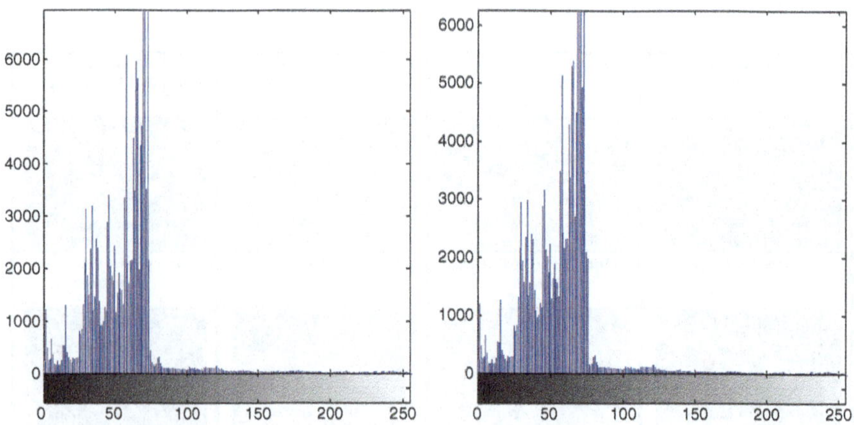

Fig. 18 Histogram of original and stego blue plane

Table 1 Results of different images against different payloads

Image	Payload (bits)	PSNR (dB)	MSE	BER (%)	SC	NAE
Lena	314,568	51.13	0.5006	0	+1	0
	629,142	48.10	1.003	0	+1	0
	943,716	46.3	1.5	0	+1	0
	1,258,290	45.11	2.001	0	+1	0
	1,572,864	44.14	2.5	0	+1	0
Airplane	24,576	61.9	0.0417	0	+1	0
	196,608	53.16	0.31	0	+1	0
	393,216	50.14	0.62	0	+1	0
	786,432	47.14	1.2	0	+1	0
	1,572,864	44.12	2.52	0	+1	0
Baboon	4746	62.12	0.0398	0	+1	0
	18,984	56.13	0.1584	0	+1	0
	151,878	47.126	1.2601	0	+1	0
	303,750	44.2	2.5	0	+1	0
Pool	234,396	51.146	0.49	0	+1	0
	468,792	48.08	1.011	0	+1	0
	703,188	46.29	1.52	0	+1	0
	937,584	45.063	2.02	0	+1	0
	1,171,980	44.12	2.5	0	+1	0

Table 2 Comparison of the proposed scheme with TPVD and PVM

Cover image	Tri-way PVD method (TPVD)		Pixel value modification (PVM)		Proposed modulus 4 method	
	Payload (Bytes)	PSNR (dB)	Payload (Bytes)	PSNR (dB)	Payload (Bytes)	PSNR (dB)
Lena	75,836	38.89	83,654	45.0926	1,17,964	45.11
Airplane	76,352	38.70	81,851	43.9345	98,304	47.14

results obtained and presented are proof that the proposed scheme is superior in comparison with others. Our technique can provide high capacity and better imperceptibility than that provided by the tri-way PVD method and pixel value modification (PVM) method. The proposed scheme offers a payload of 6 bits per pixel with an average PSNR of 44.1 dB.

5 Conclusion

The work presented in this chapter is concerned with the implementation of data hiding systems using color images as the cover medium. Modular arithmetic has been used to embed the secret data in cover images. Color images because of their three constituent bit planes provide a good cover media for hiding secret data. The proposed scheme uses MOD-4 operation for embedding and extraction. Various image indices like mean square error (MSE), peak signal-to-noise ratio (PSNR), structural content (SC), normalized absolute error (NAE), and normalized cross-correlation (NCC) have been used to evaluate the system. It was observed that the proposed scheme supports a payload of about 2bpp and provides a good quality of stego images as an average PSNR value of 46 dB is obtained. We have also carried out the histogram analysis of the proposed scheme. The histogram study of the stego images shows that the eavesdroppers cannot crack the system by mere inspection. The proposed work is fragile to various attacks, and in the future, it can be extended by designing secure and robust algorithms with relatively better data hiding capacity so that the resultant stego image could withstand any active attack from an adversary.

References

1. Parah SA, Sheikh JA, Hafiz AM, Bhat GM (2014) Data hiding in scrambled images: a new double layer security data hiding technique. Comput Electr Eng 40(1):70–82
2. Amna S et al (2018) Joint Crypto-Stego scheme for enhanced image protection with nearest-centroid clustering. IEEE Access 6:16189–16206
3. Hurrah NN, Parah SA, Loan NA, Sheikh JA, Elhoseny M, Muhammad K (2019) Dual watermarking framework for privacy protection and content authentication of multimedia. Future Gener Comput Syst 94:654–673
4. Jansi KR, Muthusamy SH (2020) Steganographic approach to enhance secure data communication using nonograms. Adv Intell Syst Comput 1056. 10.1007/978-981-15-019
5. Giri KJ, Bashir R, Bhat JI (2019) A discrete wavelet based watermarking scheme for authentication of medical images. Int J E-Health Med Commun. https://doi.org/10.4018/IJEHMC.2019100103
6. Hossein MG, Nodehi A, Enayatifar R (2019) An overview of encryption algorithms in color images. Sig Process 164:163–185
7. Parah SA, Sheikh JA, Bhat GM (2015) Hiding in encrypted images: a three tier security data hiding system. Multidimension Syst Sig Process 28:549–572
8. Giri KJ, Bashir R (2017) Digital watermarking: a potential solution for multimedia authentication. In: Dey N, Santhi V (eds) Intelligent techniques in signal processing for multimedia security. Studies in computational intelligence, vol 660. Springer
9. Muhuri PK, Ashraf Z, Goel S (2020) A novel image steganographic method based on integer wavelet transformation and particle swarm optimization. Appl Soft Comput. doi: https://doi.org/https://doi.org/10.1016/j.asoc.2020.106257
10. Giri KJ, Bashir R (2018) A block based watermarking approach for color images using discrete wavelet transformation. Int J Inf. Tecnol 10:139–146
11. Kaw JA, Loan NA, Parah SA, Muhammad K, Sheikh JA, Bhat GM (2019) A reversible and secure patient information hiding system for IoT driven e-health. Int J Inf Manage 45:262–275

12. Gull S, Loan NA, Parah SA, Sheikh JA, Bhat GM (2018) An efficient watermarking technique for tamper detection and localization of medical images. J Ambient Intell Human Comput 11:1799–1808. https://doi.org/10.1007/s12652-018-1158-8
13. Chan CK, Cheng LM (2004) Hiding data in images by simple LSB substitution. Pattern Recogn 37:469–474
14. Hengfu Y, Sun X, Sun G (2009) A high-capacity image data hiding scheme using adaptive LSB substitution. Radioengineering 18(4):509–516
15. Parah SA, Sheikh JA, Bhat GM (2012) High capacity data embedding using joint Intermediate Significant Bit (ISB) and Least Significant Bit (LSB) technique. J Inf Eng Appl. ISSN (Paper), pp 2224–5782
16. Saeed S, Akhaee MA (2013) One-third probability embedding: a new±1 histogram compensating image least significant bit steganography scheme. IET Image Proc 8(2):78–89
17. Joao T, Carneiro R, Bernardino FM (2016) Word-Hunt: a LSB steganography method with low expected number of modifications per pixel. IEEE Latin Am Trans 14(2):1058–1064
18. Khan M, Ahmad J, Rehman N, Jan Z, Sajjad M (2017) CISSKA-LSB: color image steganography using stego key-directed adaptive LSB substitution method. Multimedia Tools Appl 76(6):8597–8626
19. Khan M, Ahmad J, Rho S, Baik SW (2017) Image steganography for authenticity of visual contents in social networks. Multimedia Tools Appl 76(18):18985–19004
20. Wu DC, Tsai WH (2003) A steganographic method for images by pixel-value differencing. Pattern Recogn Lett 24(9–10):1613–1626
21. Mehdi H, Wahab AWA, Anuar NB, Salleh R, Noor RM (2015) Pixel value differencing steganography techniques: analysis and open challenge. In: 2015 IEEE international conference on consumer electronics-Taiwan. IEEE, pp 21–22
22. Mehdi H, Wahab AWA, Ho TS, Javed N, Jung KH (2017) A data hiding scheme using parity-bit pixel value differencing and improved rightmost digit replacement. Signal Process Image Commun 50:44–57
23. Shuyuan S, Huang L, Tian Q (2015) A novel data hiding for color images based on pixel value difference and modulus function. Multimedia Tools Appl 74(3):707–728
24. Gandharba S (2016) Adaptive pixel value differencing steganography using both vertical and horizontal edges. Multimedia Tools Appl 75(21):13541–13556
25. Marín G, Ismael R, Montes-Venegas HA, Marcial-Romero JR, Hernández-Servín JA, Ita GD (2016) An optimization approach to the TWPVD method for digital image steganography. Mexican Conference on Pattern Recognition. Springer, Cham, pp 125–134
26. Xin L, Qin Z, Ding L (2017) Data embedding in digital images using critical functions. Signal Process Image Commun 58:146–156
27. Chung KW, Wang CC, Hou HC (2016) Signed digit data hiding scheme. Inf Process Lett 116(2):183–191
28. Mingwei T, Song W, Chen X, Hu J (2015) An image information hiding using adaptation and radix. Optik 126(23):4136–4141
29. Wei-Sung C, Liao YK, Lin YT, Wang CM (2016) A novel general multiple-base data embedding algorithm. Inf Sci 358:164–190
30. Jeanne C (2014) A PVD-based data hiding method with histogram preserving using pixel pair matching. Sig Process Image Commun 29(3):375–384
31. Khan M, Ahmad J, Farman H, Jan Z, Sajjad Z, Baik SW (2015) A Secure method for color image steganography using gray-level modification and multi-level encryption. TIIS 9(5):1938–1962
32. Iyad FJ, Darabkh KA, Al-Zubi RT, Al Na'mneh RA (2016) Efficient reversible data hiding using multiple predictors. Comput J 59(3):423–438
33. Zhibin P, Hu S, Ma X, Wang J (2015) Reversible data hiding based on local histogram shifting with multilayer embedding. J Vis Commun Image Represent 31:64–74
34. Nai-Kuei C, Su CY, Shih CY, Chen YT (2016) Reversible watermarking for medical images using histogram shifting with location map reduction. In: 2016 IEEE international conference on industrial technology (ICIT). IEEE, pp 792–797

35. Hayat AD, Al-Ani A (2016) A steganography embedding method based on edge identification and XOR coding. Expert Syst Appl 46:293–306
36. Prasenjit D, Kar N (2015) ILSB: indicator-based LSB steganography. intelligent computing, communication and devices. Springer, New Delhi, pp 489–495

Chapter 8
Survey on Reversible Watermarking Techniques for Medical Images

Kaiser J. Giri, Zubair Jeelani, Javaid Iqbal Bhat, and Rumaan Bashir

1 Introduction

Digital watermarking refers to concealing information in digital multimedia data like digital images for the purposes of authentication [1–4], copyright protection, and fingerprinting. Many digital watermarking techniques have been proposed using both spatial [5] and frequency domains [6–9]. Traditionally, embedding the secret information (*watermark)* into the multimedia data (*cover data*) is done in such a way that distortion of the cover data due to watermarking is almost negligible perceptually. Traditionally, the multimedia data (cover data) is distorted when the secret information (watermark) is embedded in it. But, the distortion caused by embedding the watermark is perceptually negligible. Digital watermarking techniques that result in distortion of the cover data are also referred to as *lossy* or *irreversible* watermarking techniques. A wide range of digital watermarking techniques have been proposed over the course of the last few decades for ownership protection of digital media like images, audio, video, and natural language processing software [10]. Digital content, mostly multimedia, is produced in vast numbers per unit time continuously. Easy access to this data has become quite easy because of the availability of fast Internet. Security and protection of digital content on the other hand can be challenging because people can acquire access to digital content produced by

K. J. Giri · Z. Jeelani (✉) · J. I. Bhat · R. Bashir
Department of Computer Science, School of Engineering and Technology, Islamic University of Science and Technology, Awantipora, Jammu and Kashmir, India
e-mail: zubairj347@gmail.com

K. J. Giri
e-mail: kaiser.giri@islamicuniversity.edu.in

J. I. Bhat
e-mail: javaidonnet@gmail.com

R. Bashir
e-mail: rumaan.bashir@islamicuniversity.edu.in

K. J. Giri et al. (eds.), *Multimedia Security*, Algorithms for Intelligent Systems,
https://doi.org/10.1007/978-981-15-8711-5_9

someone else, modify it partially, and ultimately publish the content as their own. Digital Rights Management (DRM) refers to a set of technologies that mitigates such challenges by enforcing the usage of digital media in compliance with the established privileges and thus helps in the protection of digital media [11]. *Digital watermarking* techniques are the most successful and widely used in the range of techniques available in DRM.

Digital watermarking can ensure data protection and can help to know if the received watermarked content has been tampered with or not by extracting and analyzing the watermark. But, the watermarking process can damage the cover image content in the process, and thus the exact recovery of the original cover data may not be possible [12]. In some applications, however, even the slightest distortion in the cover work is intolerable. For example, one cannot tolerate the risk of distorting the sensitive data pertaining to deep space exploration, military investigation, and recognition, and medical diagnosis, by performing traditional digital watermarking on the cover data, because, even the slightest distortion of such sensitive data can lead to misleading inferences. For example, if the medical image test report of a patient is distorted using a lossy watermarking embedding process, these small modifications may affect the interpretation significantly, ultimately resulting in wrong diagnosis by a doctor.

The Electronic Health Records (EHRs) contain important medical history, diagnosis, medications, etc. of a patient. This information is sensitive and is used by medical doctors and practitioners and occasionally by health insurance companies. These EHRs are often embedded into the respective medical multimedia imagery (like, radiographs, mammograms etc.) of patients in the form of digital watermarks [11]. The use of irreversible digital watermarking can distort the medical imagery of the patient over time as a result of repeated extraction and embedding of the EHR data. As such, the cumulative distortion of the medical imagery may result in incorrect medical inferences from such images. Such scenarios motivated researchers to modify the irreversible digital watermarking, such that it becomes possible to obtain back the original cover data content without any distortion after extracting the watermark from the cover data.

Reversible watermarking, also known as *lossless* or *invertible* watermarking, allows full extraction of the embedded *watermark signal* along with the complete reconstruction of the *cover signal*. Reversible watermarking techniques have gained considerable attention for the last few years because of the demand for application of these techniques in healthcare, military communication, astronomical sciences, and digital forensics. Reversible watermarking, therefore, can be used in scenarios where the authenticity of the cover object has to be established, and the exact original content is required after extraction of the watermark. There is a tradeoff between *size of the watermark* that can be embedded in the cover object and the *perceptual quality of the watermarked cover object*. For irreversible watermarking, one must be careful about the size of the watermark because while embedding a watermark, the resulting distortions to the cover object cannot be reversed. Reversible watermarking techniques can withstand comparatively larger distortions, and thus accommodating larger size watermarks, because of their capability to obtain the original cover content after the watermark has been extracted.

Effectiveness of any watermarking technique can be evaluated using metrics to evaluate the *perceptual quality* of the cover data. An effective watermarking technique attempts to maintain the *quality* of the cover data as much as possible after the watermark is embedded in the cover object. The second performance metric is the *maximum size* (number of bits) of the watermark that can be embedded within a cover object without distorting it perceptually. Third is the *ability* of the watermark data *to withstand different attacks*. A robust watermarking scheme makes sure that the watermark data is least modified even if the cover data undergoes some kind of modification.

The rest of this chapter is organized as follows. Section 2 presents the general framework of the digital watermarking, applications, and different attacks that affect the watermarked content as well as the authenticity of the cover object. In Sect. 3, the classification of digital watermarking based on the purpose of watermark embedding is presented. Section 4 present the classification of reversible watermarking techniques based on the tools and algorithms used to embed and extract the watermark. Section 5 presents the different performance metrics that can be used to evaluate the performance of the watermarking techniques with regard to imperceptibility or for measuring the similarity between the watermarked and the original cover object. Section 6 is concerned with the discussion of state-of-the-art reversible watermarking techniques for medical imagery. Finally, Sect. 7 concludes the discussions of the presented study.

2 General Framework for Digital Watermarking

In this section, we use terms *cover object* (C) to refer to the data in which the data needs to be embedded. The data embedded in the *cover object* is referred to as the *watermark* (W). *Watermarked cover object* (C') refers to the *cover object* with the *watermark* embedded into it. The *key* (κ) is used to embed watermark (W) into (C). The *watermarked cover object* (C') might come under malicious or non-malicious attacks during transmission over a network. C' distorted by any such attack is modified by the attack and is referred to as $A(C')$. Here, $A()$ is the generic attack function. The attack function $A()$ can modify the *embedded watermark* (W) in addition to C'; we denote the *distorted watermark* as $A(W)$.

The general framework for digital watermarking is illustrated in Fig. 1. The different modules of a watermarking system are discussed below.

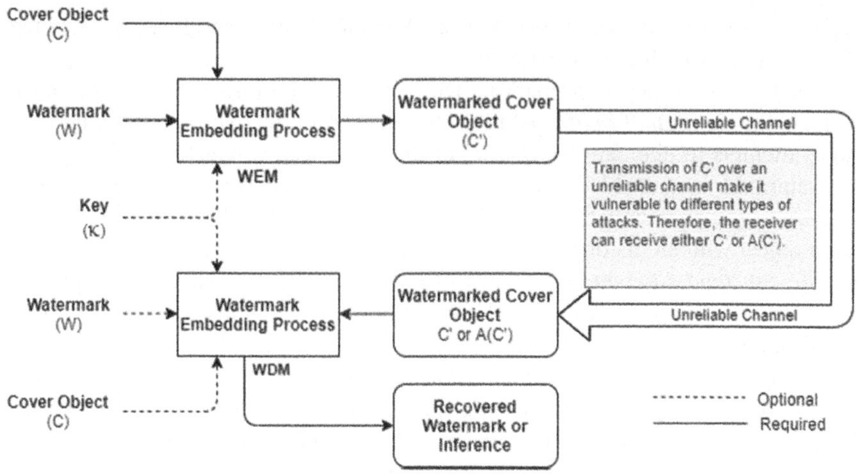

Fig. 1 Generic framework for digital watermarking

2.1 *Watermark Embedding Module (WEM)*

Input—Input to the WEM is the triple (C, W, κ).

Output—Output is the *watermarked cover object* (C').

WEM takes the triple (C, W, κ) as input with the objective to successfully embed W inside C using κ to facilitate the detection and recovery of W by the receiving party. Different attacks might distort the contents of C' during transmission before it is ultimately received by the intended receiver. Therefore, the receiver can either receive C' or $A(C')$.

2.2 *Watermark Detection/Extraction Module (WDM)*

Input—Watermarked cover object (C') or distorted watermarked cover object $A(C')$, key (κ), C and/or W depends on the technique.

Output—The output of WDM depends on the purpose of the technique. It can be extracted watermark (W) (or $A(W)$), or an inference that is obtained after processing C' or $A(C)$.

WDM usually takes the triple (C', W, κ) as input. The WDM then uses the key (κ) to extract the watermark from C'. Once the watermark is extracted, it is compared with the original watermark (W). The amount of distortion in the extracted watermark is a subjective matter and depends on the application. Based on the inferences drawn from the comparison of these watermarks, the authenticity of the cover object is

Table 1 Comparison of *fragile, semi-fragile* and *robust* watermarking

Watermark	Robustness	Applications
Fragile	• Modification of a single bit destroys the watermark • Very less robust	• Data integrity verification
Semi-fragile	• Withstand minor content modifications (compression, enhancement etc.) • More robust than fragile watermarks	• Multimedia data verification • Selective authentication
Robust	• Withstand unintentional modifications as well as more severe intentional modifications like geometrical attacks • Robust than both fragile and semi-fragile watermarks	• Ownership protection • Digital media authentication

determined. The output of the WDM is either the extracted watermark and/or the inferences drawn from the detection/extraction process.

3 Classification of Digital Watermarking

Digital watermarking characteristics can be used to classify digital watermarking in many different ways. We can have blind versus non-blind, perceptible versus imperceptible, private versus public, spatial domain versus frequency domain-based and robust versus fragile. An important characteristic of the hidden watermark is its robustness to withstand unintentional attacks that happen as a result of legitimate manipulation like slight compression as well as intentional attacks with the intention to distort and destroy the embedded watermark. In the context of the content put in this chapter, we discuss the digital watermarking classification into fragile, semi-fragile, and robust classes. The comparison of different types of watermarks with their purpose is given in Table 1. The fragile, semi-fragile, and robust watermarks are embedded in a cover object to fulfill different purposes as discussed below.

3.1 Fragile Watermarking

In *fragile* watermarking, the inserted watermark disappears or breaks if the watermarked cover object undergoes any kind of modification, thus confirming that the integrity of the data has been compromised. Fragile watermarking is, as such, suitable for digital content authentication [13, 14] that mostly include multimedia data. Fragile watermarks are therefore adopted to detect unauthorized modification to the data [15, 16]. Fragile watermarks establish complete integrity verification because even the slightest modification of the watermarked image alters or destroys the fragile watermark [17]. Fragile watermarking techniques usually employ computation of a secure keyed hash for the entire cover object to create the watermark. Therefore, even modification of a single bit modification in the cover object destroys the watermarking indicating that the integrity of the content has been compromised. A fragile watermarking technique capable of tamper detection and localization in medical/general images is proposed by Gull et al. [18].

3.2 Semi-fragile Watermarking

In *semi-fragile* watermarking, the watermark is able to survive to minor unintentional modifications, but at the same time allowing some image-processing operations [19]. Semi-fragile watermarking techniques, therefore, have the capability to differentiate between slight image processing operations or slight compression, or slight amount of noise and illegitimate content distortion. Therefore, unlike fragile watermarks, the semi-fragile watermarks are not extremely sensitive to content modification. Semi-fragile watermarks can allow slight JPEG compressions, for example, that modify the cover object content. This feature of semi-fragile watermarking techniques makes them suitable for applications where a certain degree of lossy compression has to be tolerated; thus, declare the cover image as acceptable even after slight modifications introduced as a result of compression.

3.3 Robust Watermarking

Robust watermarking represents a more restricted set of techniques within the set of semi-fragile watermarking techniques. Robust watermarking techniques, in addition to unintentional normal image-processing operations, are able to survive malicious attacks, like geometrical transformations, done intentionally to modify the cover data [20, 21]. The objective of any robust watermarking technique is to achieve maximum possible robustness against any intentional or unintentional modification of the watermarked data [11]. Robust watermarks are widely used for copyright protection that helps to prove the rightful ownership.

Unintentional attacks include normal image-processing operations including slight compression, brightness, and contrast adjustments. Malicious and intentional attacks aim to removal or alter the watermark and/or embed any other information in the cover image with an intention to destroy the watermark. Robust watermarking techniques are designed to withstand geometrical transformations including rotation, shearing, scaling, skew, partial cropping, etc. These techniques are also designed to withstand delay attacks, lossy compression with low-quality factors, noise attacks, etc.

4 Classification of Reversible Digital Watermarking

In Sect. 3, we discussed the classification of watermarking techniques into fragile, semi-fragile, and robust classes. Most of the existing reversible watermarking schemes are fragile in nature. General framework of reversible watermarking for medical images is given in Fig. 2.

Different classifications of reversible digital watermarking have been proposed in literature [11, 12, 22]. Figure 3 shows the classification of reversible watermarking techniques as proposed by Naskar and Chakraborty [11], and Fig. 4 is the classification of reversible watermarking techniques as proposed by Khan et al. 12]. Work

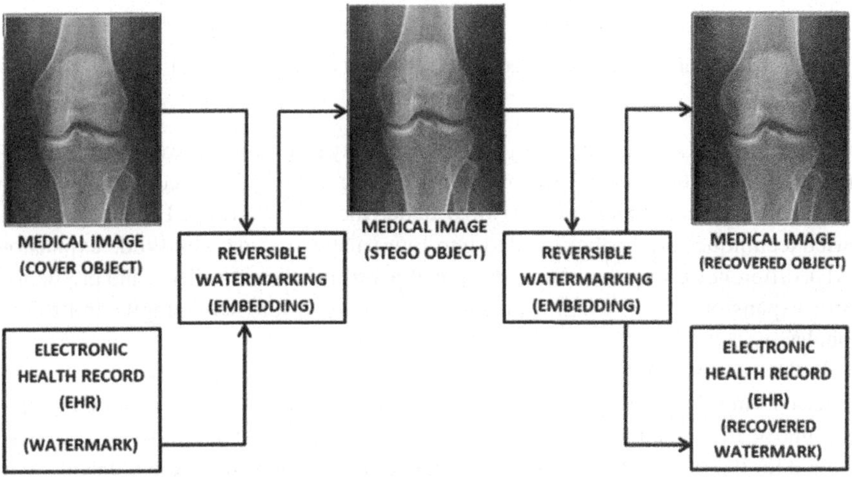

Fig. 2 Reversible watermarking for medical images

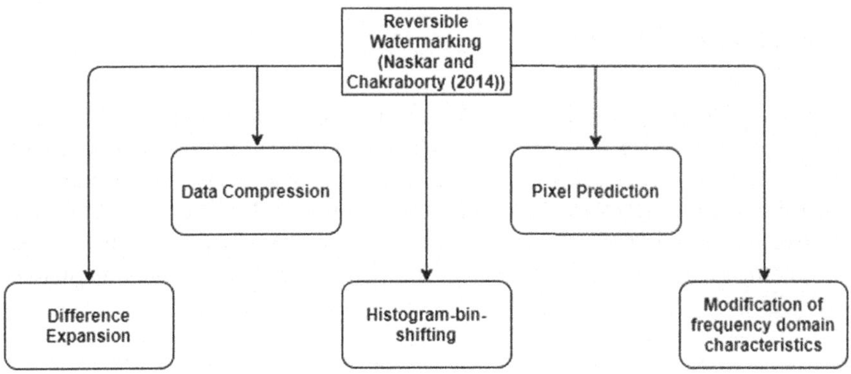

Fig. 3 Classification of reversible watermarking proposed in [11]

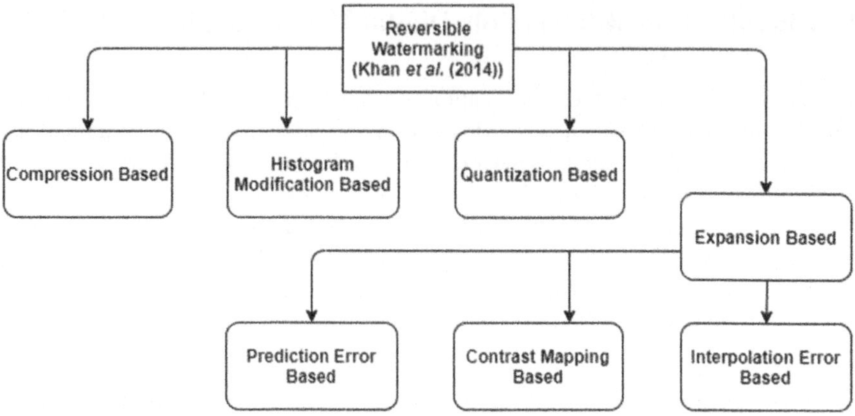

Fig. 4 Classification of reversible watermarking proposed by Khan et al. [12]

presented in [22] adds to this classification hierarchy by dividing the reversible watermarking techniques into *fragile* and *robust*. Error expansion-based and histogram shifting-based techniques are classified under the *fragile* group. Error expansion-based techniques are further divided into four subclasses, viz. amplitude expansion (AE), difference expansion (DE), interpolation error expansion (IEE), and prediction error expansion (PEE). The *robust* group is classified into semi-fragile and robust subclasses.

In this section, we provide a brief discussion of the general classes of reversible watermarking techniques identified by these studies. We, however, avoid a detailed account of the techniques because we intend to provide only the necessary background for the reversible watermarking techniques for medical imagery discussed in Sect. 6.

5 Evaluation Metrics

Maintaining the high *perceptual quality* of the cover image is one of the characteristics of an effective watermarking technique. Different metrics can be used to analyze the *perceptual quality* of the medical images objectively; they include peak signal-to-noise ratio (PSNR) and Structural Similarity (SSIM) index are the two most effective and widely used metrics for such analysis. These metrics use the original and watermarked cover signals to evaluate the perceptual quality of the watermarked cover signal.

5.1 Peak Signal-To-Noise Ratio (PSNR)

The original cover image f and the watermarked cover image g, both of size $M \times N$, are used to calculate PSNR as shown in Eq. (1).

$$\mathrm{PSNR}(f, g) = 10 \log_{10}(255^2/\mathrm{MSE}(f, g)) \tag{1}$$

where

$$\mathrm{MSE}(f, g) = \frac{1}{\mathrm{MN}} \sum_{i=1}^{M} \sum_{j=1}^{N} (f(i, j) - g(i, j))^2 \tag{2}$$

As the mean squared error (MSE) approaches zero, the PSNR value approaches infinity; this implies a higher PSNR value provides a higher perceptual quality of the watermarked cover image. A small value of PSNR, on the other hand, implies intensity values in f and g have larger difference and, thus lower image quality.

5.2 Structural Similarity (SSIM) Index

Developed by Wang et al. [23], the SSIM index is a well-known image quality metric used to measure similarity between two images. SSIM index uses combination of three important factors to model the image distortion—loss of correlation, luminance distortion, and contrast distortion. The SSIM index is defined [24] as shown in Eq. (3).

$$\mathrm{SSIM}(f, g) = l(f, g)c(f, g)s(f, g) \tag{3}$$

where

$$\begin{aligned} l(f, g) &= \frac{2\mu_f \mu_g + c_1}{\mu_f^2 + \mu_g^2 + c_1} \\ c(f, g) &= \frac{2\sigma_f \sigma_g + c_2}{\sigma_f^2 + \sigma_g^2 + c_2} \\ s(f, g) &= \frac{\sigma_{fg} + c_3}{\sigma_f \sigma_g + c_3} \end{aligned} \tag{4}$$

In Eq. (4), the first term l(f, g) measures the similarity between mean luminance (μ_f and μ_g) of f and g. This factor is maximal and equal to 1 only if $\mu_f = \mu_g$. The second term $c(f,g)$ measures the similarity of contrast in two images. Here, contrast is measured using standard deviation σ_f and σ_g. This term is maximal and equal to 1 only if $\sigma_f = \sigma_g$. The third term $s(f, g)$ measures the correlation coefficient between the two images f and g. Here, σ_{fg} is covariance between the two images. Positive values of SSIM index are in [0, 1]. A value of 0 means no correlation between images, and 1 means that $f = g$. In order to avoid a null denominator, positive constants c_1, c_2, and c_3 are introduced in these terms.

6 State-of-the-Art Reversible Watermarking for Medical Imagery

One of the most important applications of reversible watermarking is in the field of healthcare. Medical images are used to examine a variety of medical conditions in patients. This section presents a review of different state-of-the-art reversible watermarking proposed in the last five years. A quick summary of these techniques is presented in Table 2.

Kumar and Natarajan [25] proposed a hybrid local prediction error-based reversible watermarking using difference expansion. In this approach, the cover medical images are first divided into non-overlapping blocks. Border pixels in each block are predicted using the median edge detection (MED) prediction. Non-border pixels are predicted using the least square prediction, and the prediction error is expanded. The secret message is embedded in the cover object corresponding to the prediction error using difference expansion. The predictor itself is also embedded into the cover object to recover the secret message without sending any additional information. We first discuss the *prediction error-based difference expansion,* and next the *prediction methods* used are discussed. Finally, the *hybrid local prediction-based reversible watermarking technique* is discussed.

Error-based difference expansion: The difference expansion (DE) is advantageous for reversible watermarking because there is no need for lossless compression on the original values. Different methods [26, 27] use DE to achieve effective reversible watermarking. In case of prediction error-based difference expansion, the pixels in the cover image are predicted by using adjacent pixels in the image [28–32].

In the proposed method, first the predicted value $x_p(i, j)$ of the actual pixel value $x(i,j)$ in the cover image is predicted using a predictor. The prediction error $e(i, j)$ is calculated using Eq. (5).

$$e(i, j) = x(i, j) - x_p(i, j) \tag{5}$$

The prediction error $e(i, j)$ should be controlled by defining a threshold to control the possible distortion introduced by the prediction error. The approach, therefore, defines a threshold, $T = 10$ to keep the distortion low. If the prediction error is within the threshold limits, then the bit of the secret message is embedded in the cover image and the pixel is transformed using Eq. (6).

$$x_t(i, j) = x(i, j) + e(i, j) + b, \quad \text{if} \quad e(i, j) < T \tag{6}$$

Table 2 Summary of the reversible watermarking techniques for medical images

Technique	Year	Summary	Dataset
Kumar and Natarajan [25]	2015	**Hybrid local prediction error** • Use difference expansion • Median edge detector predict border pixels of blocks • Lease square prediction predict block pixels	**Dataset size** Dataset of six standard medical images (MRI, molecular, ultrasound, CT liver, MRI brain, cells)
Roček et al. [34]	2016	**Identify ROI and RONI areas in the medical image and use** • Reversible watermarking • Zero watermarking • RONI watermarking	**Dataset size** 6000 test medical images [35] **Image size** 512 × 512 px **Modalities** Angioscopy, computed radiography, computed tomography, digital radiography, mammography, magnetic resonance, radio fluoroscopy, ultrasound and X-ray angiography **Depth** 8–16-bit, (*depending on the modality*)
Arsalan et al. [40]	2016	**Companding function** For reducing embedding distortion **Integer wavelet transform** For embedding domain for achieving reversibility **Histogram processing** For avoiding underflow/overflow conditions **Genetic programming** For intelligent wavelet coefficient selection	**Dataset size** Lena, Baboon, Barbara, and Gold hill standard test images 4 standard medical images The size of each image is of size 512 × 512 MATLAB-based GPLab toolbox has been utilized in order to develop a suitable GP expression
Priya et al. [54]	2017	**Hybrid transform based** • Integer wavelet transform • Discrete Gould transform Achieve reversible watermarking for medical images without any additional key information	**Dataset size** Standard test images Medical images from Anbu Hospital, Kumbakonam, Tamil Nadu, India **Image size** 512 × 512 px

(continued)

Table 2 (continued)

Technique	Year	Summary	Dataset
Mehta and Kelkar [55]	2017	**Identify ROI and RONI areas in the medical image and use** Prediction expansion error-based reversible watermarking considering ROI **Median edge detector** For pixel value prediction **Arithmetic coding** For compression of location map	Standard medical images are used **Depth** 8-bit gray scale medical images
Abbasi and Memon [58]	2018	**Identify ROI and RONI areas in the medical image and use** • **Fragile watermarking**—Embed a binary pattern in the ROI • **Robust watermarking**—Embed a composite watermark in the RONI • **Least significant bit method**—used in embedding process • **Integer wavelet transform**—used to transform medical image	Standard medical images are used **Depth** 8-bit grayscale medical images
Liu et al. [63]	2019	**Identify ROI and RONI areas in the medical image and use** • Recursive dither module • Slantlet transform • Singular value decomposition	**Dataset Size** 200 medical images **Modalities** 40 CT images, 40 MRI images, 40 ultrasound images, 40 X-ray images, 40 fundus images **Watermark type** Image **Watermark size** 32 × 32
Dutta et al. [68]	2019	Block-based robust reversible watermarking technique	**Dataset size** 59,000 [70] **Modalities** Radiological grayscale images

If $|e(i,j)| \geq T$, then the secret message bit is not embedded; however, the pixel values are shifted to make the absolute value of the prediction error larger than the embedded ones at the detection side. If, $e(i, j) \geq T$, then $x_t(i, j) = x(i, j) + T$; and if $e(i, j) < -T$, then $x_t(i, j) = x(i, j) - (T - 1)$.

Prediction Methods: The different prediction methods used include *median-based prediction* that uses the four immediate neighbors of the current pixel and calculates the median on these pixel values to predict the pixel. *Median edge detection (MED)* uses the median of three casual pixels in the neighborhood of the current pixel to predict the pixel. The *gradient adjusted prediction (GAP) estimator* uses seven casual pixels around the current pixel and weights them based on the computed local gradients of the cover image [33]. The *least square prediction* is one of the most successful regression models and is used locally to predict pixels by dividing the image into non-overlapping blocks of size 8 × 8.

The technique divides the image into $B \times B$ (8 × 8) non-overlapping blocks including the border pixels as well. The border pixels are predicted using the MED predictor, and non-border pixels are predicted using the local least square prediction. Two predictors are used; therefore, the term hybrid local prediction is used.

Reported Results: This technique uses a dataset of six medical images of size 256 × 256 with a bit depth of eight bits and test the pixel predictions using MED, median, GAP, local and hybrid prediction methods. Please note that the prediction error is low and close to 0; then, the predicted image is similar to the cover image.

First this technique analyzes the effect of the prediction error Threshold (T) on corresponding embedded capacity. The T values from 1 to 10 inclusive are used with the six test images used in the experiments. The embedding capacity (bits per pixel—bpp) is shown to increase with increase in the T values from 1 to 10. Moreover, the proposed hybrid local predictor is shown to perform better in terms of bpp on all threshold values for all the test images.

Next, in order to establish a relationship between embedding capacity and image quality, experiments are carried out to understand the effect of varying T on PSNR. As established in the study earlier that the embedding capacity increases as we increase the value of T. More embedding capacity introduces more distortion in the cover medical image. The PSNR values reported for all the test images are shown to decrease with increase in T. Again, the hybrid local predictor is shown to perform slightly better than other predictors at all values of T.

Further, imperceptibility (invisibility of watermark in the cover image) analysis is done using the correlation coefficient analysis. The correlation coefficient is computed between the cover image and watermarked image by varying bits per pixel (*bpp*). The proposed hybrid local predictor outperforms other predictors under study.

Roček et al. [34] proposed a new approach to fully-reversible watermarking in medical images. This method combines the traditional approaches, i.e., reversible, zero, and Region of Non-Interest (RONI) watermarking, thus combining the advantages of these three approaches to make the proposed system much effective in terms of performance metrics like PSNR and SSIM. The philosophy of this approach is to identify the advantages and disadvantages of the traditional approaches used in this method and to keep the advantages in the proposed system while as eliminating the disadvantages.

The technique starts by dividing the input image into two parts, viz. Region of Non-Interest (RONI) and Region of Interest (ROI) parts. The detection of RONI and ROI regions is also an important problem. The RONI detection on a sample database of more than 60,000 medical images is done in [35]. The study carried out in [36] examined a representative set of medical images and concluded that all these images constitute the RONI region of 11% or more.

In this method, the RONI of 10% is used and the ROI region is secured using the zero watermarking. *Secret Share* refers to the information required to extract the watermark from the secured ROI region. *Secret Share* information is embedded in the RONI region itself using reversible watermarking. So, the steps to extract the watermark include: extraction of *Secret Share* information from RONI using reversible watermarking; calculation of *Public Share* information from ROI using zero watermarking; combining these into the final watermark.

Experiments are performed using a dataset of 6000 images selected from the education database containing around 60,000 medical images [35]. The images selected for the experiments are 512×512 pixels in size with nine modalities: angioscopy, computed radiography, computed tomography, digital radiography, mammography, magnetic resonance, radio fluoroscopy, ultrasound, and X-ray angiography. Test medical images have a color depth of 8–16-bits, depending on the specific image modality.

First, the effect of the percentage of RONI on PSNR and SSIM is analyzed in this study. Increasing RONI from 10 to 20% decreases RONI as well as SSIM slightly. PSNR for the test images with RONI of 10% is reported to be equal to 107.6% which is quite high. Average SSIM values reported at RONI of 10% is 0.999974 and an average SSIM of 0.999955 at RONI of 20%. These SSIM values are close to 1 indicating very high quality of watermarked image with RONI ranging between 10 and 20%. A comparative analysis of the performance of this method is done by comparing it to the methods proposed in [37–39]. All these techniques use an RONI greater than 70%, whereas the proposed method can work even with an RONI of 10%. As such, for comparison purposes the proposed method uses RONI ranging between 10 and 90%. The image quality in terms of PSNR for the proposed method remains between 81 and 75 for RONI ranging between 10 and 50%. Therefore, the proposed method results in small distortions by keeping RONI percentage low in the image.

Arsalan et al. [40] proposed an intelligent reversible watermarking technique for protecting medical images and patient-related information in health care (referred to as 'IRW–Med' in this work). IRW-Med reduces embedding distortion by using the companding function. Integer wavelet transform (IWT) is used in the embedding procedure and to achieve reversibility. Possible underflow and overflow conditions are the proposed system checked by using histogram processing. Genetic programming (GP) is employed for intelligent wavelet coefficient selection. The proposed technique achieves high larger capacity and imperceptibility.

The IRW-Med is divided into two stages meant for embedding and extraction of the watermark respectively. GP is used to calculate the best candidate expression (α) in the training phase, that is used for embedding the watermark in the testing phase. The watermark to be embedded is the Electronic Patient Record (EPR) and the payload. Histogram modification is done to avoid overflow and underflow conditions by compressing the histogram from both ends [41]. The image with histogram modification is transformed into the frequency domain by applying second generation Cohen–Daubechies–Feauveau (CDF) IWT. The transformed image of size M $\times$ N is divided into four sub-bands, i.e., approximation (LL), horizontal (HL), vertical (LH), and diagonal-detailed (HH) sub-bands [42]. In order to avoid perceptual distortion, the watermark embedding is not done in the LL sub-band. Next, GP [43] is used to intelligently select the coefficient for the embedding process. All the sub-bands except LL are divided into smaller blocks with dimensions $m \times n$, and an optimal/near-optimal GP expression is generated for the whole image. The GP expression value is different for different blocks because these values are dependent on the local features of the particular block. This GP expression does an intelligent coefficient selection for companding. Next, the watermarking is done using companding. At the receiver end, the IWT is applied on the watermarked image to obtain the image in wavelet domain. The extraction process is then performed on the four sub-bands thus obtained.

This technique is compared to other state-of-the-art techniques [41, 44–54] for its comparative performance. First, the relation between image distortion and the embedding capacity is established by doing PSNR-bpp analysis. The proposed technique provides better PSNR values (less distortion) for bpp values greater or equal to 0.4 when compared to other techniques. The GP, with a payload of 0.1 bpp, is shown to achieve best PSNR values after 18 generations of learning.

Priya et al. [55] proposed a hybrid transform-based reversible watermarking method for medical images in telemedicine applications. The proposed technique employ integer wavelet transform (IWT) and discrete Gould transform (DGT) to achieve reversible watermarking for medical images without any additional key information. The proposed method is divided into the watermark embedding and extraction phases.

In the watermark embedding phase, the watermark information (Electronic Patient Record (EPR)) like patient id, name, age, and other medical details like blood sugar level are embedded in a medical image. The following steps are taken to embed the EPR within the medical image. First, the IWT is applied to the original cover medical image (S) to produce the transformed image (St). The application of IWT to$S$ divides it into four sub-bands; the low frequency (LL) and high frequency (LH,HL, andHH) sub-bands. Next, any of the high-frequency sub-band of size ($H \times W$) of the transformed medical image (St) is divided into non-overlapping square blocks (B) of size$L \times$ L, where, $L = 2n$ and $1 \leq n \leq$ size of the wavelet coefficient sub-band. Then, GDT is applied to each block of size $L \times L$ in St to produce the hybrid transformed medical image (Stg). For a block of size $L \times L$, the size of the resulting DGT is also equal to $L \times L$. Finally, the watermark information (EPR) in each hybrid block by modifying the pixel at location (L, L). In both DGT and inverse DGT, the pixel at location (1, 1) remains unchanged. The watermark bit is embedded in the pixel at position (*L*, *L*). In the watermark extraction phase, the IWT followed by DGT is applied to the watermarked image at the receiver site. The original image is reconstructed by finding put the pixel S′4. Once the exact transformed pixel is obtained, the inverse DGT is applied followed by inverse IWT to get the original medical image.

Different experiments have been performed, and the respective results suggest that the proposed method achieves maximum imperceptibility, high robustness, and capacity without the requirement of any additional information for embedding or extraction.

Mehta and Kelkar [56] propose a reversible watermarking technique based on prediction error expansion considering region of interest for medical images. The focus of this study is to compare the performance of the classical prediction expansion error (PEE)-based and prediction error expansion-based reversible watermarking methods considering region of interest.

This method employs the classical PEE in the first stage and subsequently uses a location map to preserve the region of interest (ROI). Median Edge Detector (MED) is employed for pixel value prediction. A location map is constructed keeping in view the underflow and overflow conditions by applying a suitable threshold. The location map is compressed and embedded in the core object with ROI preservation. The ROI in this method is detected using Ostu's method [57]. The extraction of the payload is again achieved using classical PEE that employs a predictor similar to the one used in embedding phase. The decompressed location map thus obtained is used to predict the values for the locations that were used for embedding earlier.

A set of grayscale medical test images are used to evaluate the performance of this technique. Arithmetic coding [58] compression is used for compressing the location map in the experiments. The proposed PEE considering ROI is shown to produce better PSNR values when compared to the classical PEE for both watermarked as well as retrieved images. The proposed method also provides better capacity when compared to classical PEE method.

Abbasi and Memon [59] proposed a reversible watermarking technique for protecting medical image databases. This method is proposed in order to address the issues of confidentiality and authenticity in medical information. This method divides the input medical image into Region of Interest (ROI) and Region of Non-Interest (RONI) areas. Two different watermarks are embedded in these areas. First watermark is fragile in nature and is a binary pattern that is embedded in ROI. Second watermark is composite and is embedded in the RONI. The watermarking is performed using the least significant bit (LSB) method.

The lossless data hiding is achieved by the compression of data to accommodate the payload in the wavelet coefficients employing the LSB technique. The poor bias of zeros and ones provides for less embedding capacity [60]. Therefore, the bit planes are compressed in a lossless manner in the frequency domain. The payload is embedded in the space saved by the compression. Medical images like other images of natural scenery contain some amount of redundancy. This method uses this redundancy to hide and embed the watermark. Integer wavelet transform (IWT) is used in this technique because of the ability to produce larger bias. Second-generation transforms, such as, IWT, are desirable to remove redundancy and therefore facilitate providing more capacity. IWT does integer to integer mapping with CDF(2, 2) format. The proposed technique uses the CDF(2, 2) format to provide better imperceptibility when compared to the lifting scheme given in [61]. Also, the bias between zeros and ones is better for frequency coefficients when compared to spatial domain [62]. Arithmetic encoding is used for the compression of bit frames of the wavelet coefficients. The first watermark is fragile in which we generate the binary pattern to ascertain the integrity of the image. The second watermark is a robust watermark and is composite. It composed of four different types of watermarks, (a) patient records, (b) doctor's ID, (c) hospital logo, and (d) LSB information. Segmentation technique proposed in [63] is employed to divide the cover medical image into ROI and RONI areas. The experimental results reveal high perceptual quality of watermarked images ascertained by the PSNR metric. The ROI and RONI regions are also empirically shown to vary with different slices. Therefore, the payload embedding capacity varies as well. The PSNR and payload capacity are shown to be inversely related as expected.

Liu et al. [64] propose a novel reversible watermarking technique for protecting the integrity of medical images and their authentication. This technique employs recursive dither modulation (RDM) in order to avoid biases while making diagnosis decisions. Further, Slantlet transform and singular value decomposition is used to provide a reliable mechanism for image authentication. The ROI and RONI areas are identified and divided for producing the integrity data. Tamper detection information is generated for the medical image using a hash function. The ROI area is divided into non-overlapping blocks of size 16×16. Cyclic redundancy check (CRC) for every block is used to generate the tamper localization information of the ROI area. The tamper recovery information of the ROI is obtained by using IWT coefficient-based method proposed in [65]. Size of the tamper recovery information of the ROI is reduced by employing block truncation coding (BTC). The data pertaining to authentication, tamper detection of the medical image, ROI recovery information and tamper localization is embedded in the medical image.

Experiments are performed to evaluate the performance of this method with subjective as well as objective perspectives wherever possible. First, the experiments are performed to evaluate the proposed method for watermark imperceptibility in terms of SSIM and PSNR. The mean PSNR and mean SSIM values are better compared to many other techniques proposed in [65–68] indicating superior watermark imperceptibility. Next, the robustness of the proposed watermarking technique is evaluated in terms of bit error rate (BER) and normalized cross-correlation (NCC) between original watermarks and corresponding extracted watermarks. Results suggest that the proposed watermarking technique is robust and show better robustness when compared to techniques proposed in [65–68].

Tanima et al. [69] proposed a robust reversible watermarking scheme for grayscale images. This technique is designed to protect the medical images to avoid misdiagnosis of the disease by the analysis of such watermarked images. This scheme, at the same time, tries to provide robustness against different types of errors during transmission so that the watermark inside the image with stands such errors and help to prove the copyrights of the medical image if needed. The technique provides high embedding capacity while trying to keep the cover medical image distortion as less as possible.

The technique is focused on performing reversible watermarking for grayscale images. The proposed technique takes a medical grayscale image as input and divides the image into non-overlapping 4×4 pixel blocks. The correlation between the adjacent pixels in each block is very high. The adjacent pixels in a block are paired to form non overlapping pairs in each block. A watermark sequence is embedded in these non-overlapping pairs invisibly. This technique uses a binary location map to keep track of watermarked bits embedded in the medical image. The extraction procedure is just the reverse of the embedding procedure. The watermarked image is used as the input and is divided into 4×4 non-overlapping pixel blocks. The watermarked pixels are recovered to their original values based on their location inside the location map. Different transformations like blurring, geometric transformation, etc., can easily distort or destroy least significant bit in a pixel [70]. The proposed technique is shown to be robust against such undesirable transformations while recovering the pixels. The embedded watermark is extracted using the same location map.

A free and open-access medical image database having a dataset [71] of grayscale radiological images of over 12,000 patients is used in the simulations. The experiments are performed to analyze the effect of salt-and-pepper noise and Gaussian noise on the recovered image. The PSNR values obtained for all the recovered images are greater than 60 indicating superior image quality under these attacks. The bit error rate (BER) refers to the frequency of bit errors at the time of detecting a multi-bit watermark message [72]. Robustness of this technique under salt-and-pepper and Gaussian noise is shown in terms of BER.

7 Conclusion

This article presents a survey of the recent reversible watermarking techniques for medical images proposed in last five years. Two possible classifications of reversible watermarking techniques are proposed in [11, 11]. Next, the survey of eight state-of-the-art reversible watermarking techniques is presented; these techniques are based on difference expansion, pixel prediction, histogram shifting, etc. and are covered in the discussed classifications. Reversible watermarking techniques for medical images are used to embed electronic health records (EHR) of patients in the respective medical images to protect the patient privacy, image copyrights, and authentication. An important objective of the reversible watermarking is to extract the watermark without any distortion while maintaining the high perceptual quality of the medical images to avoid any misdiagnosis from the analysis of these medical images. All the techniques surveyed take due care to avoid perceptual distortion of the images. Most of the techniques achieve this perceptual quality by identifying ROI and RONI regions in the cover medical image. The surveyed techniques also attempt to enhance the embedding capacity by adopting hybrid watermarking techniques.

References

1. Hurrah NN, Parah SA, Loan NA, Sheikh JA, Elhoseny M, Muhammad K (2019) Dual watermarking framework for privacy protection and content authentication of multimedia. Fut Gener Comput Syst 94:654–673
2. Parah SA, Sheikh JA, Bhat GM (2014) Fragility evaluation of intermediate significant bit embedding (ISBE) based digital image watermarking scheme for content authentication. In: 2014 international conference on advances in electronics computers and communications, Oct 2014
3. Giri KJ, Bashir R (2016) Digital watermarking: a potential solution for multimedia authentication. Stud Comput Intell 93–112
4. Giri KJ, Bashir R, Bhat JI (2019) A discrete wavelet based watermarking scheme for authentication of medical Images. Int J E-Health Med Commun 10(4):30–38
5. Gull S, Parah SA (2019) Color image authentication using dual watermarks. In: 2019 fifth international conference on image information processing (ICIIP), Nov 2019
6. Parah SA, Sheikh JA, Loan NA, Ahad F, Bhat GM (2017) Utilizing neighborhood coefficient correlation: a new image watermarking technique robust to singular and hybrid attacks. Multidimension Syst Sig Process 29(3):1095–1117
7. Giri KJ, Peer MA, Nagabhushan P (2014) A channel wise color image watermarking scheme based on discrete wavelet transformation. In: 2014 International conference on computing for sustainable global development (INDIACom), Mar 2014
8. Giri, K.J., Ahmad Peer, M., Nagabhushan, P.: A robust color image watermarking scheme using discrete wavelet transformation. Int. J. Image Graph. Signal Process. **7**(1), 47–52 (2014)
9. Giri KJ, Bashir R (2017) A block based watermarking approach for color images using discrete wavelet transformation. Int J Inf Technol 10(2):139–146
10. Iftikhar S, Kamran M, Anwar Z (2015) A survey on reversible watermarking techniques for relational databases. Secur Commun Netw 8(15):2580–2603
11. Naskar R, Chakraborty RS (2014) Reversible digital watermarking: theory and practices. Synth Lect Inf Secur Privacy Trust 5(1):1–130
12. Khan A, Siddiqa A, Munib S, Malik SA (2014) A recent survey of reversible watermarking techniques. Inf Sci 279:251–272
13. Tian J (2002) Wavelet-based reversible watermarking for authentication. In: Security and watermarking of multimedia contents IV, vol 4675. International Society for Optics and Photonics (2002)
14. Yan Y, Sun Y, Wang HQ, Lin Y, Chen HY (2013) Reversible authentication scheme of images based on lifting wavelet transform. Appl Mech Mater 278–280:1379–1382
15. Shih, F.Y.: Adjusted-purpose digital watermarking. Dig. Watermark. Steganogr. 105–119 (2017)
16. Shih FY (2017) Robust high-capacity digital watermarking. Dig Watermark Steganogr 121–136 (2017)
17. Shih FY (2017) Digital watermarking and steganography: fundamentals and techniques. CRC Press (2017)
18. Gull S, Loan NA, Parah SA, Sheikh JA, Bhat GM (2018) An efficient watermarking technique for tamper detection and localization of medical images. J Amb Intell Hum Comput 11(5):1799–1808
19. Shih FY (2017) Genetic algorithm-based digital watermarking. Dig Watermark Steganogr 77–103 (2017)
20. Shih FY (2017) Classification in digital watermarking. Dig Watermark Steganogr 9–14 (2017)

21. Shih FY (2017) Watermarking attacks and tools. Dig Watermark Steganogr 51–63 (2017)
22. Menendez-Ortiz A, Feregrino-Uribe C, Hasimoto-Beltran R, Garcia-Hernandez JJ (2019) A survey on reversible watermarking for multimedia content: a robustness overview. IEEE Access 7:132662–132681
23. Wang Z, Bovik AC, Sheikh HR, Simoncelli EP (2004) Image quality assessment: from error visibility to structural similarity. IEEE Trans Image Process 13(4):600–612
24. Hore A, Ziou D (2010) Image quality metrics: PSNR vs. SSIM. In: 2010 20th international conference on pattern recognition, Aug 2010
25. Vinoth Kumar C, Natarajan V (2016) Hybrid local prediction error-based difference expansion reversible watermarking for medical images. Comput Electr Eng 53:333–345 (2016)
26. Lee S, Suh Y, Ho Y (2006) Reversiblee image authentication based on watermarking. In: 2006 IEEE international conference on multimedia and Expo, July 2006
27. Li X, Zhang W, Gui X, Yang B (2013) A novel reversible data hiding scheme based on two-dimensional difference-histogram modification. IEEE Trans Inf Forensics Secur 8(7):1091–1100
28. Hong W, Chen T-S, Shiu C-W (2009) Reversible data hiding for high quality images using modification of prediction errors. J Syst Softw 82(11):1833–1842
29. Ou B, Zhao Y, Ni R (2012) Reversible watermarking using optional prediction error histogram modification. Neurocomputing 93:67–76
30. Leung HY, Cheng LM, Liu F, Fu QK (2013) Adaptive reversible data hiding based on block median preservation and modification of prediction errors. J Syst Softw 86(8):2204–2219
31. Chen X, Sun X, Sun H, Zhou Z, Zhang J (2013) Reversible watermarking method based on asymmetric-histogram shifting of prediction errors. J Syst Softw 86(10):2620–2626
32. Luo H, Yu F-X, Huang Z-L, Chen H, Lu Z-M (2012) Reversible data hiding based on hybrid prediction and interleaving histogram modification with single seed pixel recovery. SIViP 8(5):813–818
33. Dragoi I-C, Coltuc D (2014) Local-Prediction-Based Difference Expansion Reversible Watermarking. IEEE Trans Image Process 23(4):1779–1790
34. Roček A, Slavíček K, Dostál O, Javorník M (2016) A new approach to fully-reversible watermarking in medical imaging with breakthrough visibility parameters. Biomed Sig Process Control 29:44–52
35. Roček A, Slavíček K, Dostál O, Javorník M (2016) RONI size and another attributes of representative sample of medical images in common hospital operation, related to securing by watermarking methods. In: International conference on image processing, production and computer science (ICIPCS-2016), Mar 26–27, 2016 London (UK), Mar 2016
36. Masek J, Burget R, Karasek J, Uher V, Guney S (2013) Evolutionary improved object detector for ultrasound images. In: 2013 36th international conference on telecommunications and signal processing (TSP), July 2013
37. Eswaraiah R, Sreenivasa Reddy E (2015) Robust medical image watermarking technique for accurate detection of tampers inside region of interest and recovering original region of interest. IET Image Process 9(8):615–625
38. Al-Qershi OM, Khoo BE (2010) ROI-based tamper detection and recovery for medical images using reversible watermarking technique. In: 2010 IEEE international conference on information theory and information security, Dec 2010
39. Al-Qershi OM, Khoo BE (2009) Authentication and data hiding using a hybrid ROI-based watermarking scheme for DICOM images. J Dig Imag 24(1):114–125
40. Arsalan M, Qureshi AS, Khan A, Rajarajan M (2017) Protection of medical images and patient related information in healthcare: using an intelligent and reversible watermarking technique. Appl Soft Comput 51:168–179
41. Xuan G, Yang C, Zhen Y, Shi YQ, Ni Z (2005) Reversible data hiding using integer wavelet transform and companding technique. Lecture notes in computer science, pp 115–124
42. Gonzalez RC, Woods RE (2007) Digital image processing, 3rd edn.
43. Xuan G, Shi YQ, Yang C, Zheng Y, Zou D, Chai P (2005) Lossless data hiding using integer wavelet transform and threshold embedding technique. In: 2005 IEEE international conference on multimedia and expo

44. Xuan G, Zhu J, Chen J, Shi YQ, Ni Z, Su W (2002) Distortionless data hiding based on integer wavelet transform. Electron Lett 38(25):1646
45. Tian J (2002) Reversible watermarking by difference expansion. In: Proceedings of workshop on multimedia and security
46. Usman I, Khan A, Ali A, Choi TS (2009) Reversible watermarking based on intelligent coefficient selection and integer wavelet transform. Int J Innov Comput Inf Control (IJICIC) 5(12):4675–4682
47. Lee S, Yoo CD, Kalker T (2007) Reversible image Watermarking based on integer-to-integer wavelet transform. IEEE Trans. Inf Forensics Secur. 2(3):321–330
48. Arsalan M, Malik SA, Khan A (2012) Intelligent reversible watermarking in integer wavelet domain for medical images. J Syst Softw 85(4):883–894
49. Xuan G, Shi Y, Ni Z (2004) Lossless data hiding using integer wavelet transform and spread spectrum. IEEE international workshop on multimedia signal processing, Siena
50. Siddiqa A, Khan A (2015) High capacity reversible image watermarking using error expansion and context-dependent embedding. Electron Lett 51(13):985–987
51. Sachnev, V, Kim HJ, Jeho Nam HJ, Suresh S, Shi YQ (2009) Reversible watermarking algorithm using sorting and prediction. IEEE Trans Circuits Syst Video Technol 19(7):989–999
52. Li X, Yang B, Zeng T (2011) Efficient reversible watermarking based on adaptive prediction-error expansion and pixel selection. IEEE Trans Image Process 20(12):3524–3533
53. Coltuc D (2011) Improved embedding for prediction-based reversible watermarking. IEEE Trans Inf Forensics Secur 6(3):873–882
54. Dragoi IC, Coltuc D, Local-prediction-based difference expansion reversible watermarking, IEEE Trans Image Process 23(4):1779–1790, Apr 2014
55. Selvam P, Balachandran S, Pitchai Iyer S, Jayabal R (2017) Hybrid transform based reversible watermarking technique for medical images in telemedicine applications. Optik 145, 655–671
56. Mehta JH, Kelkar V (2017) Comparison of reversible watermarking using prediction error expansion and prediction error expansion considering region of interest for medical images. In: 2017 2nd international conference for convergence in technology (I2CT), Apr 2017
57. Otsu N (1979) A threshold selection method from gray-level histograms. IEEE Trans Syst Man Cybern 9(1):62–66
58. Osorio RR, Bruguera JD (2000) Architectures for arithmetic coding in image compression. In: 2000 10th European signal processing conference, Tampere, pp 1–4
59. Abbasi F, Memon NA (2018) Reversible watermarking for the security of medical image databases. In: 2018 21st Saudi Computer Society National Computer Conference (NCC), Apr 2018
60. Yu-Chen H, Chun-Chi L, Chang-Ming W, Wu-Lin C, Chia-Hsien W (2013) Probability-based tamper detection scheme for BTC-compressed images based on quantization levels modification. Int J Secur Its Appl 7(3):11–32
61. Patil RD, Nigavekar AR (2013) Reversible image watermarking using lifting wavelet transform and arithmetic coding. Int J Eng Res Technol 2(2). e-ISSN: 2278–0181
62. Xuan G, Chen XJ, Zhu J, Shi YQ, Ni X, Su W (2002) Distortion less data hiding based on integer wavelet transform. IEEE Electron Lett 28(2002):1646–1648
63. Memon NA, Keerio ZA, Abbasi F (2014) Dual watermarking of CT scan medical images for content authentication and copyright protection. Springer International Publishing Switzerland. In: Shaikh FK et al (eds) IMTIC 2013, communications in computer and information science (CCIS), vol 14, pp 173–183
64. Liu X, Lou J, Fang H, Chen Y, Ouyang P, Wang Y, Zou B, Wang L (2019) A novel robust reversible watermarking scheme for protecting authenticity and integrity of medical images. IEEE Access 7:76580–76598
65. Thabit R, Khoo BE (2015a) Medical image authentication using SLT and IWT schemes. Multimedia Tools Appl 76(1):309–332
66. Maheshkar S (2017) Region-based hybrid medical image watermarking for secure telemedicine applications. Multimedia Tools Appl 76(3):3617–3647

67. Lei B, Tan E-L, Chen S, Ni D, Wang T, Lei H (2014) Reversible watermarking scheme for medical image based on differential evolution. Expert Syst Appl 41(7):3178–3188
68. Thabit R, Khoo BE (2015b) A new robust lossless data hiding scheme and its application to color medical images. Dig Sig Process 38:77–94
69. Dutta T, Bagi R, Gupta HP (2020) Robust reversible watermarking for grayscale medical images. Lecture notes in networks and systems, pp 613–622
70. Said A, Pearlman WA (1996) An image multiresolution representation for lossless and lossy compression. IEEE Trans Image Process 5(9):1303–1310
71. Dutta T, Sur A, Nandi S (2013) MCRD: motion coherent region detection in H.264 compressed video. In: 2013 IEEE international conference on multimedia and expo (ICME), July 2013
72. Dutta T, Sur A, Nandi S (2013) A robust compressed domain video watermarking in P-frames with controlled bit rate increase. In: 2013 national conference on communications (NCC), Feb 2013

Chapter 9
Security and Privacy of E-health Data

Gazi Imtiyaz Ahmad, Jimmy Singla, and Kaiser J. Giri

1 Introduction

Information and Communication Technology (ICT) plays an important role in every field of life. It is an integration of computer systems, communication devices, audio–video displays and Internet. It can be used in almost all sectors of life such as business, education, marketing, medical, agriculture, e-governance, e-commerce, banking, transport and other fields. ICT has created a great impact on healthcare sector as well. This include health and education, hospital management system, health research and health data management. Health education is used to create awareness about communicable disease such as COVID-19, symptoms, diagnosis, preventive measure and other awareness details. This way people get information not only about the disease but also about the best hospitals and doctors to approach for treatment. Use of ICT in hospital management helps organization to manage patient details, hospital details, patient safety and satisfaction, updated information about new tools and technology, knowledge on population health and statistics. In healthcare research, ICT helps in finding the possible preventive measures to eradicate and reduce the spread of disease. Advancement in diagnosis and treatment reduce time, cost and increase the lifespan of population. New models are formed for effective quality care. An important use of ICT is the electronic storage of medial

G. I. Ahmad
Goverment Degree College Awantipora, Pulwama, Jammu and Kashmir, India
e-mail: gaziimtiyaz@gmail.com

J. Singla
School of Computer Science and Engineering, Lovely Professional University, Punjab, India
e-mail: jimmysingla285@gmail.com

K. J. Giri (✉)
Department of Computer Science, Islamic University of Science and Technology Awantipora, Pulwama, Kashmir, Jammu and Kashmir, India
e-mail: kaiser.giri@islamicuniversity.edu.in

K. J. Giri et al. (eds.), *Multimedia Security*, Algorithms for Intelligent Systems,
https://doi.org/10.1007/978-981-15-8711-5_10

data which include patient details, disease information, doctor's consultation and other useful information which can be used anywhere and anytime. The use of ICT in health care has also resulted in new developments and innovations. Development of medical devices and equipments and other areas of healthcare systems such as biotechnology and pharmaceuticals during the recent years have made significant contributions in improving the overall health scenario of the people around the globe. Tele-health which uses digital and communication technologies to access healthcare services remotely has also gained much advancement during the recent years. Storage and retrieving of healthcare data, their use and sharing among medical entities for knowledge and decision making involves both hardware and software. Use of ICT-enabled devices, digital technologies, increased number of medical devices, access to healthcare services, and increased communication among healthcare entities such as doctor–doctor, doctor–patient, and doctor–nurse and vice versa has resulted in accumulation of large volume of medial or clinical data which include text, audio, video, and still images. This multimedia information in healthcare systems needs to be securely accessed, stored, transferred and communicated to the relevant persons without loss or infringement.

E-health involves both health services and health-related information which is delivered through the means of information and communication-related technologies. It is an emerging field which comprises of medical informatics, public health and business. E-health does not only involve technological development in the field of medicine and pharmaceuticals, it acts as a way of thinking and participation of healthcare experts around the world for better healthcare facilities at the doorsteps and at cheaper rates, to improve the healthcare locally and globally using modern tools and technologies [1]. The "e" in e-health not only stands for electronic, but it also means efficient use of resources, enhancement of quality, encourage to educate people to participate and access health-related information at ease while maintaining ethical values, equality and equity. It includes a number of subdomains of digital health, viz. electronic Health Records (EHR), electronic Medical Records (EMR), tele-health, Telemedicine and other healthcare systems, patient health records, virtual healthcare systems, mobile healthcare systems and recently big data systems used in digital healthcare systems [2]. In today's new era, healthcare systems and services in medicine and technology are essential for realizing the healthcare needs individually as well as globally. Therefore to provide better healthcare to the masses, new healthcare systems are being designed which are both efficient and effective [3] (Fig. 1).

It is estimated that the amount of healthcare data generated globally was 500 petabytes in 2012, and 25,000 petabytes are expected in 2020 [5]. The rapidly growing unstructured medical data is attributed to the digitization of images, laboratory reports and results, laboratory and image sensors and other forms of multimedia data which needs to be captured, analyzed, stored and visualized. This data is continuously growing resulting in accumulation of large volume of data. The data may be medical and healthcare information like personal information, medical records, genomics and biometric sensor readings, radiology images, etc. All this multimedia data is used, manipulated and stored by healthcare systems in the form of cloud computing.

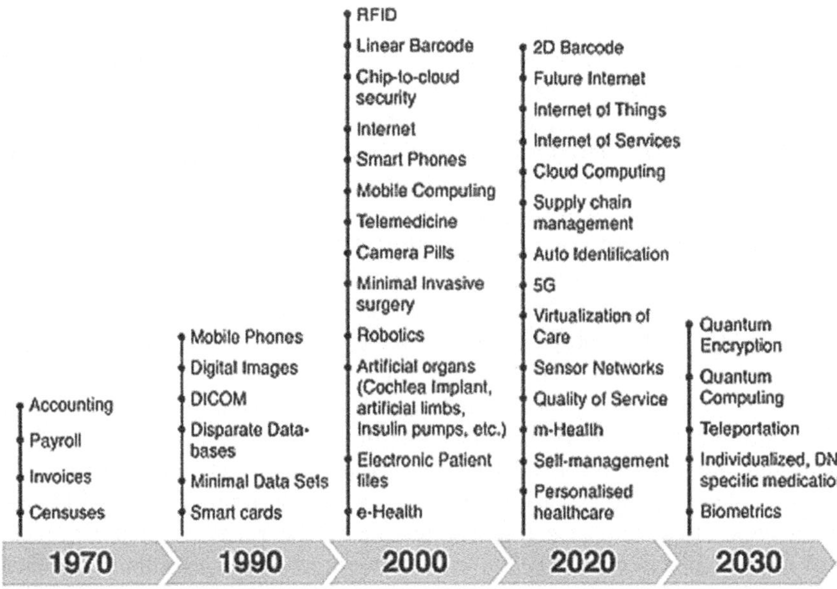

Fig. 1 Digital health evolution [4]

Healthcare systems are adopting new tools and techniques to improve the delivery of health services and assist both patients and healthcare workers. A number of applications are used by the stakeholders to monitor the performance of medical facility for financial efficiency and treatment success rates. Government organizations are also using IT solutions to monitor the safety and quality of healthcare organizations. Through Internet, mobile and wireless technologies patients also monitor their vitals and communicate with healthcare personnel through various healthcare applications. However, digital information have also posed many security risks in healthcare industry. Therefore, a reliable and robust Information Security Service for healthcare service providers is necessary for better and secure health-related facilities. In recent years, cyber criminals have taken interest in electronic health records due to higher black market rates than bank accounts and credit cards, it is the responsibility of healthcare service providers not only to protect health care data but also prevent it from any unauthorized assaults. In contrast to the financial institutions where a strong data protection mechanism is in place by way of a two-factor authentication, the healthcare systems do have weak protection of medical records. The medical facilities must protect data from unlawful activities such as hacking, and penetration of virus, employee actions due to illiteracy or purposely stealing of medical records. From data collection to data storage and dissemination of information through electronic channels, a reliable user authentication to access electronic medical records must be adopted for securing this vital information.

A number of approaches are in place to protect electronic health information from unauthorized access. Availability, accountability and controlled access are the key components of modern healthcare information systems. Digital watermarking is used for integrity of medical images during distribution of these images to various stakeholders. The authors in [6] proposed methods for securing medical images by way of watermarking techniques in transform domain using discrete cosine transform. Two approaches have been proposed. In the first technique, both Region of Interest and Region of Non-Interest have been used to embed digital watermark and electronic patient record (EPR). In second algorithm, only the Region of Interest is used to embed digital watermark and EPR. These techniques not only prevent singular attacks but also hybrid attacks in a robust manner. A high-capacity and reversible data hiding technique capable of temper detection and localization of medical images has been proposed in [7]. For obtaining cover image, original image is scaled using interpolation techniques. The cover image so obtained is divided into n x n non-overlapping blocks, and in each block, pixels are categorized into seed pixels and non-seed pixels. Embedding of electronic patient record (EPR) is carried out in non-seed pixels while as the seed pixels are kept unchanged to facilitate reversibility during extraction process. For smooth transmission of EPR and for detection of any temper during transmission, a watermark coupled with block checksum is embedded to EPR. For embedding process, the algorithm uses intermediate significant bit substitution (ISBS) in order to protect it from attacks during transmission. Experiment results showed that the proposed system detects all types of temper and geometric attacks which make the proposed system ideal for content authentication of EPR in an e-healthcare system.

Authors in [8] proposed an efficient Electronic Health Record Scheme for medical images used in an Internet of Things (IoT)-driven healthcare systems. The proposed system consists of transmitter and receiver phases. In the transmission phase, the EPR along with cover image is transmitted after proper preprocessing of cover image that includes pixel repetition method (PRM). Data embedding processing is done using modular arithmetic which results in a single image called stego-image. At the receiving end, the extraction of stego-images results in the extraction of EPR and original image. The proposed system was experimented on various types of medical images, and the results showed that the proposed scheme is computationally efficient, having high capacity and more secure than the contemporary models.

A data hiding technique based on pixel repetition method (PRM) and hybrid edge detection for medical and general images has been proposed in [9]. The proposed technique is semi-reversible and has high capacity. To enhance and scale up the smaller images, PRM is used, and to include all information of the images including complete information of their edges, the hybrid edge detection technique is used. The embedding process involves least significant and intermediate significant bit substitution along with RC4 encryption technique to impose additional security levels. The technique was experimented on various medical images and compared with state-of-the-art techniques to show that the proposed system is capable of handling high payload.

An embedding scheme for hiding EPR in colored medical images for IoT-driven healthcare systems and health monitoring systems which are both secured and of high capacity has been proposed in [10]. The color image is divided into three color spaces of red, green and blue corresponding to three equal length messages of the message vector containing EPR for embedding. Two address vectors are generated to embed EPR in the cover image. A reverse process is applied at the receiving end to extract the watermark and to ensure that the image is secure during transmission. A number of experiments have been conducted to ensure that the proposed system is not vulnerable against attacks such as noise, filtration and rotation during transmission.

A digital image watermarking algorithm for copyright protection of medical images has been proposed in [11]. The watermark is encrypted to increase the security level. The cover image is divided into 8 × 8 blocks, and Arnold transform followed by DCT is applied on each block. The encrypted watermark is embedded in cover image using diagonal coefficients of singular matrices using SVD operation. The singular matrices are obtained by selecting middle frequency coefficients from a DCT block. The performance of the proposed system was tested under various image-processing operations line noise addition, filtering and geometric manipulations. The results showed that the proposed system is highly robust to both singular and simultaneous attacks. Therefore, the proposed system can be used for application where user privacy and copyright protection are important.

A digital watermarking scheme to hide electronic patient record and logo by using magic rectangle encryption and bilinear interpolation is proposed in [12]. The proposed method is capable to detect tempering caused during transmission, highly secure and provides a high payload. In addition, reversible data-hiding techniques make the proposed scheme secure, efficient and can be used to improve the healthcare system applications.

A scheme based on the principle of histogram bin shifting for secure, high embedding capacity and to ensure reversibility of digital medical images has been proposed in [13]. For extra security in the embedded electronic patient record, RC6 encryption techniques has been used prior to embedding. The proposed scheme was analyzed on various types of medical images, and it was revealed that the system provides higher values of PSNR for a very high payload. The proposed techniques can be useful in telemedicine and tele-health applications.

Similar watermarking techniques for protection of copyright information in digital images based on discrete wavelet transformation have also been proposed in [14–16]. These methods used the concept of nearest neighborhood relationship for embedding of EPR images/signature images as watermark in a color medical image. The edge and sharp detail information of an image which corresponds to high-frequency components are used for watermark embedding. The embedding process uses pseudo-random key in a distributed fashion across the neighboring pixels of an image. The experimental results showed improved robustness and transparency.

For securing EPR in a medical image and authentication of medical images, a discrete wavelet transformation (DWT) scheme is proposed in [17]. For taking appropriate portion in a medical image for embedding purpose and to examine the distribution of frequency across the image, the image is subjected to discrete wavelet transformation. Watermark pixels are embedded into multiple pixels to increase the possibility of pixels during watermark extraction. A series of experiments have been carried on the proposed system, and it was revealed that the proposed system is highly resistant to noise and other types of attacks.

The study is based on the research studies of recent years and information has been collected from Google Scholar, IEEE Explore, Research Gate, ACM digital Library, Science Direct and other relevant web repositories. The quantitative as well as qualitative research papers have been referred. The study shows how much work has been done in the field of security and privacy of e-health content.

2 Background

From the last decade, it has been observed that there is a trend of moving from traditional paper-based health records (PBHR) to the electronic health records (EHR). Primarily, the efforts for this shifting have been done by independent organization. However, due to its manifold benefits, government and health organizations have also now started shifting from PBR to electronic records particularly in developing countries like India. The benefits include cost-effectiveness, quality improvement and encouragement of evidence-based medical and health record storage and their transmission [18]. High-quality health data is being produced and accumulated daily. However, little or no developments have been made in policies involving privacy issues raised in shifting from paper-based health record systems to integrated and EHR systems [19]. Also advancements in ICT have led to the privacy and security issues of patient's health data. The fundamental goals of privacy and security of electronic health records are integrity, availability and confidentiality [20]. The protection and safety of personal health related information is of vital importance in the health sector; therefore, it is imperative that these measures and goals are implemented in letter and spirit. It must be ensured that information must be accessible to the authorized persons only. Moreover, information must be accurate and should not have been tempered or modified in an unauthorized manner. This information is also important for its timely and effective delivery mechanism. The healthcare systems must be operational in normal, constrained and in natural disasters. Health information is regarded as the most confidential type of personal information. Therefore, protection of this confidential information is need of the hour. Secure information must be available and accessible to the authorized persons, which is therefore necessary in healthcare systems. It should also be defined that how much information should be communicated to the concerned persons on a given time. The primary threats for electronic health records are hackers, viruses and worms. Besides, use of ICT also exposes Information and Communication tools to these kinds of threats.

According to India Digital Health Net (IDHN), 1.3 billion people in India are seeking healthcare through public and private service providers. Most of the interactions between patient and doctors are captured on handwritten papers. However, efforts are being made to switch to the automation of health records for its delivery in an efficient and effective way. With the introduction of Ayushman Bharat to provide health insurance coverage and improve access to primary healthcare, the digitization process of health records has speed up. Ministry of Health and Family Welfare in 2018 has introduced "Digital Information Security in Healthcare Act" for establishment of national and regional electronic health authorities to ensure and monitor that proper healthcare data is collected and stored.

But, despite these efforts and availability of Health Information Systems along with their wider capabilities, most of the hospitals in India do not communicate healthcare data among hospitals electronically rather use the traditional and existing ways of sharing of patient health information.

2.1 Applications of Electronic Health Records (EHR)

Use of electronic health records and electronic medical records has gained momentum during the recent years throughout the world. There are a number of benefits of electronic health records (EHR) over traditional paper-based system (TPBS) for smooth functioning of healthcare systems.

1. Through electronic health records, doctors and health workers are able to provide better diagnosis and proper medical care as patient's data is readily available and easy accessible.
2. Through easy accessibility and transmission of health information between patient and health workers by using electronic health records, continuity of care is ensured. Moreover, timely information to patients such as immunization, follow-up care, preventive measures, etc., is also ensured.
3. EHR also helps in efficient time management as only one-time registration at a health center is needed. In addition, EHR also saves money by way of preserving patient's previous history and diagnosis services.
4. Due to increased visibility and accessibility of patient's previous history and statistics, EHR provides better accountability for health providers, besides improving traceability in medical malpractice and ethics related incidents.
5. EHR through automated reporting and real-time situation assessment can also help in identification of diseases such as COVID-19 which are threat to the public health. EHR can also help in collate data related to chronic diseases through available clinical data. Through EHR, we can collect huge amount of medical data for research purpose.

2.2 Privacy in Electronic Health Systems

The collection, storage and dissemination of healthcare information through ICT-enabled devices have made it easy to access this information for future reference, but has put forth many challenges in terms of providing protection of privacy and security of the healthcare and medical information. Privacy refers to individual's right to share his/her personal health information and how the information stored and communicated [21]. It is the ability of individuals to exercise control on their personal data held by others. It is concerned with the compilation, preservation, usage and disclosure of an individual's personal information [22]. Privacy requires confidentiality, intimacy, anonymity and invisibility to outsiders [23]. The privacy of the individual must be respected before his/her information is shared or published.

In countries like USA and Australia, information privacy is legalized in the context of personal information held within public and private sector organizations. In India, right to privacy has been made a fundamental right and to protect e-health records, Ministry of Health and family Welfare under Digital Information Security in Healthcare Act (DISHA), constituted a National Digital health Authority to promote and adopt e-health standards, enforce privacy and security safeguards for e-health data standards and exchange of e-health records. The security and privacy requirements include identification, authentication and authorization which come under the ambit of access control. Identification involves to identify the authorized users, authentication means that the data access request is authentic and valid, and the communication is between the authentic users only [24]. Authentication involves choosing of valid data and authentic user for communication and dissemination of information and restriction of data thereof. Therefore, a solid access control mechanism should be in place for privacy in e-health data. The data must be accessible to only authorized users. This feature called availability means that data should be accessed anywhere anytime by the authorized users without disruption of service [25]. Dependability ensures that the e-health data is accessible and retrievable in constraint environments such as failure of a node or entire network without causing threat to patient's life [26–28]. Last but not least, there must be some sort of flexible mechanism so that in case of an emergency, specific data is accessible to unauthorized participant in order to save life of a patient [29].

3 Security in Electronic Health Systems

While privacy is the ability to protect sensitive healthcare information of a person, security is concerned with protection against unauthorized access with integrity and availability [30]. Security of electronic e-health data is important because unauthorized access by intruders can affect the overall performance of healthcare systems. Threats and attacks at various phases of e-health data are.

1. *At Data Collection Phase*: This phase involves collection of data from various sources in different formats. In order to protect data at this phase, patient privacy must be preserved and data must be collected from trusted sources, besides ensuring that data and information systems are protected from unauthorized access. Possible threats include alteration, obliteration, unauthorized use or stealing of information. Common attacks include jamming, data collision, data flooding and spoofing.

2. *At Data Transmission Phase*: At this phase, necessary transformation, filtration and classification take place for analysis and to improve the quality of data by way of removing noise and other unnecessary details. Therefore, extra caution is needed at this phase for safety of data. Access-level security and access control are maintained at this phase as security measures [31]. Common threats include spying, information alteration, communication interruption, etc. Therefore, possible attacks like snooping of the patient's healthcare information, alteration of communication between stakeholders known as man-in-the-middle attack, modification of message attacks, Hello flood attack and wormhole attack usually occur at this phase.
3. *At Data Storage Phase*: At this phase, collected data is stored in the form of directories, databases and other form of repositories. In order to enforce security of data, it should remain protected and isolated by enforcing access control mechanism and defining other security measure like anonymization approach, permutation and data partitioning [30]. At this phase, various types of threats and attacks may occur such as changing patient's medical information, social engineering attacks, removable distribution media attacks, altering configuration of servers and connecting systems etc. Also software vulnerabilities caused by the hackers, viruses, spyware, trojans etc., are also common at this phase [32].

4 Security Models Proposed in E-healthcare

For proper diagnosis and efficient treatment, the patient's history and treatment details need to be shared among various concerned persons through a number of communicative devices. The dissemination of patient's data to a number of users poses potential threat to privacy and security of personal information of an individual. Therefore, security models for e-health systems are needed for protection of patient's data and for proper access controls.[33, 34]. The important issue in this regard is the protection of data at collection phase, storage and transfer phase and at access control level. Therefore, confidentiality, reliability, responsibility, ease of use and access control are the fundamentals of security issues when dealing with e-health data [35]. Traditional encryption techniques provide simpler access control solutions but are not applicable to the complex e-health systems. In addition, encryption techniques require computational overhead, and modifications lead to sensitivity of personal health information [36].

(a) Security Models at Data Collection Phase

The basic component of a security model is the secure data collection and its integration with other repositories. At this phase, various electronic health records are verified in terms of authenticity, confidentiality and atomicity. A number of security models proposed for this phase include:

The authors in [37] presented a modular data collection and access control system, so that quick and context-aware access control decisions according to contextual-based security policies are taken in extensive healthcare systems. The modular design besides simplifing the access control policies addressed well-defined medical situations and allows for system configuration. This allows the system to incorporate the system functionalities and operations in various deployment scenarios and situations. The approach ensures fast reaction in critical situations, thereby provides user safety and simpler end-user policy composition. The system extends traditional RBAC model to assign and distribute access control policies and to store medical context in terms of location, time and health information about a patient which influence the access control decisions on patient's medical condition. The model provides access control to sensitive data to the medical staff in critical conditions. However, during a critical situation, the model does not detect an unauthorized access which poses a threat to the overall system.

Authors in [38] proposed a lightweight security model in limited recourse constraints. The authors described body sensor network (BSN) architecture in the VITRUVIUS and introduced potential threats to its security. The authors examined a number of security protocols like LLSP, TinySec, MiniSecand RC4-based methods along with Skipjack, AES, and RC4 base cipher algorithms for development of their own lightweight model. It has been found that RC4 and Skipjack ciphers provide efficient message confidentiality of healthcare data.

A constraint model in wireless sensor networks (WSNs) and body sensor networks (BSNs) for access control mechanism with limited speed and storage was discussed in [39]. The model exhibits flexibility and efficient access control decisions on requests. Poder2 framework along with additional extensions for unexpected events generated by privileged user was used for the development of the model. The model was discussed in detail for adaptive access control, viz. its design, development, implementation and policy testing.

Authors in [40] presented a three-tier security framework for authentication and pair-wise key establishment for data collection from sensor nodes using mobile sinks. Two key pools were used in the framework for connection establishment and mobile sink to access sensor network. An improved security performance of authentication between stationary nodes and sensor nodes was exhibited by the model. However, since the model uses basic key pre-distribution scheme, it becomes easy for an attacker to gain a number of keys by way of catching a small fraction of nodes and thus gains control of the network by deploying a virtual mobile sink preloaded with some compromised keys which is a big disadvantage of the model.

A biometric-based security framework for data authentication within WBAN was proposed in [41]. Since each patient has his/her own specific biometric information, in order to ensure that a patient's healthcare information is secured from assimilation with other, the model uses Sender's electrocardiogram signals feature as a key. This also resulted in improvement in terms of efficiency and effectiveness while reducing the computational complexity over cryptographic key distribution. However, the model uses authentication process based on sensors which results in restriction of process with limited resources.

(b) **Security Models at Data Transmission Phase**

Presentation, processing and sharing of electronic health data is an important and critical phase in e-healthcare application as a number of devices such as WSNs and other terminal devices play an important role in data clustering and communication. With the advancement of technology especially in medical health, one can easily access his/her health status and health-related information anywhere and anytime. However, during transmission of this important information, it should be protected from unauthorized access.

A framework which is secure for authentication and data communication was presented in [42]. The model includes efficient protocol architecture for e-health services using channel security, data and device security. SSL and HTTP layer was used for providing channel security where as data and device security was provided by SOAP layer constructed over HTTP. An adaptive authentication mechanism for various scenarios along with application of various encryption algorithms for secure e-health services was also presented in the model.

To support secure and authorized access along with data access control, a framework was proposed in [43]. A two-stage key access along with zero knowledge protocol was used in the model. Minimum access is granted and extracted in each access request based on user rights. A two-step combination of public key encryption and derive unique key per-transaction (DUKPT) is used for secure connection between different entities. Experimental results showed that the proposed method uses normal response time while accepting a number of concurrent authentication requests.

Authors in [44] presented a novel and secure access control mechanism "Mask-Certificate Attribute Based Encryption (MCABE)" for cloud integrated body sensor networks. To mask the plain text, a specific signature is developed. This masked data is then outsourced securely to cloud servers. To grant privilege at the receiving end, an authorization certificate composed on signature and other related privilege items is developed. For each data receiver, a unique value is chosen to mask the certificate to differentiate between the receivers and to ensure security of the data. Experimental results showed that the proposed scheme has less computation cost and storage cost compared with other common models. However, since the model uses encryption techniques, it requires more computation time and to achieve state-of-the-art security and access control mechanism is very difficult.

A lightweight security framework "SecourHealth" designed specifically for highly sensitive data collection and transmission applications was presented in [45]. The models presented a number of security services for both stored as well as in-transit data by using lightweight mechanism of TLS/SSL for securing data exchange without needing any extra security layer. The framework includes security features such as authentication, data confidentiality and lack of connectivity.

(c) **Security Models at Data Storage and Access-Level phase**

Electronic Health Record is one of the uses of information technology in health sector. It is fully computerized, easily accessible and secured storage of health information. The information is stored in portable storage device. The data is stored in database so that user can access it easily and in a secure way. For the access of medical information, the user must have authority. Authority can be provided by different ways like username/password, digital certificate with digital card. It is more difficult to remember password, username and taking digital card; therefore, concept of service-oriented architecture (SOA) is introduced and implemented. In this architecture, virtual healthcare service is implemented. Another easy way of identification is biometrics, where the fingerprint can be applied. Most commonly used access control model is RBAC, and most commonly used authentication model is digital signature with public key. In EHR, access control model can have hybrid or mix according to requirement of organization.

A secure framework using Advanced Encryption Standard (AES) encryption algorithm for to encrypt patient's medical information was proposed in [46]. The model provides an approach which enables a user to maintain data in a secured cloud environment. It was ensured that the patient's medical information is accessed and used by known and authorized persons only that makes the model highly secure.

Authors in [47] proposed a comprehensive access control approach with extensions for private data protection for e-health-related service. The model proposed was projected suitable for e-healthcare systems in term of access control mechanism. Various decision factors such as authorization, obligation and other conditions, in addition to continuity properties, were analyzed for usage control. The decision factors determined which person should be allowed to access the information. The model also helps in solving shortcomings occurring during accessing information. The model also proposed approaches for controlling purpose data in a dynamic environment.

A new scheme "PEACE" was proposed in [48] for patient-centric access control along with privacy and security-related issues by using attribute-based encryption techniques. The security model is based on different privacy levels. The proposed model accumulates medical information in a centralized storage area, thereby reducing the maintaining cost of service providers. User privacy is achieved by way of data integrity. However, when data is accessed concurrently by a number of users, the overall performance of the system reduces.

An access control system using privacy for e-health-related services was proposed in [49]. Access control models such as DAC, MAC and RBAC were combined to develop an new model for healthcare service providers along with various privilege settings. Experiments conducted showed that the model works well under different policy settings and data access methods.

Authors in [50] presented a novel healthcare information system which is secure and patient centric. The proposed model used proxy re-encryption protocol for sharing and access control in cloud computing. The main components of the proposed model include sharing of patients health related information among various stakeholders, defining various access control policies, storage of data in a cloud, validation of access requests and auditing the stored encrypted data. Various experiments were conducted on the proposed model, and it was revealed that the system is efficient and resistant against possible attacks and malicious behaviors.

Hierarchical Identity-Based Encryption Scheme (HIBE) used to develop a generic framework was proposed in [51]. The model was developed using Role-Based Access

Control (RBAC). The data is first encrypted using HIBE and then forwarded to a storage server. The user privileges are granted using RBAC. The experimental results showed that the proposed system keeps patient's data confidential and secure. However, the system does not offer accurate access control requirements and thereby having encryption drawbacks.

5 Conclusion

Everyday large volume of healthcare data is generated and used by professionals to save, heal and enhance the lives of patients. Despite best efforts from healthcare organization and increasing strictness of health insurance portability and accountability compliances, huge number of electronic health records are being lost or stolen in increasing year on year. Therefore, proper privacy and security mechanisms should be employed for protection of valuable healthcare information and unauthorized access. It is very important to implement e-healthcare solutions to improve the existing healthcare delivery system. In order to have efficient healthcare solutions, it is necessary to put in place the required privacy and security methods for protecting data and allied machinery from security breach and vulnerability. The government and policymakers should develop a comprehensive model for e-health system that would secure electronic health records from attacks and vulnerability. Privacy regulations should be in place for healthcare services and practices so that various stakeholders particularly patients should feel highly protected and motivated while sharing their health-related information. Every person involved at various stages of an e-health system is equally responsible to secure electronic health records in order to harvest the promises of Digital Health Information System for better health outcomes, smarter spending and healthier people. Therefore, it is the responsibility of developers and service providers that while designing and developing e-health systems, technology; social and legal aspects must be taken into consideration. Implementing proper protection strategies and solutions in electronic healthcare systems will facilitate medical fraternity to fulfill the monitoring and reporting regulations and share data securely. In this chapter, privacy-related issues of e-health data were discussed. Security models at different phases have also been discussed. In future, a detailed survey regarding securing and privacy of e-health data shall be carried.

References

1. Allen A (2000) Morphing telemedicine-telecare-telehealth-ehealth. Telemed Today, Special
2. Eysenbach G, Jadad AR (2001) Evidence-based patient choice and consumer health informatics in the Internet age. J Med Internet Res 3(2):e19
3. Maksimović, M., Vujović, V.: Internet of things based e-health systems: ideas, expectations and concerns. In: Handbook of large-scale distributed computing in smart healthcare, pp. 241–280. Springer, Cham (2017)

4. Thümmler, C.: Digital health. In: Requirements engineering for digital health, pp. 1–23. Springer, Cham (2015)
5. Feldman, B., Martin, E.M., Skotnes, T.: Big data in healthcare, hype and hope, Dr. Bonnie 360 (2012)
6. Parah SA, Sheikh JA, Ahad F, Loan NA, Bhat GM (2017) Information hiding in medical images: a robust medical image watermarking system for E-healthcare. Multimedia Tools Appl 76(8):10599–10633
7. Parah SA, Ahad F, Sheikh JA, Loan NA, Bhat GM (2017) A New Reversible and high capacity data hiding technique for E-healthcare applications. Multimedia Tools Appl 76(3):3943–3975
8. Parah SA, Sheikh JA, Akhoon JA, Loan NA (2020) Electronic health record hiding in images for smart city applications: a computationally efficient and reversible information hid
9. Loan NA, Parah SA, Sheikh JA, Akhoon JA, Bhat GM (2017) Hiding Electronic Patient Record (EPR) in medical images: a high capacity and computationally efficient technique for e-healthcare applications. J Biomed Inform 73:125–136
10. Parah SA, Sheikh JA, Ahad F, Bhat GM (2018) High capacity and secure electronic patient record (EPR) embedding in color images for IoT driven healthcare systems. In: Internet of things and big data analytics toward next-generation intelligence. Springer, Cham, pp 409–437
11. Hurrah NN, Parah SA, Sheikh JA (2019) A secure medical image watermarking technique for e-healthcare applications. In: Handbook of multimedia information security: techniques and applications. Springer, Cham, pp 119–141
12. Parah SA, Bashir A, Manzoor M, Gulzar A, Firdous M, Loan NA, Sheikh JA (2019) Secure and reversible data hiding scheme for healthcare system using magic rectangle and a new interpolation technique. In: Healthcare data analytics and management. Academic Press, pp 267–309
13. Parah SA, Digoo T, Hamdani GA, Shah AA, Khan I, Khan O, Sheikh JA (2019) A reversible and secure electronic patient record embedding technique using histogram bin shifting and RC6 encryption. In: Healthcare data analytics and management. Academic Press, pp 245–266
14. Giri KJ, Peer MA, Nagabhushan P (2014) A channel wise color image watermarking scheme based on discrete wavelet transformation. In: 2014 international conference on computing for sustainable global development (INDIACom). IEEE, Mar 2014, pp 758–762
15. Giri KJ, Bashir R (2017) Digital watermarking: a potential solution for multimedia authentication. In: Intelligent techniques in signal processing for multimedia security. Springer, Cham, pp 93–112
16. Giri KJ, Bashir R (2018) A block based watermarking approach for color images using discrete wavelet transformation. Int J Inf Technol 10(2):139–146
17. Giri KJ, Bashir R, Bhat JI (2019) A Discrete wavelet based watermarking scheme for authentication of medical images. Int J E-Health Med Commun (IJEHMC) 10(4):30–38
18. Fernández-Alemán JL, Señor IC, Lozoya PÁO, Toval A (2013) Security and privacy in electronic health records: a systematic literature review. J Biomed Inform 46(3):541–562
19. Rothstein MA (2007) Health privacy in the electronic age. J Leg Med 28(4):487–501
20. Farzandipour M, Sadoughi F, Ahmadi M, Karimi I (2011) Security requirements and solutions in electronic health records: lessons learned from a comparative study. J Med Syst 34(4):629–642. Int J Med Inform 80(2):e26–e31
21. Brodnik M, Rinehart-Thompson L, Reynolds R (2012) Fundamentals of law for health informatics and information management professionals (Chap. 1). AHIMA Press, Chicago
22. Sahama T, Simpson L, Lane B (2013) Security and privacy in eHealth: is it possible? In: 2013 IEEE 15th international conference on e-health networking, applications and services (Healthcom 2013). IEEE, Oct 2013, pp 249–253
23. Wilkowska W, Ziefle M (2012) Privacy and data security in E-health: Requirements from the user's perspective. Health Inf. J. 18(3):191–201
24. Wang J, Zhang Z, Xu K, Yin Y, Guo P (2013) A research on security and privacy issues for patient related data in medical organization system. Int J Secur Its Appl 7(4):287–298
25. Kumar P, Lee HJ (2012) Security issues in healthcare applications using wireless medical sensor networks: a survey. Sensors 12(1):55–91

26. Zubaydi F, Saleh A, Aloul F, Sagahyroon A (2015) Security of mobile health (mHealth) systems. In: 2015 IEEE 15th international conference on bioinformatics and bioengineering (BIBE). IEEE, Nov 2015, pp 1–5
27. Nagaty KA (2014) Mobile health care on a secured hybrid cloud. J Sel Areas Health Inf 4(2):1–9
28. Kotz D (2011) A threat taxonomy for mHealth privacy. In: 2011 third international conference on communication systems and networks (COMSNETS 2011). IEEE, Jan 2011, pp 1–6
29. Mare S, Sorber J, Shin M, Cornelius C, Kotz D (2011) Adapt-lite: privacy-aware, secure, and efficient mhealth sensing. In: Proceedings of the 10th annual ACM workshop on privacy in the electronic society, Oct 2011, pp 137–142
30. Masood I, Wang Y, Daud A, Aljohani NR, Dawood H (2018) Towards smart healthcare: patient data privacy and security in sensor-cloud infrastructure. Wireless Commun Mobile Comput
31. Abouelmehdi K, Beni-Hessane A, Khaloufi H (2018) Big healthcare data: preserving security and privacy. J Big Data 5(1):1
32. General Dynamics Health Solutions white paper UK: Securing Big Health Data"©2015. https://gdhealth.com/globalassets/health-solutions/documents/brochures/securing-big-health-data_-white-paper_UK.pdf
33. Wellington K (2013) Cyberattacks on medical devices and hospital networks: legal gaps and regulatory solutions. Santa Clara High Tech LJ 30:139
34. Sun L, Wang H (2011) A purpose based usage access control model for e-healthcare services. In: 2011 international conference on data and knowledge engineering (ICDKE). IEEE (2011)
35. Altamimi A (2016) SecFHIR: a security specification model for fast healthcare interoperability resources. Int J Adv Comput Sci Appl (IJACSA) **7**(6)
36. Thilakanathan D, Calvo RA, Chen S, Nepal S, Glozier N (2016) Facilitating secure sharing of personal health data in the cloud. JMIR Med Inf 4(2):e15
37. Garcia-Morchon O, Wehrle K (2010) Efficient and context-aware access control for pervasive medical sensor networks. In: 2010 8th IEEE international conference on pervasive computing and communications workshops (PERCOM Workshops). IEEE, Mar 2010, pp 322–327
38. Amini S, Verhoeven R, Lukkien J, Chen S (2011) Toward a security model for a body sensor platform. In: 2011 IEEE international conference on consumer electronics (ICCE). IEEE, Jan 2011, pp 143–144
39. Maw HA, Xiao H, Christianson B (2013) An adaptive access control model for medical data in wireless sensor networks. In: 2013 IEEE 15th international conference on e-health networking, applications and services (Healthcom 2013). IEEE, Oct 2013, pp 303–309
40. Linciya T, Anandkumar KM (2013) Enhanced three tier security architecture for WSN against mobile sink replication attacks using mutual authentication scheme. Int J Wireless Mobile Netw 5(2):81
41. Rasheed A, Mahapatra RN (2010) The three-tier security scheme in wireless sensor networks with mobile sinks. IEEE Trans Parallel Distrib Syst 23(5):958–965
42. Ramli SN, Ahmad R Abdollah MF, Dutkiewicz E (2013) A biometric-based security for data authentication in wireless body area network (wban). In: 2013 15th international conference on advanced communications technology (ICACT), pp. 998–1001. IEEE Jan 2013
43. Boonyarattaphan A, Bai Y, Chung S (2009) A security framework for e-health service authentication and e-health data transmission. In: 2009 9th international symposium on communications and information technology. IEEE Sept 2009, pp 1213–1218
44. Kahani N, Elgazzar K, Cordy JR (2016) Authentication and access control in e-health systems in the cloud. In: 2016 IEEE 2nd international conference on big data security on cloud (BigDataSecurity), IEEE international conference on high performance and smart computing (HPSC), and IEEE international conference on intelligent data and security (IDS). IEEE, Apr 2016, pp 13–23
45. Guan Z, Yang T, Du X (2015) Achieving secure and efficient data access control for cloud-integrated body sensor networks. Int J Distrib Sens Netw 11(8):101287
46. Simplicio MA, Iwaya LH, Barros BM, Carvalho TC, Näslund M (2014) SecourHealth: a delay-tolerant security framework for mobile health data collection. IEEE J. Biomed. Health Inf. 19(2):761–772

47. Sunagar V, Biradar C (2014) Securing public health records in cloud computing patient centric and fine grained data access control in multi owner settings. Int J Sci Appl Inf Technol (IJSAIT) 3(4):18–21
48. Sun L, Wang H (2011) A purpose based usage access control model for e-healthcare services. In: 2011 International conference on data and knowledge engineering (ICDKE), pp 41–46. IEEE, Sept 2011
49. Barua M, Liang X, Lu R, Shen X (2011) Peace: an efficient and secure patient-centric access control scheme for ehealth care system. In: 2011 IEEE conference on computer communications workshops (INFOCOM WKSHPS). IEEE pp 970–975
50. Gajanayake R, Iannella R, Sahama TR (2012) Privacy oriented access control for electronic health records. In: Data usage management on the web workshop at the worldwide web conference. ACM, Mar 2012
51. Barua M, Lu R, Shen X (2013) SPS: secure personal health information sharing with patient-centric access control in cloud computing. In: 2013 IEEE global communications conference (GLOBECOM). IEEE Dec 2013, pp 647–652

Part III
Security in IoT and Big Data-Driven Enterprise

Chapter 10
Implementation of Intrusion Detection System for Internet of Things Using Machine Learning Techniques

Jatinder Manhas and Shallu Kotwal

1 Introduction

Kevin Ashton first coined the term Internet of Things (IoT) in the year 1999 [1]. IoT is a network that connects physical devices of various domains like home automation, industrial process, vehicles, healthcare, weather monitoring, etc. to Internet. IoT networks consist of various sensors, actuators, softwares and Internet connectivity to collect and exchange data. The IoT networks are not only restricted to conventionally interconnected devices, i.e., desktops, laptops, smartphones and tablets, but it also includes unconnected computers over the Internet. Technological advancement allows these devices to be embedded with sensors which allow these to communicate with each other, and these can also be controlled and monitored remotely. Some of IoTs' most common applications are in logistics, industrial processes, community security, house automation, ecological monitoring and healthcare. This highly interconnected environment apart from its benefits also brings various cyber-security threats along with it. This makes the critical infrastructures like power stations, gas pipelines, transportation, etc. more prone to various cyber-attacks which pose serious consequences on the nation as a whole. Traditional protection measures for IoT networks might not work competently because of the various standards and communication stacks implicated, inadequate computational power and highly interconnected devices. That's why it's important to build unique security solutions for IoT networks to get all the benefits it offers.

Some ongoing IoT security enhancement projects include methods for encryption and data validation, right to use IoT network, and compliance of defense and privacy guidelines [2]. However, even these mechanisms are not sufficient enough to prevent the IoT networks from getting exposed to the multiple attacks that have been devised

J. Manhas (✉) · S. Kotwal
Department of Computer Science and IT, University of Jammu, Bhaderwah Campus, Jammu and Kashmir, India
e-mail: manhas.jatinder@gmail.com

K. J. Giri et al. (eds.), *Multimedia Security*, Algorithms for Intelligent Systems,
https://doi.org/10.1007/978-981-15-8711-5_11

for these networks. Therefore, another line of protection is required, designed to detect attackers. Intrusion detection systems (IDSs) are one of the ways in which we can effectively achieve this goal.

IDS is used mainly to secure the existing basic networks but to protect IoT networks with IDS pose difficulties due to IoT characteristics. In traditional network, IDS agents are deployed in nodes with high processing power, but nodes in IoT network are resource constrained.

Some of popular IoT vulnerabilities are:

- Sybil threat
- Man-in-the-middle vulnerability
- Hello flood threat
- Denial-of-service(DoS) threat
- Wormhole attack
- Sinkhole attack
- Selective forwarding vulnerabilities
- Zero-day attacks.

Intrusion detection refers to the mechanism of monitoring network traffic for identifying the suspect activity by scanning the network for harmful activity or infringement of policies [1]. On the identification of any suspicious activity, it generates alerts and reports to the administrator or it is distributed centrally to the security information and event management (SIEM) system. This approach uses multi-source outputs and warning filtering methods to distinguish malicious commotion from false alarms. Intrusion detection system can be categorized primarily as network IDS and host IDS and often detects the attack using signature-based approach and anomaly-based approach [2]. Machine learning techniques have been used in recent years to introduce IDS, and these techniques have proven efficient for intrusion detection. In this paper, the authors used various machine learning techniques to implement IDS. The main purpose of the current study is to evaluate the efficiency of machine learning techniques in intrusion detection to determine the effective one (Fig. 1).

This research aims to study various security aspects of IoT networks. IoT networks should provide their users with confidentiality, integrity and availability, but they are vulnerable to a variety of attacks such as sinkhole attacks, wormhole attacks, hello flood attacks, sybil attacks, etc. Manufacturers provide security patches to authenticate, monitor access, etc. through firmware updates, but IoT networks are still susceptible to numerous attacks due to resource constraints, low power consumption and less computing capacity. So we need another line of defense which can protect IoT networks from various attacks. IDS is one of such defense techniques against attacks. We propose to develop IDS for IoT networks that can use various machine learning techniques to detect major attacks.

- **Types of Intrusion Detection system**:
  IDS can primarily be divided into five categories

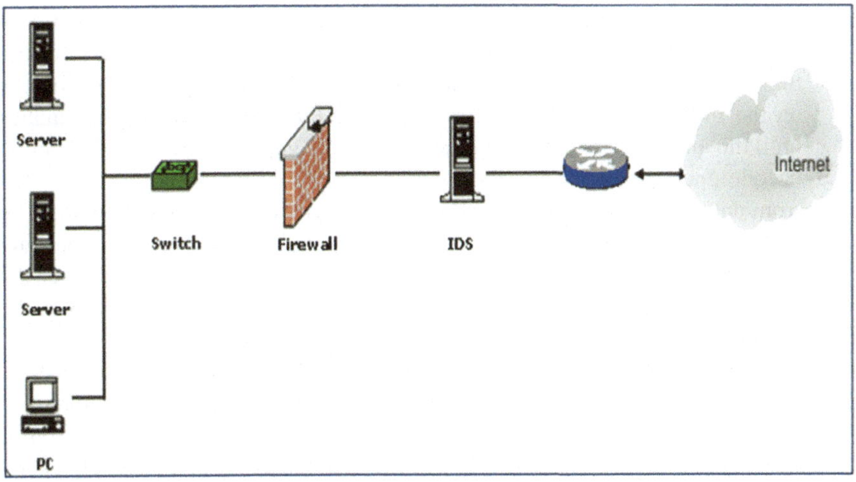

Fig. 1 General framework of IDS [3]

- **Network Intrusion Detection System (NIDS)**: NIDS refers to IDS framework which is set up to track traffic from all devices over the network at a point within the network designed. This tracks traffic passing through the intact subnet and compares to the inventory of known attacks the signals transmitted over the subnet. The alert can be sent to the administrator before the detection of any vulnerability or unusual action. Installation of NIDS over the subnet on the location of the firewalls observes whether anyone attempts to break the firewall, which is one of the examples of NIDS. This type of IDS monitors and analyzes packets in the network to detect intrusion in the network (Fig. 2).

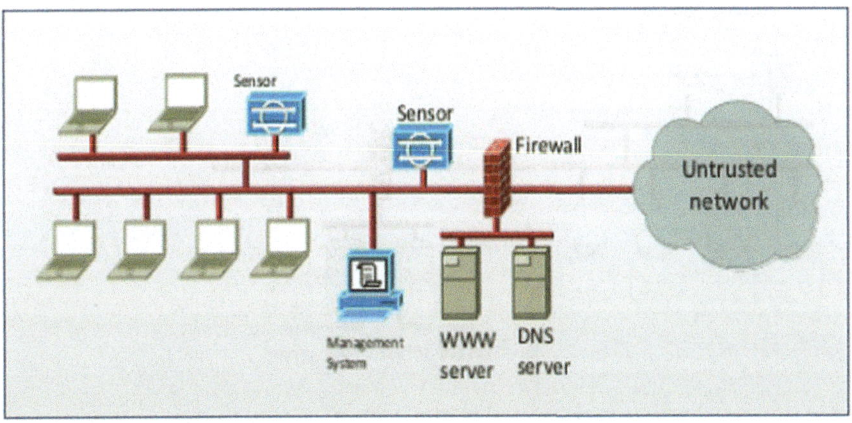

Fig. 2 Network-based IDS [4]

NIDS tracks network operation. An administrator installs sensors over network devices like routers and firewalls into the NIDS. Such sensors gather information and report anomalies to a central control server running an NIDS console on the individual devices or workstations. NIDS cannot identify anomalies until the anomaly creates a large-network traffic gap. Therefore, NIDS cannot decode encrypted data. In other words, it can only track and evaluate risks from data transmitted in plaintext or non-encrypted data on the network. The preference of where the sensors are to be mounted depends on what you want to calculate. For instance, the sensor can perceive all of the traffic over the Internet side of the firewall. On the inside side of the firewall, however, the sensor can only make out traffic going through the firewall.

- **Host Intrusion Detection System (HIDS)**: HIDS presents an IDS that operate over network computers or separate hosts. A HIDS only tracks incoming and outgoing system packets and warns the administrator regarding detection of unusual or malicious behavior. To evaluate it to the above screenshot, it takes a screenshot of previously on hand device files. When the files for the analytical program have been changed or removed, an alert message will be forwarded to the administrator for investigation. An example of HIDS involves vital task machines that are not supposed to alter their configuration. This IDS monitors the activity on individual computer or host on it is installed. It checks operating system files, logs, system calls, etc. for intrusion detection (Figs. 3 and 4).
- **Protocol-based Intrusion Detection System (PIDS)**: PIDS reflects IDS that will be permanently placed at the front end of a file. It monitors and interprets the set of rules between a server and a user/device. It helps to secure the web server by tracking the HTTPS protocol stream on a regular basis and accepting

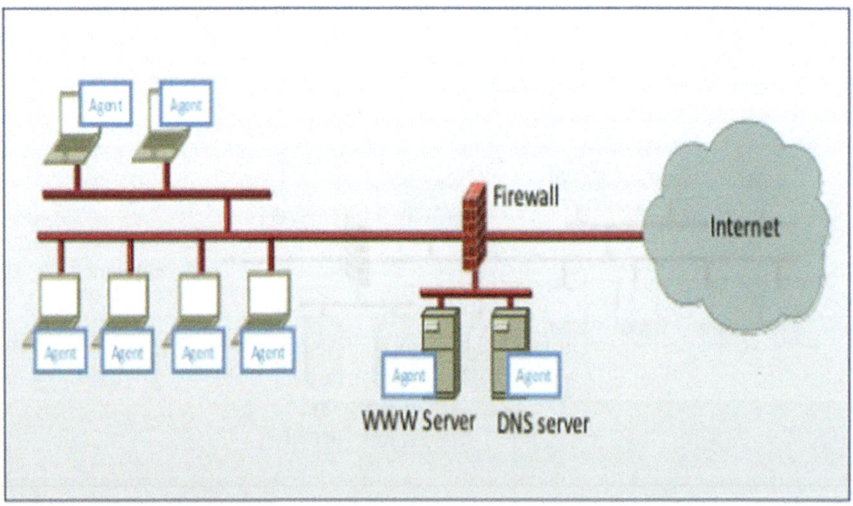

Fig. 3 Host-based IDS [4]

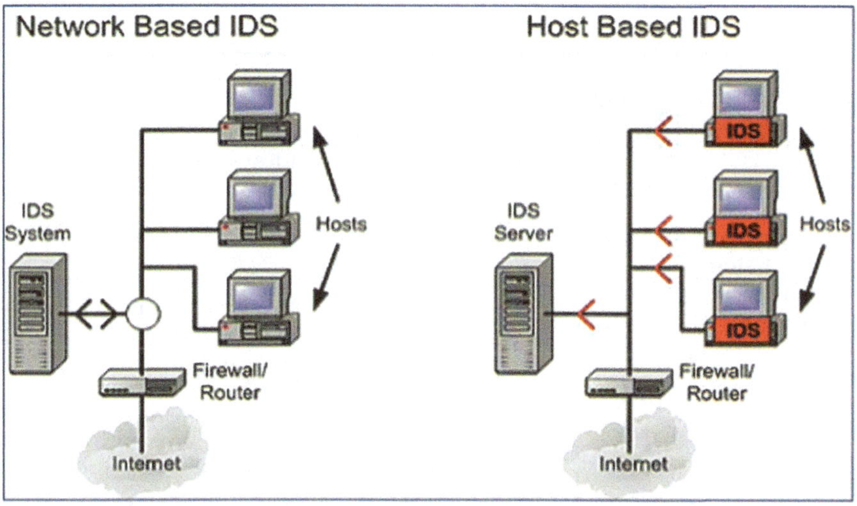

Fig. 4 NIDS versus HIDS [5]

the appropriate request. HTTPS is unencrypted, and this device will have to live in this browser before immediately accessing its web display layer, between using the HTTPS. A PIDS should try and implement the correct use of the protocol at a basic level. The PIDS can learn or be taught appropriate protocol constructs at a more advanced stage, and thus better detect anomalous behavior.

- **Application Protocol-based Intrusion Detection System (APIDS)**: APIDS refers to system or agent that usually resides within a network of servers. This detects the intrusions through monitoring and touch inference on various collections of rules for use. For example, this will actually monitor the SQL protocol toward the middleware layer as it transacts through the database of the web server.
- **Hybrid Intrusion Detection System**: Hybrid IDS refers to IDS that is rendered by integrating two or more methods of intrusion detection systems. Within the hybrid intrusion detection method, the host agent or application information is combined with network data to create a comprehensive view of the network environment. Compared with traditional intrusion detection system, the hybrid intrusion detection system is more effective. Prelude provides an example of a hybrid IDS (Fig. 5).

  The hybrid IDS shown in the figure is composed of six components, i.e., data collection software, database of signatures, analyzer, detector of anomalies, generator of signatures and software of countermeasures.

  The module features many data acquisition sensors. Sensors are either mounted on a single host, or on a different network segment. Sensors built on individual hosts detect packets entering and leaving the host. Sensors installed on a specific network segment will read packets as they move in and out of each segment.

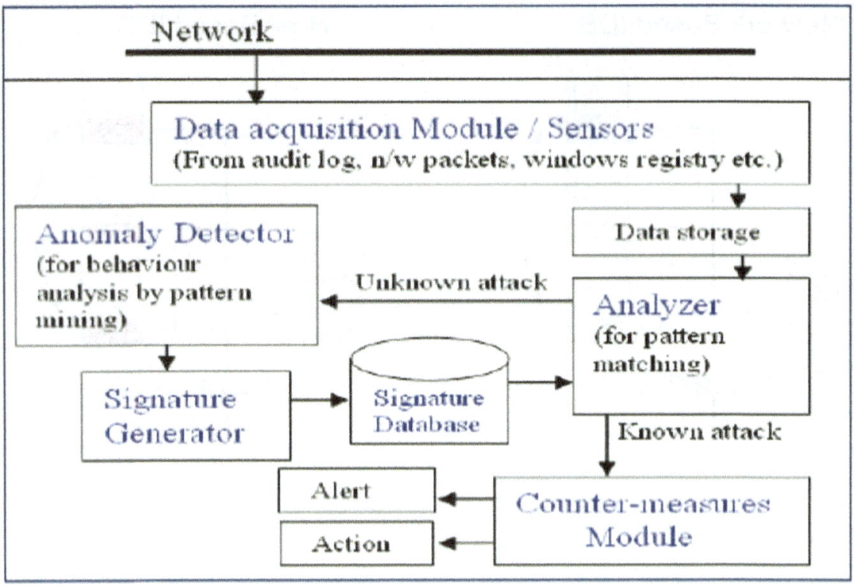

Fig. 5 Architecture of hybrid IDS [6]

Using pattern matching algorithm, the analyzer module evaluates the packets it identifies with the signatures or laws of normal activity patterns stored in the signature database. If any match is identified by the analyzer, then it sends the correct alert message for known attack to the countermeasure module. It inserts the log-file entry about the warning–triggering incident as well. If the analyzer finds no match, then it sends data to the anomaly detector using pattern mining technique to come across anomaly. On detection of any threat, the appropriate message will be sent to the signature generator. When the countermeasure module retrieves the alert message of a known threat from analyzer, it notifies the administrator in one of several ways previously configured by the administrator. One of the examples involves the module which shows a pop-up window, or forwards an e-mail to the person listed. This module can be programmed to take action in addition to the automatic response sent to the administrator when receiving a warning message. Typical behavior includes forwarding alert message to the administrator, dropping of the packet devoid of transmitting an error message to the root device, and resetting of the IDS to discontinue and reboot the network, thus preventing especially severe threats. Network administrator over and over again uses this module to evaluate the alert message and obtain apposite action such as dropping a packet, closing a connection, etc. The administrator will anticipate the need to fine-tune the database of signatures to account for situations that appear to be intrusions to the IDS but are in fact legitimate traffic.

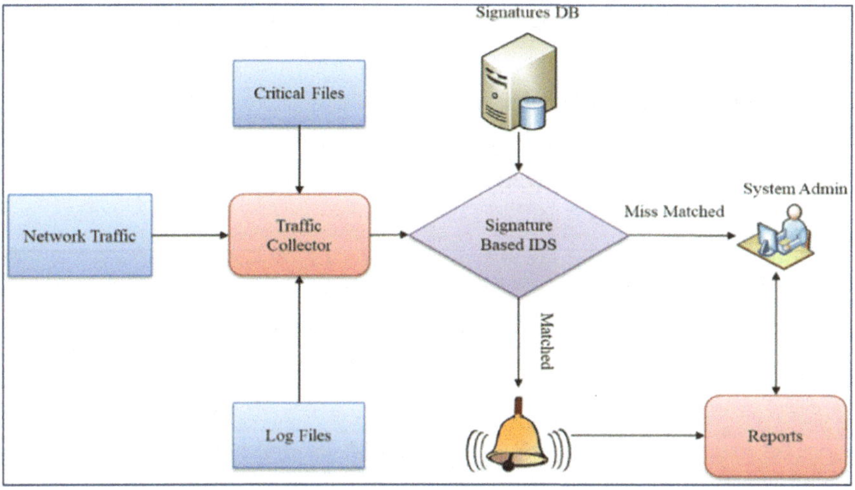

Fig. 6 Signature-based IDS [7]

- **Types of detection techniques**:
  - **Signature-Based IDS**: Signature-based IDS includes detecting net traffic threats based on general trends, i.e., number of bytes, number of 1 s, number of 0 s, etc. This also requires malware detection, based on the already reported malicious instruction sequence. The patterns that have been observed at IDS are called signatures. Through this process, the signature of recognized threats is inserted into the database; and for identification of the identified attacks, numerous link records are evaluated against the previously stored signatures. This utilizes a collection of known vulnerabilities or known attack patterns. This includes launching a SYN flood attack through tools on servers by just entering the device's IP address to strike the threat (Fig. 6).
  - **Anomaly-Based IDS**: Anomaly-based IDS has been introduced to become aware of the unexplained malware attacks. This methodology includes implementation of machine learning in order to create a model of confident commotion and everything to come with said depiction get evaluated and declared apprehensive unless it is defined in model. This type of IDS is based on the fact that attacker behavior is different from the normal traffic. This approach models the system's normal behavior, and any deviation from that normal behavior is marked as intrusion. Machine-based learning method has strong generalized ability compared to the signature-based IDS approach because device and hardware configuration train those methods. It determines, first, normal or abnormal behavior. A zero-day vulnerability is generally described as one the vendor doesn't know about. However, administrators describe a zero-day vulnerability as one for any application in which the developer did not issue a fix. In other words, the provider may be aware of the vulnerability, but a patch to close the vulnerability has not already been written, checked and

published. In both instances, there are the insecurity and vulnerable devices (Fig. 7).

Features of signature based IDS:

- Very fast in detecting recognized threats.
- Unfamiliar threats (zero-day attacks) do not get identified.

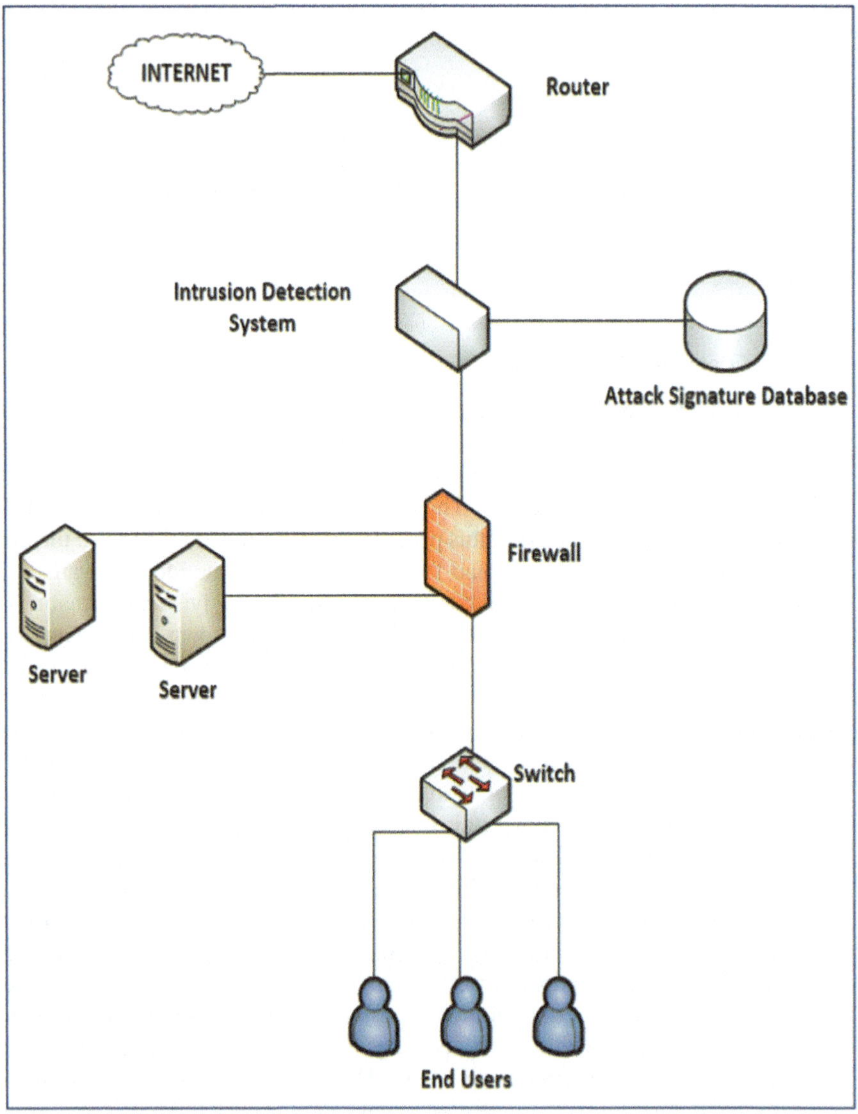

Fig. 7 Anamoly-based IDS [8]

- Very good accuracy.
- Low false positives.
- Requires updation of signature database frequently.

Features of Anomaly-Based IDS:

- Less efficient in detecting known attacks.
- Unknown attacks can be detected.
- High occurrence of false positives.
- Challenge is in modeling the normal behavior.

- **IDS versus Firewalls**:

While IDS and firewall are both synonymous with network protection, IDS differs from a firewall as a firewall searches for intrusions externally to prevent them from occurring. Firewalls block network access to prevent interference and do not alert if a network attack occurs. An IDS spots a potential intrusion after it happens and then sends an alert while both contributing to network security, and there is a dissimilarity between an intrusion detection system and a firewall in that a firewall looks outwardly for intrusion to prevent it. Firewalls block admittance to the network for interference prevention and do not signal an attack from within the network. An IDS assesses a possible intrusion after it has occurred and signals a warning. An IDS also checks for threats that arise from inside a network. It is generally done by analyzing network communications, recognizing specific device threat heuristics and patterns (often known as signatures) and taking action to alert operators. A device that terminates connections is called an intrusion prevention system, which is another form of application layer for the firewall (Fig. 8).

As anomaly-based IDS possesses the potential to detect zero-day vulnerabilities, more work is attracting this sort of IDS. Modeling an anomaly-based IDS, researchers

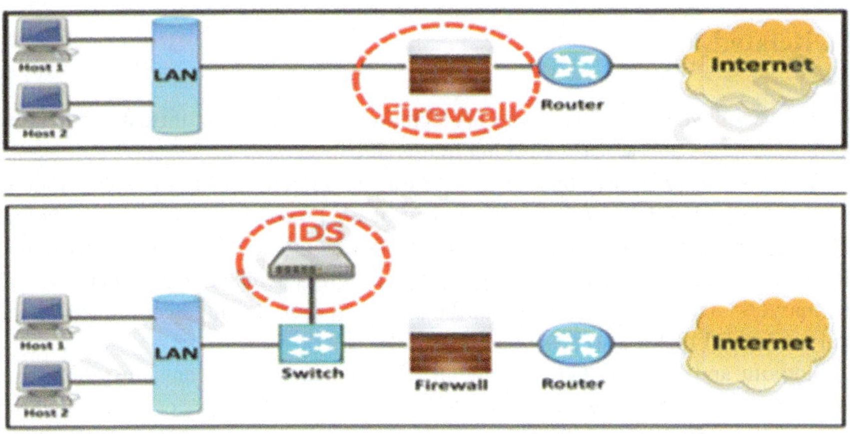

Fig. 8 IDS versus firewall [9]

primarily take advantage of classification and clustering techniques for machine learning.

- **Types of classification techniques used in intrusion detection**:
  - K-nearest neighbor
  - Artificial neural network
  - Decision tree
  - Support vector machine
  - Naïve Bayes
- **Types of clustering techniques used in intrusion detection**
  - K-means
  - K-medoids
  - Expectation maximization

Many researchers have used ensemble learners like random forest, AdaBoost, etc. to detect intrusion. Some researchers also proposed the use of deep learning and reinforcement learning in intrusion detection.

Optimal IDS would meet the following requirements for IoTs:

- The implementation does not build new vulnerabilities in the current program.
- It should require little resources and not impair system performance.
- It should not modify the existing infrastructure.
- It should be lightweight and less complex.
- It should have less false positives and false negatives.
- It should detect known as well as unknown attacks.
- Its performance should not be affected by the mobility of the nodes.
- It should work in real time environment without human intervention.

2 Related Work

Researchers have proposed out various techniques to implement intrusion detection system for years and are still working in this direction.

Guofei et al. [10] developed a system for the detection of botnet. The authors provided a general IDS system independent of the protocol and structure of botnet C&C, and no apriori knowledge of the botnets was needed. The authors have also introduced and tested the BotMiner prototype program using several actual traces of the network. The results showed that the proposed model can detect botnets in the real world, and that it also has a very low FPR.

Siddiqui and Naahid [11] implemented IDS by using K-means clustering. The authors concentrated on creating a relationship between the types of attacks and the protocols used by hackers using the Oracle 10 g data miner program.

Lahre et al. [12] also analyzed various approaches for intrusion detection systems on KDDcup'99 dataset. Ambedkar et al. [13] presented three-layered architecture for the detection of probe attacks by using KDDcup99 dataset. The proposed architecture was implemented by applying Naïve Bayes, support vector machine and multilayer perceptron.

Indre et al. [14] proposed a framework for identification and prevention of malevolent links based on machine learning techniques. The proposed framework worked by combining features to extract correlations from packet history based on the rate of flags and link states of REJ, SYN and ACK, with features of HTTP extracted from URI and RESTful form. The proposed methodology achieved the precision of 98.4% for binary classification problem under study.

Anthi et al. [15] developed IDS using machine learning techniques for IoT network named "Pulse." This network consists of two conventional machines, one for network traffic processing and the other for data injection attacks.

Xin et al. [16] implement specific methods of machine learning and deep learning for cyber-security. The authors clarified the machine learning (ML) and deep learning (DL) similarities and differences. Differences in ML and DL have been mainly due to data dependencies, hardware dependencies, feature processing, etc. They explained how the researchers used different machine learning methods in the literature. They also explained various datasets used by the researchers to build and test their model.

Ambedkar et al. [17] carried out a detailed study on probe vulnerabilities through implementation of numerous common machine learning techniques, namely Naïve Bayes, SVM, multilayer perceptron, decision trees, etc. The authors used KDDcup99 dataset for model construction. In the study, they projected three-layered design for identification of probe attacks. PCA has been implemented for dimensionality reduction.

Kamlesh et al. [18] applied KDDcup'99 dataset to analyze different approaches for intrusion identification systems. IDS is one of the network's key security elements. IDS involves network control and analysis. Detection of intrusion includes the identification of a collection of malicious actions which compromise information resources. Conventional intrusion detection approaches employ comprehensive signature information for documented attacks. The networking boom has finally come of age over the last six years.

Zarpelao et al. [19] conducted a comprehensive review of IDS literature for IoT networks. Various IDS placement approaches such as clustered, dispersed and hybrid were addressed. We also addressed methods of IDS identification, such as signature, anomaly-based, specification-based and hybrid-based approaches. They addressed numerous security threats to IDS, such as traditional attacks, routing attacks, man-in-the-middle and service denial. They also explored various testing methods such as theoretical, analytical and simulation. They discussed various issues, concerns and directions for IDS implementation in IoT networks in the future.

INTI IDS was projected by Cervantes et al. [20] for discovery and segregation of sinkhole threats. INTI incorporates anomaly-based principles for tracking the swap of packets between nodes and specification-based methods to obtain node evaluation, i.e., confidence and credibility. For both fixed and mobile settings, INTI is tested and compared with SVELTE IDS and found to have more precision and fewer false positives and less false negative.

Sharma et al. [21] present comprehensive literature survey of VANETs from the year 2004–2017. They lay emphasis on the security aspects of VANETs as human lives are at stake in such networks. They presented various possible attacks in VANETs and categorize the attacks on the basis of action of the attacks. They presented challenges in implementing security policies in VANETs because of their high mobility and resource constraints. They also proposed honeypot-based intrusion detection system for VANETs and discussed future direction for research in this area.

Mishra et al. [22] offered thorough review and investigation of the various intrusion detection techniques used in machine learning. They touched on every aspect of intrusion detection and the techniques of machine learning to incorporate intrusion detection. In introducing intrusion detection, they addressed the problems encountered by previous researchers. They provided comprehensive descriptions of the datasets KDD'99 and UNSW-NB. They defined which dataset function is critical in the detection of a specific type of attack. They also addressed the issue of using machine learning to detect low-frequency attacks. They divided their approach into four parts: one classifier with all the features, one classifier with limited features, multiple classifiers with all features and multiple classifiers with limited features. They proposed using deep learning and reinforcement learning to detect intrusion.

Sharma et al. [23] dealt with the value of the VANET security system. They optimized the IDS with dolphin swarm algorithm based on SVM machine learning algorithm. Echolocation, information exchange and labor-sharing are biological characteristics of dolphins. Combined with swarm intelligence, these biological features can be used to optimize multi-cluster head anomaly-based IDS. This system's performance is compared with different security frameworks using false positives, detection rate, detection time, etc.

3 Methodology

The experiments were performed in Python using Scikit-learn library to test the performance of machine learning techniques for implementing intrusion detection program. The experimental research was conducted on a benchmark dataset, i.e., KDD'99 dataset [24], consisting of 7 million records each with 42 features of local-area-network (LAN) raw TCP dump data simulating a typical U.S. Air Force LAN with numerous assaults peppered. This dataset deals mainly with four types of attacks, namely service denial (DoS), R2L, U2R and probing. For training ML-based IDS, the dataset is split into train–test set in the 70:30 ratio. The overall methodology of the current work is presented in Fig. 9.

Five machine learning techniques, namely k-nearest neighbor (KNN), multilayer perceptron (MLP), decision tree, Naïve Bayes and support vector machine (SVM), were used to implement the IDS. The data collected were preprocessed to convert nominal data into numerical data, deal with outliers and impute missing values using statistical techniques. The performance of undertaken machine learning-based IDS

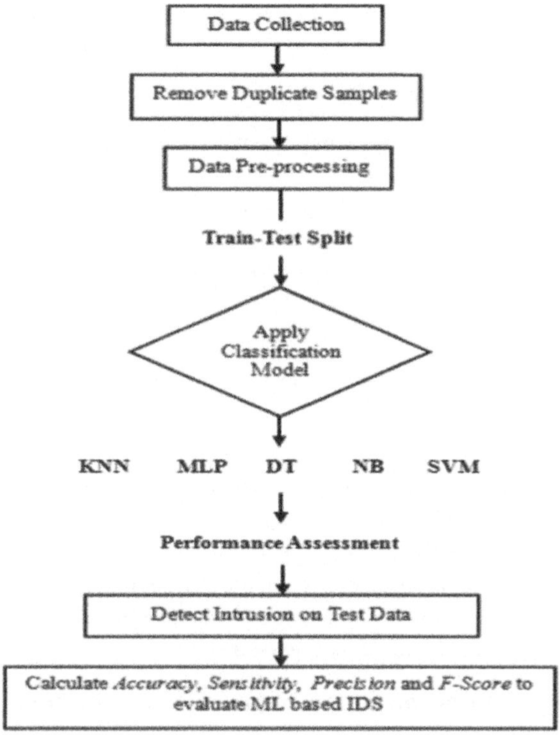

Fig. 9 Flowchart for the current study

has been evaluated on the basis of four measures, i.e., *accuracy*, *sensitivity*, *precision* and *F-score.*

A. Implementation via K-Nearest Neighbor

KNN is a supervised technique in machine learning, applied to classification and regression tasks. This algorithm uses "feature similarity" to predict the values of new data points, meaning that the new data point will be assigned a value based on how close it matches the points in the training set [25]. KNN's fundamental work is reflected in the following steps:

Step 1—Start
It needs dataset for implementing this algorithm. Therefore, it requires loading the training data as well as testing data during the first step of KNN.
Step 2—Select the optimum value of K, i.e., the nearest data points. K can be any integer.
Step 3—Do the following for every single point in the test set:

- **3.1**—Estimate the distance between test data and each row of training data using either of the methods like Euclidean, Manhattan or Hamming distance. Euclidean is the most widely used method for distance calculation.
- **3.2**—Sort them in upward order, based on the distance value.

- **3.3**—From the sorted list, pick the top rows K.
- **3.4**—Assign label to the test point based on most frequent class of these rows.

Step 4—End

It determines the similarity pattern by evaluating the distance metrics that include Minkowski, Euclidean and Hamming. This distance metric performs calculation of the least distance between the test data to all data points in order to trace out the k-nearest neighbors. The test data output belongs to majority of k-nearest neighbors' classes. Most researchers used Euclidean distance metric to determine minimum distance represented as below:

$$E(A, B) = \sqrt{(a_1 - b_1)^2 + (a_2 - b_2)^2 + \cdots + (a_m - b_m)^2}$$

where

A defines the data sample for training set and ***B*** represents the test data point with ***m*** input attributes in the dataset [26].

In the current study, KNN-based IDS has been trained with *42 input parameters* and *1 output parameter* from the dataset under study and implemented with *tenfold cross-validation* and *Euclidean* distance metric. The optimum number of nearest neighbors (k) was found to be 5.

B. **Implementation via Support Vector Machine**

SVM is also the supervised methodology in machine learning, applicable to classification and regression tasks. It works by defining an optimal hyper plane M-dimensional which separates the data into two groups. There are several potential hyperplanes that could be chosen to distinguish the two data point groups. The main goal is to find a plane with the highest range, that is to say the highest distance between both classes' data points. Maximizing the distance from the margins offers some support for classifying potential data points with greater confidence. This optimal hyperplane aims to optimize the gap between the closest observations, thus generating the limitations for the attack and normal groups [27]. Figure 10 presents an illustration of hyperplanes and support vectors.

The linear SVM score function can be calculated as

$$g(x) = \sum (x_i \cdot y)\beta_i + b$$

where

$\boldsymbol{x}$ defines the test sample whose class to be predicted, $\boldsymbol{x_i}$ refers to all the support vectors in the training set with $\boldsymbol{M}$ input parameters, the vector $\boldsymbol{\beta}$ comprises the coefficients to define an orthogonal vector to the hyperplane and $\boldsymbol{b}$ reflects the bias term [28].

For the current study, SVM-based IDS has been trained with *tenfold cross-validation*, *linear kernel* function and *sign* function for score transformation. The

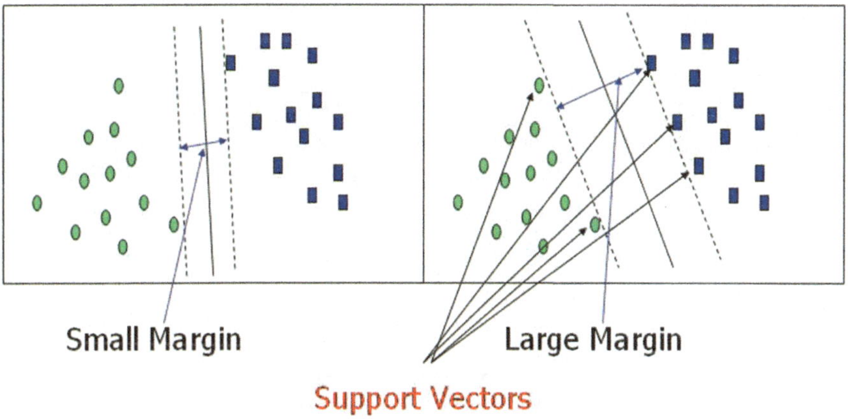

Fig. 10 Illustration of support vectors

system takes *42 input attributes* and *1 output attribute* for classifying the network connections as *normal* or *attack*.

C. **Implementation via MLP**

MLP represents a neural network that comprises two or more perceptrons. It consists of three components, i.e., single *input layer* for receiving input parameters, one or more *hidden layers* to process the inputs, and single *output layer* for producing the final output on the test data [29].

MLP refers to perceptron coupling two or more perceptrons which are coupled in many layers for solving multifaceted troubles. Following diagram presents the general framework of the MLP. For single neuron in neural network, the output of neuron can be computed mathematically as:

$$Y = W * X + B$$

where

X represents the input vector that contains values for input parameters
B contains bias value
W presents the weight associated with each input (Fig. 11).

The working of MLP can be interpreted in mathematical terms as:

$$y = \varphi\left(\sum_{i=1}^{n} \omega_i x_i + b\right) = \varphi\left(\mathbf{W}^T \mathbf{X} + b\right)$$

where

$\mathbf{w}$ refers to weights of associated inputs in vector form

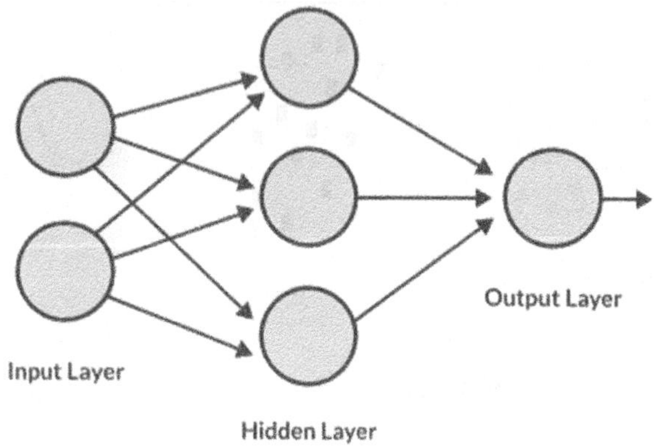

Fig. 11 General structure of MLP

x represents input parameters in vector form
b refers to bias value of neuron
Φ refers to activation function to be applied to output of neuron for transforming the output in the range [−1, 1].

In the current study, MLP-based IDS has been implemented with *relu* activation function. The classifier has been trained for *100 epochs* with *tenfold cross-validation.*

D. **Implementation via Decision Tree**

Decision tree is an algorithm used to classify or predict the data using machine learning. It is represented as a tree-like structure, where each internal node represents the feature, each branch refers to the rule of decision, and each leaf node is associated with the label of output class.

DT is "learned" by splitting the dataset into subsets, based on internal node attribute values. For each derived subset, this division method continues in a recursive fashion known as recursive partitioning. This recursive method is terminated when the subset of nodes has the same target variable value; or when process splitting adds value to the predictions. It classifies data points by sorting the data from the root node to the terminal node down the list, which classifies the test data point [30].

The basic idea behind every decision tree algorithm is as follows:

1. Implement attribute selection measures (ASM) to pick the correct attribute to break the data.
2. Link a decision node to it, and divide the dataset into smaller subsets.
3. Starts tree-building by repeating this cycle recursively for every child until one of the conditions matches:

   - Every tuples belongs to the same attribute value.
   - There is left no more remaining attributes.
   - No more instances are left (Figs. 12 and 13).

Measurement of selection of attributes is a form of heuristic approach to select in the best way possible the dividing criteria for partition data. The most common

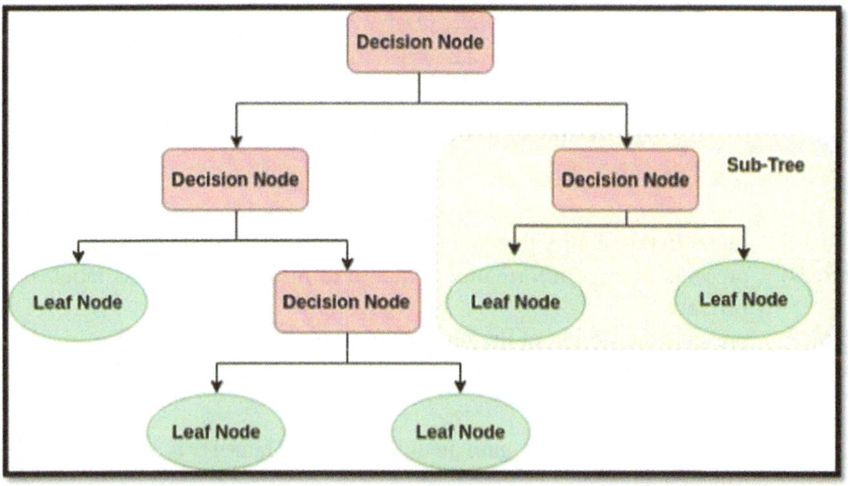

Fig. 12 General architecture of decision tree [31]

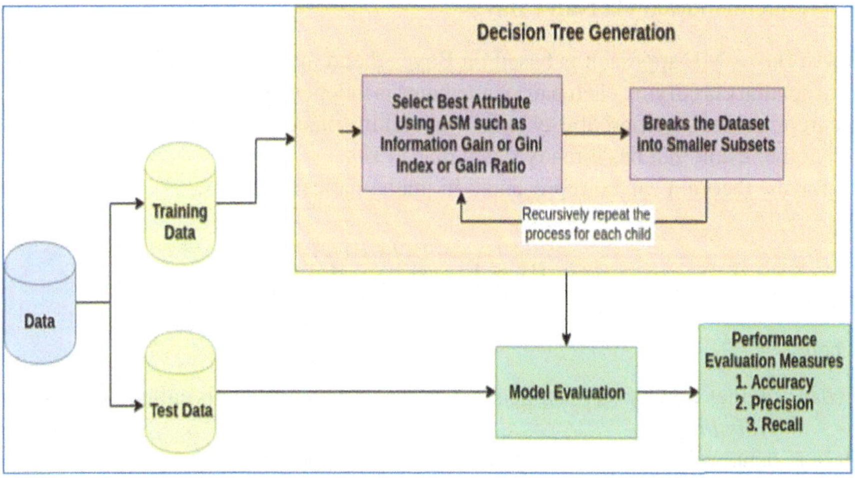

Fig. 13 Train–test split in decision tree [32]

selection metrics are *information gain, Gain ratio* and *Gini Index*. These metrics are presented mathematically as under:

$$\text{Info}(D) = -\sum_{i=1}^{m} pi \log_2 pi$$

$$\text{Info}_A(D) = \sum_{j=1}^{V} \frac{|D_j|}{D} \times \text{Info}_A(D_j)$$

$$\text{Gain}(A) = \text{Info}(D) - \text{Info}_A(D)$$

where

pi refers to probability that an instance *D* would go to category C_i

Info(*D*) presents average information required to make out the class label of *D* instance

$|D_j|/|D|$ presents the weight of the *j*th partition

$\text{Info}_A(D)$ is the expected information required to categorize instance *D* based on the partitioning by *A*.

Attribute *A* with the highest gain in information, Gain(*A*), is selected as the dividing attribute at node *N*.

In the current study, DT-based IDS has been implemented with *ID*3 algorithm, *tenfold cross-validation* and *entropy* for information gain.

E. **Implementation via Naive Bayes**

Naive Bayes ML approach is based on Bayes theorem. This approach works on the concept that classifying each pair of attributes is independent from each other. Bayes' theorem assesses the probability of an event taking place, considering the probability of another event that has already taken place [33].

Bayes' theorem can be represented in mathematical terms as:

$$P\left(\frac{I}{J}\right) = \frac{P(I).P\left(\frac{J}{I}\right)}{P(J)}$$

where

$P(I|J)$ defines how often *I* happens given that *J* happens and is called as *posterior probability*

$P(J|I)$ denotes how often *J* happens given that *I* happens

$P(I)$ is called *prior probability,*

$\frac{P(\frac{J}{I})}{P(J)}$ denotes *likelihood ratio.*

For the current work, Naive Bayes-based IDS has been implemented with *multinomial* approach and *tenfold cross-validation.*

4 Result and Analysis

After performing the experiments, it has been found out that machine learning techniques can be effectively used for implementation of IDS. The performance of these

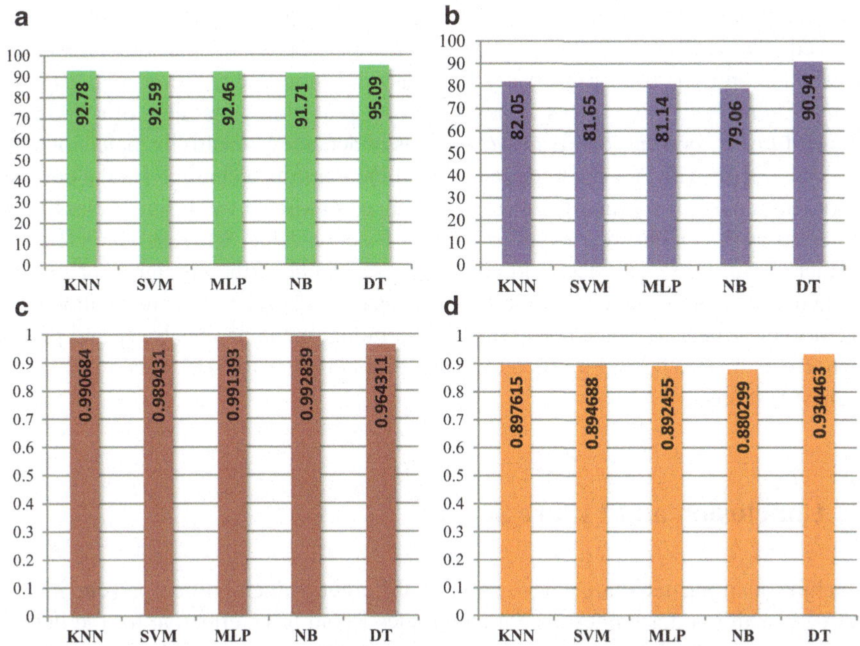

Fig. 14 **a** Accuracy, **b** sensitivity, **c** precision and **d** F-score predicted by different machine learning techniques for IDS

Table 1 Experimental results

Parameter ML algorithm	Accuracy (%)	Sensitivity (%)	Precision	F-score
KNN	92.78	82.05	0.990684	0.897615
SVM	92.59	81.65	0.989431	0.894688
MLP	92.46	81.14	0.991393	0.892455
DT	95.09	90.64	0.964311	0.934463
NB	91.71	79.06	0.992839	0.880299

techniques have been computed by four metrics, i.e., *accuracy*, *sensitivity*, *precision* and *F-score*. In the current experimental study, KNN achieved accuracy of 92.78% with sensitivity of 82.05%, precision of 0.9907 and F-score of 0.8976. SVM also predicted good performance with accuracy of 92.59%, sensitivity of 81.65%, precision of 0.9894 and F-score of 0.8946. MLP showed 92.46% accuracy, 81.14% sensitivity, 0.9914 precision and 0.8924 F-score. Decision tree and Naive Bayes also revealed good accuracy for intrusion detection. All the machine learning techniques taken in this work predicted accuracy and precision of more than 90%. But some techniques revealed low-sensitivity values. From Table 1, it can be analyzed that *deci-*

sion tree predicted highest accuracy with highest sensitivity of 95.09% and 90.64%, respectively, whereas *Naive Bayes* predicted lowest accuracy with lowest sensitivity of 91.71% and 79.06%, respectively. It means decision tree achieved highest *true positive rate*. But Naive Bayes achieved highest precision with value of 0.992839. Figure 14 (a–d) presents comparative analysis of accuracy, sensitivity, precision and F-score, respectively, for the undertaken machine learning techniques in the current work. Figure 14(d) represents the *F-score* values predicted by different classifiers under study. Decision tree achieved highest F-score value, whereas Naïve Bayes predicted lowest F-score value. Finally, for assessment of the efficiency of undertaken methods of machine learning, *F-score* has been considered the most important performance measure since it presents the harmonic mean of sensitivity and precision. Decision tree has been found to be the best classifier for IDS since it achieved highest F-score.

5 Conclusion and Future Scope

Machine learning techniques can identify different vulnerabilities present in the network with good accuracy. These techniques found to be effective for the detection of various attacks occurring in the network at different layers. The experimental findings indicate that methodologies in machine learning are efficiently utilized for the functioning of intrusion detection system. In the current study, various machine learning techniques have been applied for intrusion detection over the network. Decision tree has been found as the best classifier as compared to other techniques of machine learning for intrusion detection in this work. In future, machine learning techniques coupled with feature selection techniques can be explored to implement optimum IDS. Deep learning techniques can also be implemented to evaluate its behavior for intrusion detection.

References

1. Intrusion Detection System, https://searchsecurity.techtarget.com/definition/intrusion-detection-system. Accessed 1 Dec 2019
2. Intrusion Detection System (IDS), https://www.geeksforgeeks.org/intrusion-detection-system-ids/. Accessed 1 Dec 2019
3. https://krazytech.com/technical-papers/intrusion-detection-and-avoidance-system. Accessed 3 Dec 2019
4. https://gbhackers.com/intrusion-detection-system-ids-2/. Accessed 3 Dec 2019
5. https://www.quora.com/What-does-IPS-Intrusion-Prevention-System-mean. Accessed 2 Dec 2019
6. Patel K, Buddhadev B (2013) An architecture of hybrid intrusion detection system. Int J Inf Netw Secur 2:197–202. https://doi.org/10.11591/ijins.v2i2.1753
7. https://www.researchgate.net/figure/Signature-based-intrusion-detection-system_fig4_338028960. Accessed 1 Dec 2019

8. https://www.researchgate.net/figure/a-Anomaly-Based-Intrusion-Detection-System-b-Signature-Based-Intrusion-Detection-System_fig1_324189357. Accessed 1 Dec 2019
9. https://networkinterview.com/firewall-vs-ips-vs-ids/. Accessed 1 Dec 2019
10. Guofei G, Zhang J, Lee W (2008) BotSniffer: detecting Botnet command and control channels in network traffic. In: Proceedings of network and distributed system security symposium, NDSS, 2008
11. Siddiqui MK, Naahid S (2013) Analysis of kdd cup 99 dataset using clustering based data mining. Int J Database Theor Appl 6(5):23–34
12. Lahre K, Dhar T, Kashyap D, Aggrawal P (2013) Analyze different approaches for IDS using KDD 99 data set. Int J Recent Innov Trend Comput Commun 1(8):645–651
13. Ambedkar C, Kishore Babu V (2015) Detection of probe attacks using machine learning techniques. Int J Res Stud Comput Sci Eng **2**(3), 25–29
14. I. Indre and C. Lemnaru, "Detection and prevention system against cyber attacks and botnet malware for information systems and Internet of Things", IEEE 12th International Conference on Intelligent Computer Communication and Processing (ICCP), 2016
15. Anthi E, Williams L, Burnap P (2018) Pulse: an adaptive intrusion detection for the internet of things. In: Proceedings of living in the internet of things: cybersecurity of the IoT, 2018
16. Xin Y, Kong L, Liu Z, Chen Y (2018) Machine learning and deep learning methods for cyber security. IEEE Access. https://doi.org/10.1109/ACCESS.2018.2836950
17. Ambedkar Ch, Kishore Babu V (2015) Detection of probe attacks using machine learning techniques. Int J Res Stud Comput Sci Eng (IJRSCSE) 2(3):25–29
18. Lahre K, Diwan T, Kashyap SK, Agarwal P (2013) Analyze different approaches for IDS using KDD 99 data set. Int J Recent Innov Trends Comput Commun 1(8):645–651
19. Zarpelao BB, Miani RS, Kawakani CT, Alvarenga SC (2017) A survey of intrusion detection in internet of things. J Netw Comput Appl 84
20. Cervantes C, Poplade D, Nogueira M, Santos A (2015) Detection of sinkhole attacks for supporting secure routing on 6LoWPAN for Internet of Things. In: 2015 IFIP/IEEE International Symposium on Integrated Network Management (IM), pp 606–611
21. Sharma S, Kaul A (2018) A survey on Intrusion detection systems and honeypot based proactive security mechanisms in VANETs and VANET cloud. Veh Commun 12:138–164
22. Mishra P, Varadharajan V, Tupakula U, Pilli ES (2018) A detailed investigation and analysis of using machine learning techniques for intrusion detection. In: IEEE communications surveys & tutorials
23. Sharma S, Kaul A (2018) Hybrid fuzzy multi-criteria decision making based multi cluster head dolphin swarm optimized IDS for VANET. Veh Commun 12:23–38
24. KDD Cup 1999 Data, http://kdd.ics.uci.edu/databases/kddcup99/kddcup99.html. Accessed 8 Dec 2019
25. https://www.tutorialspoint.com/machine_learning_with_python/machine_learning_with_python_knn_algorithm_finding_nearest_neighbors.htm. Accessed 8 Dec 2019
26. K Nearest Neighbor Algorithm, https://people.revoledu.com/kardi/tutorial/KNN/HowTo_KNN.html. Accessed 2 Dec 2019. Accessed 7 Dec 2019
27. https://towardsdatascience.com/support-vector-machine-introduction-to-machine-learning-algorithms-934a444fca47. Accessed Online 6 Dec 2019
28. Support Vector Machine, https://towardsdatascience.com/support-vector-machine-introduction-to-machine-learning-algorithms-934a444fca47. 2 Dec 2019
29. Multilayer Perceptron, https://www.techopedia.com/definition/20879/multilayer-perceptron-mlp. Accessed 2 Dec 2019
30. Decision Tree Implementation Using Python. https://www.geeksforgeeks.org/decision-tree-implementation-python/. Accessed 21 Nov 2019
31. https://www.datacamp.com/community/tutorials/decision-tree-classification-python. Accessed 5 Dec 2019
32. https://www.geeksforgeeks.org/decision-tree/. Accessed 5 Dec 2019
33. Naive Bayes Classifier in Python, https://dzone.com/articles/naive-bayes-tutorial-naive-bayes-classifier-in-pyt. Accessed 21 Nov 2019

Chapter 11
An Efficient Approach for Phishing Detection using Machine Learning

Ekta Gandotra and Deepak Gupta

1 Introduction

Due to the availability of Internet at low price, people are shifting to online platforms instead of visiting banks, shops, etc. Attackers are taking advantage of this fact and trying to find their victims online to make money instead of taking risks to rob banks/shops, etc. They are making use of various attacks like phishing to steal the passwords, credit card details, etc., by misleading users to visit malicious and fake websites.

Phishing attack is one of the top security threats on the Internet today. Attackers tend to gather victims' confidential information using fake websites. According to Anti-Phishing Working Group [1], the trend of phishing attacks is increasing every year and 138,328 phishing pages were informed in 2018, 4th quarter. It causes a lot of financial losses. On the basis of the cases informed to Federal of Investigation [2], there occurred a loss of around \$48 million in USA in the year 2018. In addition, phishing attacks are also becoming the top delivery method of malware [3–5]. A recent report of Microsoft security intelligence [6] reported that in 2018, phishing attack was the top web attack.

In year 2013, South Korea faced a phishing attack, in which 3 banks and 3 media companies were compromised [7]. This attack involved stealing passwords and was spread to more than 32,000 machines. In December 2014 [8], the cyber attackers were able to get the data of 80 million customers of Anthem Healthcare. The recent

E. Gandotra
Department of Computer Science and Engineering, Jaypee University of Information Technology, Waknaghat, Solan, India
e-mail: ekta.gandotra@gmail.com

D. Gupta (✉)
Department of Computer Science and Engineering, Thapar Institute of Engineering & Technology (Deemed to be University), Patiala, India
e-mail: deepak.vd@gmail.com

K. J. Giri et al. (eds.), *Multimedia Security*, Algorithms for Intelligent Systems,
https://doi.org/10.1007/978-981-15-8711-5_12

worldwide lockdown due to novel coronavirus (COVID-19) epidemic has forced many employees to work from home where they do not have cybersecurity resources as they usually have in their offices. Moreover, due to the infectious nature of the virus, there is anxiety among the people over any online information related to coronavirus. The attackers are exploiting these facts to perform phishing attacks through emails or fake websites with the aim of making money. Threat Analysis Group [9] of Google has identified attacker groups who are using subject as COVID-19 to attempt malware and phishing attacks.

There exist various tools [10–12] which attackers use to duplicate a website in order to create a phishing website. Sometimes, the websites designed in such a way are identified by anti-phishing tools because of the website's elements such as copyright, anchor link, logo or keywords of the genuine websites. There are various approaches for detection of phishing websites which are having their own pros and cons. Signature-based method is the most popular method used for detecting phishing webpages. This method maintains a database of blacklisted websites which consists of a list of phishing webpages. Any newly requested webpage is checked against this database to identify whether it is phished. Two of the important blacklisted databases are PhishTank [13] and Google Safe Browsing (GSB) [14]. This approach fails to identify the newly generated phishing websites which have not been added to the database. In case of heuristic approach, the webpage is analyzed to come up with discriminating features like process creation, page redirection, etc. [15]. However, this approach is resource-intensive. Another approach used to detect phishing websites is visual-based approach [16] in which a database of images, logos, text formatting, HTML tags, etc., is created. But this method is highly time-consuming.

Recently, researchers have started to employ machine learning and computational [17] based methods for identifying phishing websites. In this method, different types of features pertaining to phished and benign websites are extracted, and subsequently, these are used for training purpose to build the classification models. The efficiency and performance of such methods depend on the classification algorithms, number of training examples, and the feature set. Keeping the classification models and the number of training examples apart, considering a large and variety of features helps in enhancing the performance of classification models. However, it takes more time for model building which hampers the timely detection of phishing websites. Therefore, an appropriate set of features should be selected to build the classification models in less time without compromising the accuracy. Keeping this point in mind, this chapter aims to present a study on the role of feature selection methods in detecting phishing webpages. Consequently, a comparative analysis is performed on the performance of machine learning algorithms without and with feature selection.

To carry out the study, this chapter is structured as follows. Section 2 presents the recent research work in classifying the phishing webpages using machine learning techniques. Section 3 explains the methodology used in carrying out this study. Section 4 analyzes and discusses the experimental results. Section 5 provides conclusion and future scope.

2 Related Work

This section presents the recent research work pertaining to classification of phishing websites using machine learning algorithms.

Tan et al. [18] presented a method named PhishWHO for the detection of phishing websites in three stages. In first stage, the keywords are extracted from the websites (using N-gram method). In second stage, these keywords are used in a search engine to find the name of target domain. Finally, in the last stage, they used a matching system for detecting the legality of the website.

An image-based approach was proposed by Chiew et al. [19] for the detection of malicious websites. They extracted the logo and fed it into Google image search engine to check the identity of the website. They compared the webpage query with the domain name returned by Google for distinguishing a phishing webpage from the benign one. Experiments were conducted to prove the usefulness of the proposed method.

In [20], the authors analyzed various features of phishing webpages and shortlisted the most important 19 features. After extracting these 19 features from the source code of webpage, they used support vector machine (SVM), random forest (RF), neural network, logistic regression, and Naïve Bayes (NB) for classification of phishing webpages and obtained around 99% accuracy.

Rao and Pais [21] proposed a method which is based on human behavior while they get exposed to malicious webpages. This human behavior of feeding the fake credentials is automated with an additional step of applying heuristic filtering. They were able to get an accuracy of 96.38% using this approach.

In [22], the authors proposed a real-time system that makes use of the features pertaining to natural language processing (NLP). They used seven different machine learning algorithms for the classification of phishing and benign webpages. They tested their proposed system on a self-created dataset. Their results showed that RF with features based on NLP provides the best accuracy while classifying the webpages.

Xiaoqing et al. [23] presented an automatic intelligent system for the detection of phishing webpages. They analyzed the features of uniform resource locator (URL) and used NB for classification. In case of suspicious webpages, it is parsed and re-classified by using SVM. Through their results, they claim that the system gives high accuracy of detection in less time.

A similar approach was used in [24]. The authors presented a method which used a combination of SVM and decision tree model. SVM is used for training purpose and decision tree is used to generate the rules for detecting phishing websites targeting the banking domain.

"Cantina + " [25] approach was used as an extension of "Cantina" [26] (based on textual content of a webpage and term frequency-inverse document frequency algorithm) where additional features were used to detect the phishing webpages. In this approach, the authors used eight features. They used search engines, document object model, and the services from the third party along with machine learning algorithms for the detection of phishing webpages.

Joshi et al. [27] proposed a system which selected important URL features and machine learning algorithms for the detection and classification of phishing webpages. They concluded that ReliefF and RF algorithms worked well than other combinations.

Wu et al. [28] presented a phishing detection tool that is developed by combining the URL of the webpage and its source code. They used Levenshtein method for computing the string similarity and SVM as a machine learning model in their proposed system for detection of phishing webpages.

Zamir et al. [29] analyzed the features of phishing data by using features selection methods. They proposed two features by combining various attributes. They used various machine learning and stacking techniques for detecting phishing webpages.

Almseidin et al. [30] presented a study where they employed various machine learning algorithms and features selection methods to improve the efficiency of their system. The experiments conducted on a phishing dataset of 48 features containing 5000 benign and 5000 phishing webpages. They concluded that RF algorithm with only 20 features gives the best accuracy.

Yerima and Alzaylaee [31] presented an approach based on deep learning for detecting phishing webpages. They used convolutional neural networks (CNN) for this purpose and evaluated on a dataset consisting of 6,157 genuine and 4,898 phishing websites. They compared their results with traditional machine learning algorithms and concluded that their approach using CNN gives much better results than conventional machine learning algorithms.

Basnet and Doleck [32] proposed a heuristic approach using URL-based features. Experiments were carried out on a dataset containing 138 features which were extracted from 31,000 malicious and 16,000 phishing webpages. These 138 features belong to four groups, namely search engine, reputation, lexical, and keyword-based. They used seven different machine learning algorithms for classification purpose. Random forest achieved the best accuracy.

Recently, due to the growth of multimedia systems [33, 34], there is an exponential growth in cyber-attacks. In order to protect the digital media, there is a need to develop the techniques to prevent their unauthorized distribution [35]. These involve digital signatures and water marking. A lot of research is being reported by the researchers on authentication of the content [36, 37].

From the literature, it is found that the research community has worked on improving the classification accuracy by considering a variety of large number of features for the classification of phishing webpages. However, it seems that there is less focus on the problem related to the building time of any classification model without compromising the accuracy. This chapter offers a study which compares the classification performance and efficiency of detecting phishing webpages using various machine learning algorithms without and with feature selection. It makes use of information gain method for selecting an appropriate set of features.

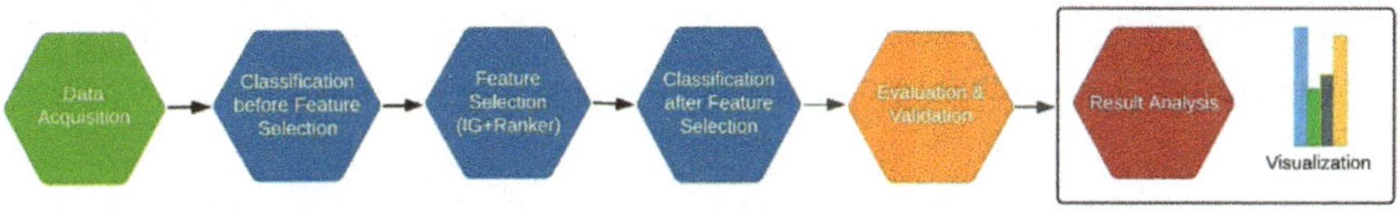

Fig. 1 Workflow of methodology used for proposed phishing detection system

3 Methodology Used

This section provides a description of the methodology used for discriminating benign webpages from the malicious ones. Figure 1 demonstrates the proposed design of phishing detection system. After downloading a suitable dataset, eight machine learning-based algorithms are employed to classify the phishing webpages. Afterward a feature selection method is deployed to choose top rank features. These selected features are then deployed to classify the webpages. Subsequently, the experimental results are compared by considering building time and performance of the models—before and after feature selection. The detailed steps of the proposed system are described in the following sub-sections.

3.1 Data Acquisition

Dataset used in this study is downloaded from [38, 39]. It is composed of 30 features (other than index and class label) extracted from 6157 legitimate and 4898 phishing webpages. Benign webpages are collected from Alexa [40] while phishing webpages are taken from PhishTank. This dataset is chosen in the present study because it is the most recent dataset available in the public domain.

3.2 Classification before Feature Selection

In this work, we have used a free data mining tool—Waikato Environment for Knowledge Analysis (WEKA), version 3.8.4 for Windows Operating System [41]. We used eight machine learning classifiers, namely IB1, NB, J48, AdaBoost, decision table (DT), bagging, RF, and sequential minimal optimization (SMO) for classifying phishing webpages. In this step, all 30 features present in the original dataset are used for constructing the classification models. A brief description of these algorithms is given below:

- IB1 is a classifier that stores the training data and starts building the classification model only when it gets the test data. Due to this reason, it is also known as lazy learner. It is based on nearest-neighbor classifier which makes use of a similarity measure, i.e., Euclidean distance to find the training example which is closest to the specified test data. Euclidean distance between two points having coordinates (x, y) and (a, b) is computed by using the equation given below.

$$\text{Dist}((x, y), (a, b)) = \sqrt{(x - a)^2 + (y - b)^2}$$

- NB method is originated from Bayes' theorem with Naïve independent conventions. It predicts the probability of an instance of a dataset belonging to a specific class. The posterior probability of a class *c*, given attribute *x*, is computed by using the equation mentioned below.

$$P(c|x) = \frac{P(x|c)P(c)}{P(x)}$$

where $P(x|c)$ represents the probability of attribute x given its class c. $P(x)$ and $P(c)$ are the prior probabilities of attribute x and class c respectively.

J48 [42] is a WEKA classification package of C4.5. It builds the decision tree based on the labeled training data using the concept of entropy.

The main aim of adaptive boosting (AdaBoost) [43] is to transform a group of weak classifiers into a strong classifier. It is a weighted grouping of weak classifiers. It assigns equal weights to all the examples in the dataset. Then the weights of wrongly classified instances are modified. At last, weighted mean of all weak classifiers is considered to take the final decision about the class of an instance.

DT is a type of classifier in which specific features are selected while performing the learning process. It is done by computing cross-validation performance of the table for different sub-sets of attributes and selecting the sub-set which is the best performer. Cross-validation error from a decision table is obtained by manipulating the class counts related to entry of the table. Best first search is usually used to search the feature space.

Bagging (bootstrap aggregation) is a meta-classifier which combines the output of different models to make the final decision more reliable by reducing the variance error [44]. It increases the predictive performance over a single model. A general way to predict the final class is to calculate the weighted average for numeric prediction problems and take a weighted vote for classification problems.

RF is an ensemble method which comprises of various decision trees. It can run effectively on a large dataset. It gives highly accurate results than a single decision tree as it reduces the overfitting [44]. Every decision tree is generated using a random selection of attributes. The class of unknown data is predicted using the aggregation (voting scheme) of predictions made by individual decision trees.

SMO [45] is an optimized version of SVM from WEKA library. It is an optimization algorithm which is used to solve the problem which arises while training SVM. The name of the problem is the quadratic problem. SVM [44] is based on computing a hyperplane with the maximal margin between the data dimensions.

3.3 Feature Selection

A feature selection method is deployed for the selection of top-ranked features which helped in discarding the irrelevant ones. Building a classification model with a lesser number, but a relevant set of features helps in improving its generalization and learning speed. In this work, we have used information gain (IG) [46] method for selecting the top features. It is defined as the extent of information obtained from a feature and is based on the degree of randomness (entropy) in the data. It is calculated by finding the difference between the entropy of data distribution after and before the split.

$$\text{Entropy} = -\sum_i P_i \log_2 P_i$$

where P_i is the class probability.

Ranker algorithm is used for ranking the attributes. Figure 2 shows the top 15 attributes after applying IG + Ranker algorithm. These top rank attributes are used for classification purpose.

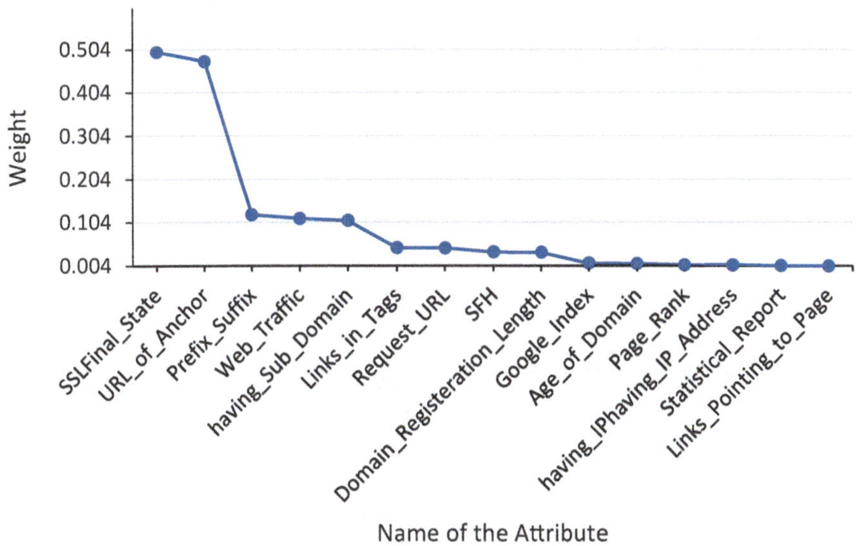

Fig. 2 Top 15 features after feature selection

3.4 Classification after Feature Selection

The selected top 15 features are then employed for building the classification models using the same set of eight algorithms as described earlier. When experiments are performed on both datasets before and after feature selection, the model building time is recorded for carrying out the comparative analysis.

3.5 Evaluation and Validation

A tenfold cross-validation technique has been used for conducting the experiments. This technique works on dividing the original dataset randomly. Firstly, the original dataset is split into ten equal parts. Thereafter, nine parts are used for training and one for testing purpose. The same process is repeated ten times with different combinations. The averaged result is used to measure the performance of algorithms. In order to evaluate machine learning algorithms, we have used various evaluation parameters. These are true positive rate (TPR), false positive rate (FPR), precision, F-measure, accuracy, and Matthews correlation coefficient (MCC) [47]. These are computed by using the fields of confusion matrix as described in Table 1.

- **TPR**: It is also known as recall and is defined as the rate of correctly identified phishing webpages.

$$\mathrm{TPR} = \frac{\mathrm{TP}}{\mathrm{TP} + \mathrm{FN}}$$

- **FPR**: It is the rate of incorrectly recognized benign webpages.

$$\mathrm{FPR} = \frac{\mathrm{FP}}{\mathrm{FP} + \mathrm{TN}}$$

- **Precision**: It is a degree of exactness.

Table 1 Confusion matrix

Actual	Classified/predicted as		
	Class	Phishing	Benign
	Phishing	True positive (TP) Phishing webpages predicted correctly as phishing	False negative (FN) Phishing webpages predicted incorrectly as benign
	Benign	False positive (FP) Benign webpages predicted incorrectly as phishing	True negative (TN) Benign webpages predicted correctly as benign

$$\text{Precision} = \frac{\text{TP}}{\text{TP} + \text{FP}}$$

- **F-Measure**: It is defined as the harmonic mean of recall and precision.

$$\text{F} - \text{Measure} = \frac{2 \times \text{Precision} \times \text{Recall}}{\text{Precision} + \text{Recall}} = \frac{2 \times \text{TP}}{2 \times \text{TP} + \text{FN} + \text{FP}}$$

- **Accuracy (%)**: It is the percentage of correctly recognized phishing and benign webpages.

$$\text{Accuracy}(\%) = \frac{\text{TP} + \text{TN}}{\text{TP} + \text{FN} + \text{TN} + \text{FP}} \times 100$$

- **MCC**: It is used to evaluate the performance of machine learning algorithms for binary classification. It measures the correlation between the actual and predicted labels and takes the values between -1 and $+1$.

$$\text{MCC} = \frac{\text{TP} \times \text{TN} - \text{FP} \times \text{FN}}{\sqrt{(\text{TP} + \text{FP})(\text{TP} + \text{FN})(\text{TN} + \text{FP})(\text{TN} + \text{FN})}}$$

4 Experimental Results

This section provides a comparative analysis of experimental results along with their pictorial visualizations. All eight classifiers are employed to classify phishing websites by considering the dataset containing 30 features. Table 2 shows the weighted average values of TPR, FPR, precision, F-measure, MCC, and accuracy for

Table 2 Classification results before feature selection (with 30 features)

Classifier	TPR	FPR	Precision	F-measure	MCC	Accuracy (%)
IB1	0.958	0.044	0.958	0.958	0.915	95.78
NB	0.907	0.098	0.907	0.907	0.811	90.7
J48	0.958	0.044	0.958	0.958	0.915	95.82
AdaBoost	0.925	0.077	0.925	0.925	0.848	92.47
Bagging	0.956	0.045	0.956	0.956	0.911	95.59
DT	0.934	0.074	0.935	0.934	0.867	93.41
RF	0.965	0.034	0.965	0.965	0.927	96.52
SMO	0.928	0.077	0.928	0.927	0.853	92.75

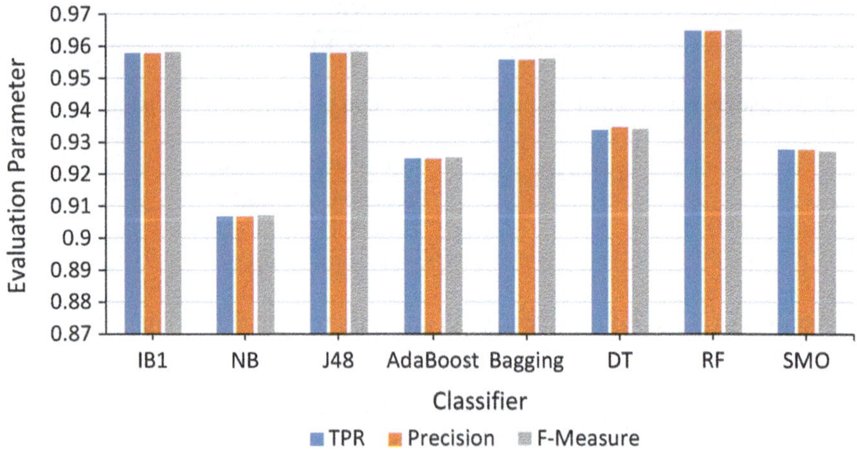

Fig. 3 Comparison of classifiers on the basis of TPR, precision, and F-measure before feature selection

Table 3 Classification results after feature selection (with 15 features)

Classifier	TPR	FPR	Precision	F-measure	MCC	Accuracy (%)
IB1	0.952	0.05	0.952	0.952	0.902	95.16
NB	0.907	0.097	0.907	0.907	0.812	90.73
J48	0.95	0.052	0.95	0.95	0.899	95.03
AdaBoost	0.925	0.077	0.925	0.925	0.848	92.47
Bagging	0.955	0.047	0.955	0.955	0.909	95.5
DT	0.931	0.076	0.932	0.931	0.861	93.13
RF	0.963	0.04	0.963	0.963	0.925	96.3
SMO	0.921	0.087	0.921	0.92	0.839	92.06

the eight machine learning algorithms using 30 features, i.e., before feature selection. It shows that RF provides the highest accuracy of 96.52% followed by J48, IB1, and bagging with 95.82%, 95.78%, and 95.59% accuracy, respectively. RF gives the maximum value of MCC, i.e., 0.928. NB provides the least accuracy of 90.7% with MCC value as 0.811. Figure 3 visualizes the comparison of various classifiers on the basis of TPR, precision, and F-measure before feature selection.

Afterward, the top 15 features are selected using IG feature selection method, and the same set of classification models are again used for the classification purpose. Table 3 shows the weighted average values of TPR, FPR, precision, F-measure, MCC, and accuracy for the eight machine learning algorithms using 15 features selected after employing a feature selection method. It shows that RF provides the highest accuracy of 96.3% followed by bagging, IB1, and J48. Once again, the minimum accuracy is provided by NB, i.e., 90.73%. RF gives the maximum value of MCC,

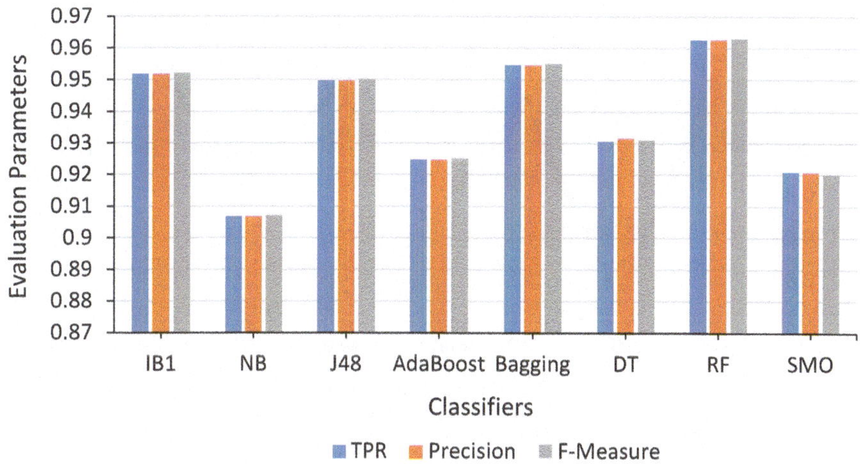

Fig. 4 Comparison of classifiers on the basis of TPR, precision, and F-measure after feature selection

i.e., 0.925. Figure 4 visualizes the comparison of various classifiers on the basis of TPR, precision, and F-measure after feature selection.

Figure 5 depicts the comparison of classifiers on the basis of accuracy before and after features selection. It shows that for almost all the machine learning algorithms considered in this work, the accuracy remains either same or there is insignificant decrement before and after feature selection. It means that the features which are shortlisted by using IG feature selection method are the most relevant ones for the

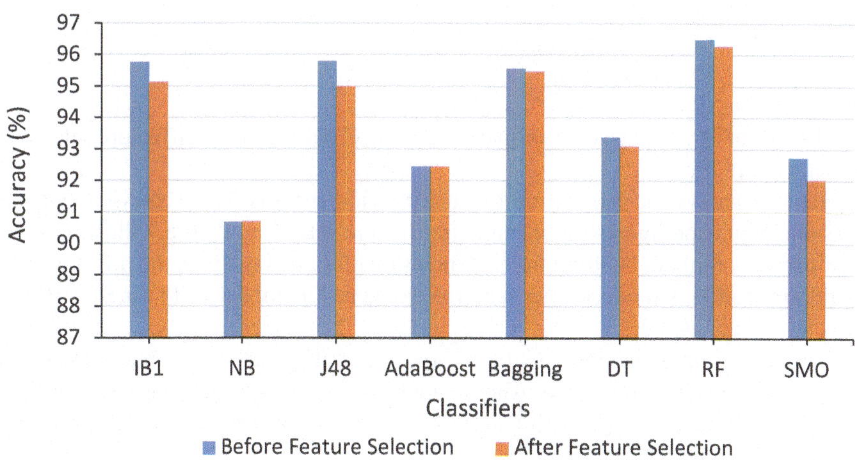

Fig. 5 Comparison of classifiers on the basis of accuracy before and after feature selection

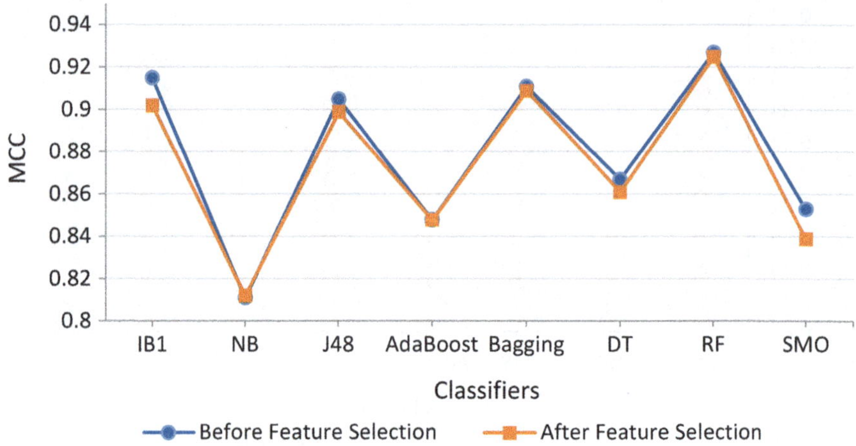

Fig. 6 Comparison of classifiers on the bases of MCC before and after feature selection

Table 4 Model building time (in sec.) before and after feature selection

Classifier	Model building time (in sec.)	
	Before feature selection	After feature selection
IB1	0	0
NB	0.28	0.06
J48	0.62	0.3
AdaBoost	1.14	0.33
Bagging	2.75	1.06
DT	5.39	2.11
RF	5.53	3.43
SMO	19.91	9.27

classification of webpages. A comparison of MCC before and after feature selection is depicted in Fig. 6.

Table 4 shows the build time of classification models while carrying out the classification experiments before and after feature selection. Figure 7 illustrates the comparison of model building time. It shows that for all classifiers, the time taken to build the models is reduced after feature selection. The model build time for IB1 is 0 as it is a lazy learner.

The improvement in model building time does not seem to be as significant since the experiments are conducted using a small dataset. If a large dataset or big data [48, 49] is considered, it would take more time for model building and thus there would be a significant improvement.

In order to detect the new phishing websites, the researchers have used machine learning algorithms. To improve the classification accuracy, they considered a large number of features. However, they do not focus on the problem related to the building

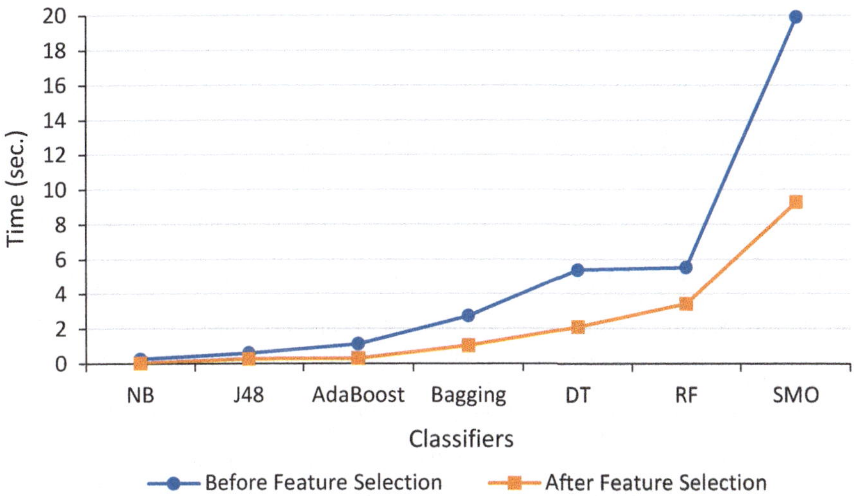

Fig. 7 Comparison of model building time before and after feature selection

time of classification models without compromising the accuracy. In this work, we used a feature selection method to select an appropriate set of features for improving the efficiency of the models without compromising with the accuracy.

5 Conclusion and Future Scope

This chapter presented an efficient system for the detection of phishing websites. It makes use of a feature selection method apart from machine learning algorithms for the purpose of classification. The performance of eight machine learning algorithms is compared before and after feature selection for the classification of webpages into phishing and benign. The experimental results clearly depict that use of a feature selection method for selecting a relevant set of features improves the building time of classification models without a significant reduction in the accuracy for the detection of phishing webpages. Further, the results show that RF gives the best accuracy for both before and after feature selection.

As future work, we intend to use an ensemble of algorithms to compare the accuracy rates with the individual machine learning algorithms. The experimental study can be further extended to understand the performance of deep learning algorithms. We may also enhance our study on large phishing datasets including IoT-based phishing attacks using big data technologies to uncover such attacks more quickly than existing methods.

References

1. Anti-Phishing Working Group (APWG) https://docs.apwg.org//reports/apwg_trends_report_q4_2018.pdf
2. IC3 Annual Report 2018 https://pdf.ic3.gov/2018_IC3Report.pdf
3. Razorthorn phishing report https://www.razorthorn.co.uk/wp-content/uploads/2017/01/Phishing-S
4. Gandotra E, Bansal D, Sofat S (2014) Malware analysis and classification: a survey. J Inf Security 56–65
5. Gupta D, Rani R (2020) Improving malware detection using big data and ensemble learning. Comput Electr Eng 106729
6. Microsoft Security Intelligence Report (2019) vol 24 https://www.microsoft.com/security
7. Logic Bomb Set Off South Korea Cyberattack. https://www.wired.com/2013/03/logic-bomb-south-korea-attack/
8. Los Angeles Times https://www.latimes.com/business/la-fi-mh-anthem-is-warning-consumers-20150306-column.html
9. Threat Analysis Group, Findings on COVID-19 and online security threats https://blog.google/technology/safety-security/threat-analysis-group/findings-covid-19-and-online-security-threats/
10. Selenium https://docs.seleniumhq.org/download/
11. Gandotra E, Bansal D, Sofat S (2016) Tools and techniques for malware analysis and classification. Int J Next-Gener Comput
12. Jsoup Java HTML Parser, with best of DOM, CSS, and jquery https://jsoup.org/
13. OpenDNS, PhishTank https://wwwphishtank.com/
14. Google Safe Browsing API—Google Code https://code.google.com/apis/safebrowsing/
15. Seifert C, Welch I, Komisarczuk P (2008) Identification of malicious web pages with static heuristics. In: 2008 Australasian Telecommunication Networks and Applications Conference, IEEE, pp 91–96
16. Jain AK, Gupta BB (2017) Phishing detection: analysis of visual similarity based approaches. Secur Commun Network
17. Gandotra E, Bansal D, Sofat S (2015) Computational techniques for predicting cyber threats. In: Intelligent computing, communication and devices, pp 247–253, Springer, New Delhi
18. Tan CL, Chiew KL, Wong K (2016) PhishWHO: phishing webpage detection via identity keywords extraction and target domain name finder. Decision Support Systems, pp 18–27
19. Chiew KL, Chang EH, Tiong WK (2015) Utilisation of website logo for phishing detection. Comput Security 16–26
20. Jain AK, Gupta BB (2018) Towards detection of phishing websites on client-side using machine learning based approach. Telecommun Syst 687–700
21. Srinivasa Rao R, Pais AR (2017) Detecting phishing websites using automation of human behavior. In: Proceedings of the 3rd ACM workshop on cyber-physical system security, ACM, pp 33–42
22. Sahingoz OK, Buber E, Demir O, Diri B (2019) Machine learning based phishing detection from URLs. Expert Syst Appl 345–357
23. Gu X, Wang H, Ni T (2013) An efficient approach to detecting phishing web. J Comput Inf Syst 5553–5560
24. Moghimi M, Varjani AY (2016) New rule-based phishing detection method. Expert systems with applications, pp 231–242
25. Xiang G, Hong J, Rose CP, Cranor L (2011) Cantina+ a feature-rich machine learning framework for detecting phishing web sites. ACM Transactions on Information and System Security (TISSEC), pp 1–28
26. Zhang Y, Hong JI, Cranor LF (2007) Cantina: a content-based approach to detecting phishing web sites. In: Proceedings of the 16th international conference on World Wide Web, ACM, (2007) pp 639–648

27. Joshi A, Pattanshetti P, Tanuja R (2019) Phishing Attack Detection using Feature Selection Techniques. In: Nutan College of Engineering & Research, International Conference on Communication and Information Processing (ICCIP)
28. Wu CY, Kuo CC, Yang CS (2019) A phishing detection system based on machine learning. In: 2019 International Conference on Intelligent Computing and its Emerging Applications (ICEA), pp 28–32
29. Zamir A, Khan HU, Iqbal T, Yousaf N, Aslam F, Anjum A, Hamdani M (2020) Phishing web site detection using diverse machine learning algorithms. The Electronic Library
30. Almseidin M, Zuraiq AA, Al-kasassbeh M, Alnidami N (2019) Phishing detection based on machine learning and feature selection methods. Int J Interactive Mobile Technol (iJIM) 171–183
31. Yerima SY, Alzaylaee MK (2020) High accuracy phishing detection based on convolutional neural networks. arXiv preprint arXiv:2004.03960
32. Basnet RB, Doleck T (2015) Towards developing a tool to detect phishing URLs: a machine learning approach. In 2015 IEEE International Conference on Computational Intelligence & Communication Technology, IEEE, pp 220–223
33. Hurrah NN, Parah SA, Loan NA, Sheikh JA, Elhoseny M, Muhammad K (2019) Dual watermarking framework for privacy protection and content authentication of multimedia. Future Gener Comput Syst 654–673
34. Parah SA, Sheikh JA, Bhat GM (2014) Fragility evaluation of intermediate significant bit embedding (ISBE) based digital image watermarking scheme for content authentication. In: 2014 International conference on advances in electronics computers and communications, IEEE pp 1–6
35. Gull S, Loan NA, Parah SA, Sheikh JA, Bhat GM (2018) An efficient watermarking technique for tamper detection and localization of medical images. J Ambient Intell Humanized Comput pp 1–10
36. Gull S, Parah SA (2019) Color image authentication using dual watermarks. In: 2019 fifth international conference on image information processing (ICIIP), pp 240–245
37. Giri KJ, Bashir R, Bhat JI (2019) A discrete wavelet based watermarking scheme for authentication of medical images. Int J E-Health Med Commun (IJEHMC), pp 30–38
38. UCI Machine Learning Repository, "Phishing Websites Dataset" https://archive.ics.uci.edu/ml/datasets/phishing+websites
39. Mohammad RM, Thabtah F, McCluskey L (2012) An assessment of features related to phishing websites using an automated technique. In 2012 International conference for internet technology and secured transactions, IEEE pp 492–497, IEEE
40. Alexa Most Popular sites. https://www.alexa.com/topsites
41. Hall M, Frank E, Holmes G, Pfahringer B, Reutemann P, Witten IH (2009) The WEKA data mining software: an update. ACM SIGKDD explorations newsletter, pp 10–18
42. Quinlan JR (2014) C4.5: Programs for Machine Learning. Elsevier
43. Schapire RE (1990) The strength of weak learnability. Machine Learning, pp 197–227
44. Witten IH, Frank E (2002) Data mining: practical machine learning tools and techniques with Java implementations. Acm Sigmod Record pp 76–77
45. Platt J (1998) Sequential minimal optimization: a fast algorithm for training support vector machines
46. Gandotra E, Bansal D, Sofat S (2016) Zero-day malware detection. In: 2016 sixth international symposium on embedded computing and system design (ISED), IEEE, pp 171–175
47. Gandotra E, Bansal D, Sofat S (2017) Malware threat assessment using fuzzy logic paradigm. Cybern Syst 29–48
48. Gupta D, Rani R (2019) A study of big data evolution and research challenges. J Inf Sci 322–340 (2019)
49. Gupta D, Rani R (2018) Big data framework for zero-day malware detection. Cybern Syst 103–121

Chapter 12
Use of Blockchain and Internet of Things for Securing Data in Healthcare Systems

Atul Banotra, Jyoti Swaroop Sharma, Swastik Gupta, Sachin Kumar Gupta, and Mamoon Rashid

1 Introduction

Blockchain and IoT are the two most important technologies that are going to have a huge influence on all the industries in the coming years. Healthcare industry will also be greatly impacted by these two technologies. The Internet of things (IoT) is the network of physical objects that contain embedded technology to communicate and sense or interact with their internal states or the external environment [1]. As the price of sensors and actuators keeps on falling, so adopting IoT platform has become easy for many industries. Various healthcare solutions involving only IoT technology are already available. But the problem with them is that they have a centralized database. So the data available might not be secure and also the organizations might charge more for the storage. But by the use of blockchain technology, these concerns are addressed, and the user is made the complete owner of his data. The data cannot be shared with anyone without his approval.

A. Banotra · J. S. Sharma · S. Gupta · S. K. Gupta
School of Electronics & Communication Engineering, Shri Mata Vaishno Devi University, Kakryal 182320, Katra, Jammu & Kashmir, India
e-mail: atulbanotraab@gmail.com

J. S. Sharma
e-mail: sharmajyotiswaroop@yahoo.com

S. Gupta
e-mail: swastik.gupta@smvdu.ac.in

S. K. Gupta
e-mail: sachin.gupta@smvdu.ac.in

M. Rashid (✉)
School of Computer Science & Engineering, Lovely Professional University, Jalandhar 144411, India
e-mail: mamoon873@gmail.com

K. J. Giri et al. (eds.), *Multimedia Security*, Algorithms for Intelligent Systems,
https://doi.org/10.1007/978-981-15-8711-5_13

1.1 Blockchain Concept

Blockchain is a distributed, collaborative, and collective electronic registry that is used to document payments through multiple devices so that the database cannot be arbitrarily changed without modifying all corresponding frames and platform complicity [2]. This helps the stakeholders to review payments and to review them cheaply. The blockchain platform is independently operated through a peer-to-peer infrastructure and a decentralized period molding database. They are independently verified by perceived self-interests operated by continuous coordination [3]. The effect is a stable process where there is minimal confusion about the stakeholder's consumer data or information protection. Blockchain use eliminates the endless methodological rigor attribute from digital currency. This guarantees that each variable of expenditure has been moved only once, addressing the enduring dual revenue problem. Blockchain was defined as a specification for the exchanging of values [4].

1.2 Internet of Things

Internet of things (IoT) is a model that is used for efficient computing and helps us in the functioning of different ways used in our daily lives. In IoT, interconnection of various devices is enabled to exchange them between them whenever interactions are done in these IoT systems. The most popular devices used in IoT are cellular phones, smart sensors, and coffee machines [5, 6]. As IoT systems help humans and various devices in easier connectivity; however, it results in large volumes of data due to their high speeds, and thus, these systems get more complex in terms of data aspects.

In IoT systems, existing network infrastructure is used for the direction integration of remotely controlled physical objects and computer-based systems [7, 8]. This technology of IoT has resulted into minimum intervention of humans for performing any job with greater efficiency and accuracy. IoT systems have power of capturing data from different kinds of sensors as well [9]. Various organizations are showing interests in IoT-based technology due to the lower costs of sensors and actuators that are used in this technology [10].

We can see IoT deployed in various streams of healthcare management. Some instances of IoT in health care are:

Clinical Management: Hospitalized patients whose physiological status requires close thought can be constantly watched using IoT-driven, non-obtrusive checking.

Patient's Remote Monitoring: IoT systems in terms of remote patient monitoring wearables help in the transmission of patient data from home premises and make it to reach out quite accurately in the premises of doctors or nurses at remote locations. These systems greatly help in the cut down of waiting times and difficulty in reaching to hospitals for those patients who are unable to do it.

Early Intervention/Prevention: Unhealthy and physically dependent people are benefitted in their daily activities by IoT-driven checking systems. A senior living alone, for example, may need a watching sensor that can distinguish a fall or other impedance in normal development and report it to emergency responders or relatives.

Wireless Sensor Networks (WSNs): WSN is a basic enabling development of IoT. It interfaces different sensor and actuator center points into a framework through remote correspondence. This fuses the framework into a bigger sum structure through a framework passage. WSN is composed with an application game plan of the IoT.

We can also see blockchain arranged in various streams of healthcare management. Some occurrences of blockchain in health care are:

Clear, Comprehensive Medical Record: Longitudinal medical records-incorporating images, illness databases, test tests, medications can be done via blockchain like inpatient, walking and wearable details, allowing providers to learn for better methods of treatment distribution.

Total Patient List: Regularly records are crossed or copied when managing information on medicinal services is required.

Adjudication of Arguments: Since blockchain takes a shot at a permission-based transaction, cases will obviously be tested where the network fits with the way by which the arrangement is implemented.

Control of Supply Chain: Blockchain-based contracts will allow organizations with medical services to track the supply-request in moving through its entire life cycle. Examples are how the trade takes place or whether the transaction is effective or if there are any delays.

The organization of this chapter is as follows: Sect. 2 provides history of blockchain. Section 3 introduces blockchain along with its functions, and Section 4 gives the literature survey. Further, Section 5 discusses the proposed architecture, and Section 6 talks about the scope and improvement of model and the chapter ends with conclusion in Sect. 7.

2 History of Blockchain

Stuart Haber and W described cryptographically protected block chain in 1991 [11]. In 1992, Merkle trees were introduced into the layout by Bayer, Haber and Stornetta, which enhanced its productivity by enabling the consolidation of many records into a single block or frame [11]. The history of innovations in block chain is shown in Table 1.

Table 1 History of Innovations in Blockchain

Reference	Innovations	Description
[12]	First Innovation	The first major discovery in blockchain was bitcoin, an invention in the digital money. Bitcoin's market valuation currently floats around $10–$20 billion dollars and is being used for transactions by millions of individuals along with a rapidly enlarging transaction processing sector
[13]	Second Innovation	The second breakthrough was named blockchain. Nearly every single main banking organization in the globe is currently conducting blockchain work, and in 2017, blockchain was projected and seen by 15 per cent of corporations
[14]	Third Innovation	The third invention was named the "smart contract," expressed in a second-generation blockchain service named Ethereum, which developed separate computer algorithms explicitly through blockchain that allowed the representation of investment vehicles, such as bonds or securities, instead of just the money-like currencies of bitcoins. The Ethereum smart contract network currently has an enterprise value of nearly $1 billion, with tens decades of ventures bound for the sector
[15]	Fourth Innovation	The fourth large advancement, the new reinvention in blockchain technology, is named "stake proof." Latest generation blockchains are protected by "job evidence," wherein resolutions are being made by the party with both the greatest overall processing capacity. Such individuals are named "miners" and in return for cryptocurrency transactions, they run large server farms to supply this protection. Such power plants are replaced by the new frameworks, exchanging themselves by complicated financial structures for a comparable and even greater measure of security
[16]	Fifth Innovation	The fifth major breakthrough is named blockchain scalable. Every correct machine in the web performs each payment in the blockchain environment. It is a sluggish operation. A distributed blockchain exacerbates the problem by calculating out how machines are required to verify every payment and then effectively divide the job, despite compromising protection. Managing anything without losing blockchain's renowned protection and reliability is a challenging issue and not an intransigent issue. A scalable blockchain is supposed to be quick and sufficiently drive the Internet of Things and then go head-to-head with both the financial biggest global transaction intermediaries (VISA and SWIFT)

3 Working of Blockchain

The earlier concept of sharing information was dependent on centralized systems. There were many disadvantages of centralized systems. In distributed systems, all the nodes participate in the information-sharing process. It has been observed that centralized systems are good, but they are not scalable and robust to failures. So need for moving toward decentralized or distributed platform was felt. Blockchain is a distributed platform where information can be shared in a trustworthy manner. A blockchain is shared ledger where transactions are permanently recorded by appending blocks. The blockchain serves as a historical record of all transactions that ever occurred from the genesis block to the latest block [17].

Blockchain can be considered a chain of blocks. Each block contains data, hash, and hash of the previous block. The data that is stored inside a block depends upon the type of blockchain; e.g., bitcoin blockchain stores details of the sender, receiver, and amount of coins [18, 19]. Hash in a block can be compared to fingerprint. It identifies the block and all its contents and is always unique. Hash of the previous block helps in creating a chain of blocks and because of this feature blockchain is so secured. The first block cannot point toward the previous block as it is the first one, and it is called the genesis block. Once a block is created, its hash is calculated. Changing anything inside the block will cause the hash to change. The following steps are followed if a transaction is done in a blockchain [20], and it is shown in terms of Fig. 1:

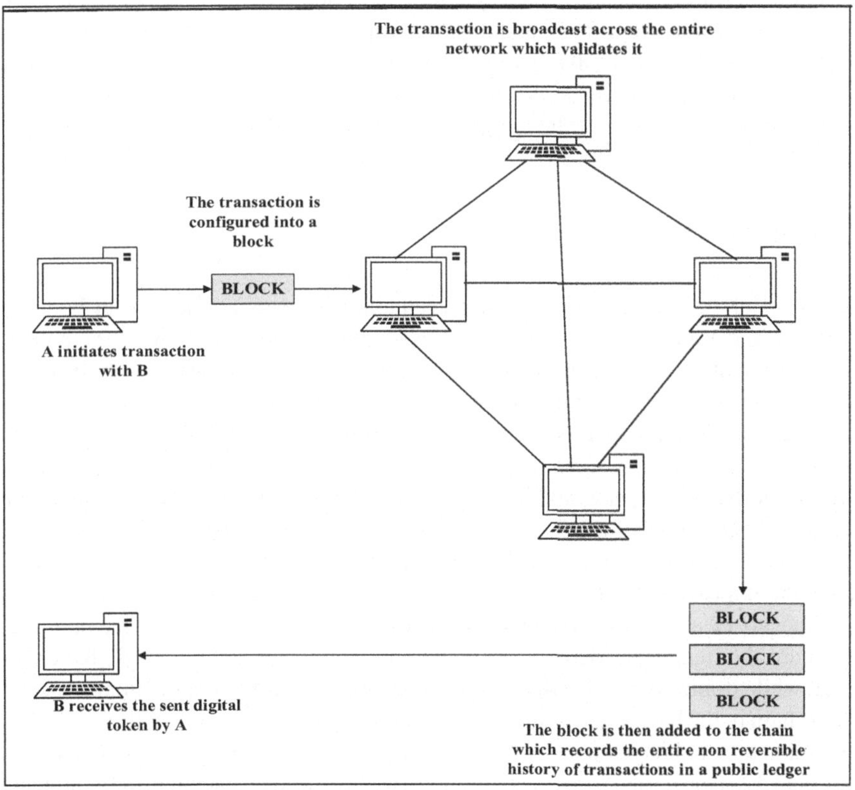

Fig. 1 Working of blockchain

- Transaction in the network can be initiated by any node.
- The first node that wants to do the transaction is the sender node. It creates a transaction and sends it to the network. The transaction message contains the cryptographic public address of the receiver, transaction details, and cryptographic digital signature of the sender.
- Every transaction is authenticated by the sender's private key until it is transmitted to the network.
- This transaction is broadcasted to every node in the network, and they validate the authenticity of the message by verifying the digital signature. The authenticated transaction is placed in the pool of pending transactions.
- The pending transactions are encapsulated in a block by one of the nodes, and it also alerts all other nodes in the network about the newly built-in block for its validation.
- A great number of computational resources are required for validation. The nodes which validate the block are called miners, and they are awarded by giving them incentives.
- The multiple systems use specific validation strategies. Bitcoin, for example, requires proof of research. The key objective of this technique is to guarantee that any transaction is legitimate and that illegitimate transactions remain impossible. After authentication of all transactions, the new block is combined with previously chained blocks comprising the blockchain.
- The present condition of the ledger is then transmitted to the network.

4 Literature Survey

Crosby et al. [21] describe blockchain technology and its application in both the financial and non-financial sector. The paper clearly explains how blockchain is helpful in creating a consensus in the digital online world. Also, various risks involved in the adoption of blockchain technology, e.g., scaling and government regulations are discussed in the paper. Miller et al. [22] define the importance of blockchain and IoT in the industrial sector and its impact in the future. It states that using these two technologies system would become more efficient. Further, it also suggests that certain regulation, legal and insurance claims have to be properly addressed. Also, there would be a number of IoT devices and these must provide scalable solutions to get the improved results.

Esposito et al. [23] characterize the ability to utilize blockchain technologies to secure the healthcare data stored inside the cloud. Various practical challenges faced while doing so are also discussed. This paper also gives out the importance of blockchain in the healthcare industry in addressing various interoperability and security features. Challenges faced while storing data in blockchain are also discussed. Dey et al. [24] have tried to present a solution to the problems faced by the healthcare industry. A blockchain-IoT model is presented where real-time data is collected by IoT devices, and after that, it is stored in a blockchain. Also, it states that blockchain being decentralized and distributed it efficiently facilitates the system. Hanley et al. [25] have presented a system that uses a blockchain and an off-chain centralized data storage so that instant access of medical records to patients and medical professionals can be given from anywhere. The benefits of usage of blockchain are highlighted in this paper that how quick communication takes place and how interoperability issues are dealt with.

The usage of this system is also discussed in [26].

Alhadhrami et al. [27] capture blockchain's role in the management and exchange of electronic health records to allow doctors, physicians, clinics, and other medical stakeholders to share data and improve interoperability. Various blockchain structures are investigated, and the solutions they provide to existing challenges are discussed. This paper also describes permission and permissionless blockchains, their architecture and their implementation in health care. Also, security and privacy challenges are discussed. In addition to the above, a parallel analysis of the current approach to blockchain and IoT related studies has been summarized in Table 2.

Table 2 Equivalence of the new monitoring policies to similar studies in the literary works

Implemented technology	Objective	Inference
Blockchain and IoT [28]	• To build a transparent, aggregated logistic network using blockchain and IoT	• The potent supply chain mechanism • Monitoring of real-time data
Blockchain [29]	• The purpose of this paper is to protect client information from infringing parties	• Elimination of third-party regulation over user data using the blockchain methodology
Blockchain and IoT [30]	• This paper explores how cloud and fog are used as hosting sites	• Because of its transparent connection, blockchain is used within IoT, for example, to handle and store sensor data in such a way that device complexity is regulated, and small-scale increments are empowered
Blockchain [31]	• This paper offers seven layers of a computational blockchain model	• Blockchain is one of the newly established parallel transport management systems (PTMS) that has been tested and entrusted with the architecture of the network
IoT data with smart contract code of Blockchain [32]	• In this paper, we propose a blockchain approach and use of Ethereum Smart Contracts to apply IoT details to a mechanized version, including no mediator	Focused on following: • Technology of building • Affairs of company • Contact between the stakeholders • Installation and checking of the general performance of the program
IoT Platform [33]	• This paper involves adaptable network structures, more modules and steps based on the Internet of Things (IoT) structure	• Conveniently combining IoT programs with usable business logic or smart contracts • Integrating the IoT phases with blockchain can lead to new avenues for successful governments to devour, simplicity and goods that depend on a faction-based economy

(continued)

Table 2 (continued)

Implemented technology	Objective	Inference
IoT and Blockchain [34]	• Solving a major issue with extremely analytical energy that involves high bandwidth overheads and processing time in blockchain technology. So that IoT is more well-matched	• Statically analyzing design in the sense of a challenge paradigm strengthens its utility in ensuring privacy and security • In comparison with that used in bitcoin, overhead capacity is significantly reduced
Blockchain [29]	• This paper extends our knowledge of blockchain technology on the payment network	It outlines its research agenda on the three functional areas of the payment network as shown below (1) Operational difficulties (2) Business sector concerns (3) Feature defects in systems

5 Proposed Methodology

The existing model in the healthcare industry that uses IoT devices is server/client models; i.e., it is centralized. The problem with a centralized system is that basic patient information, and data lies with the hospital which might charge users more for storage. Also in a centralized system in case of a single point of failure, the whole data might be lost if there is no proper backup. As the healthcare data is confidential, so there are also security concerns for the data as it would be easy for hackers to hack the data from a centralized system. Interoperability is also a big concern in centralizing healthcare models.

So we are proposing a model in which both IoT and blockchain technologies are used to overcome all these concerns. The architecture is divided into various layers as shown in Fig. 2.

•The first layer is the physical layer and consists of the patient whose data has to be recorded.

•The second layer is the device layer and consists of mobile phones (having sensors), wearables, medical devices (CT scan, X-Ray), etc. These devices are used for measuring and transmitting the data from the patient end.

•The third layer is the communication layer. This layer is responsible for the transmission of data that is recorded by devices. The device can communicate directly to cloud through API or devices can communicate using LORA, ZigBee, and Profibus; further, MQTT and COAP are used to transmit data from gateway to cloud.

•The next is cloud interface which involves platform components that are further categorized as real-time engine and blockchain engine.

The data that is sent to the cloud firstly moves to the real-time engine where raw data is collected, and data aggregation and normalization are done. This data is then fed to blockchain engine which is responsible for the storage of data. Since blockchain is a decentralized technology, data stored in it is in decentralized form. The blockchain stores the records of a patient which is measured and collected by IoT devices. The data is stored is in the form of blocks, and it is also properly secured in it. Through the API management, this data can be accessed by the various stakeholders

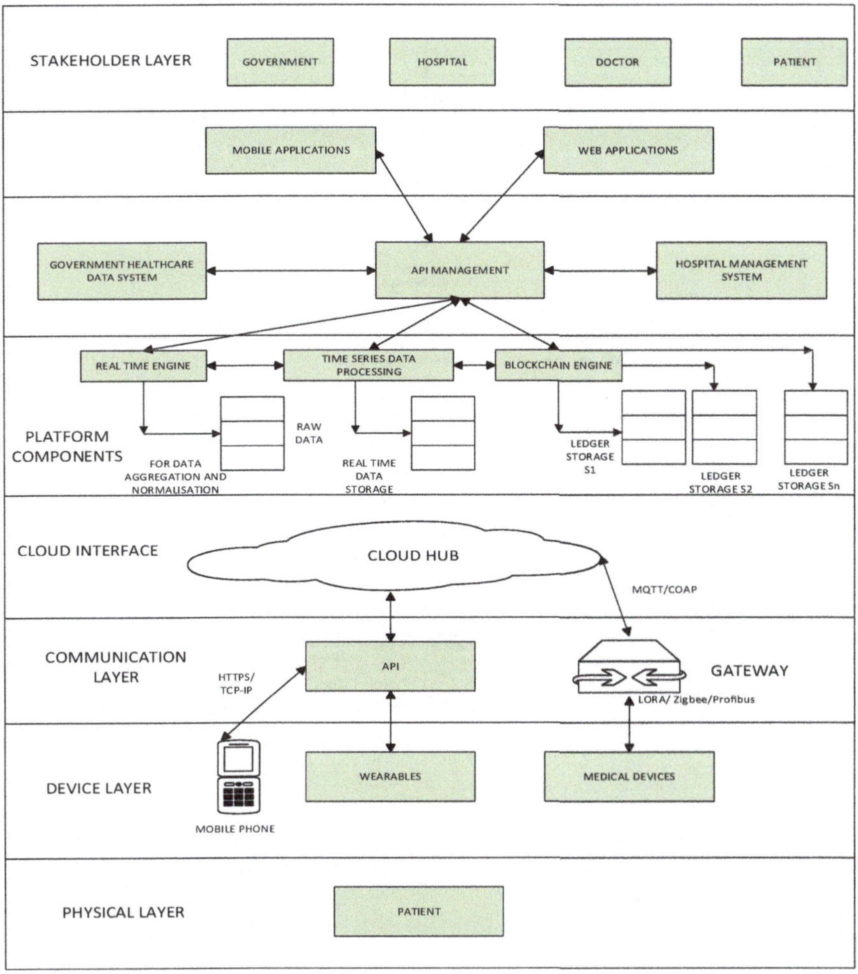

Fig. 2 Proposed layered architecture

which include government regulators, hospitals, doctors, and patients. This data can be accessed by mobile applications or web applications, and this can be done by API management shown in the architecture as shown in Fig. 2.

The flowchart shown in Fig. 3 describes the role of the following entities in healthcare management. The flowchart describes the role of entities in healthcare data storage. The storage of healthcare data starts with the query of the patient about the request for consultation in the smart contract. Hospital authority acts on the smart contract. As the doctor is available, a consultation request is accepted by the hospital authority. After that patient diagnosis is done by the doctor who will examine if the patient is fine. If the patient is fine, then it will terminate the request. If the doctor feels the patient needs further treatment, then he would be sent to the diagnostic department

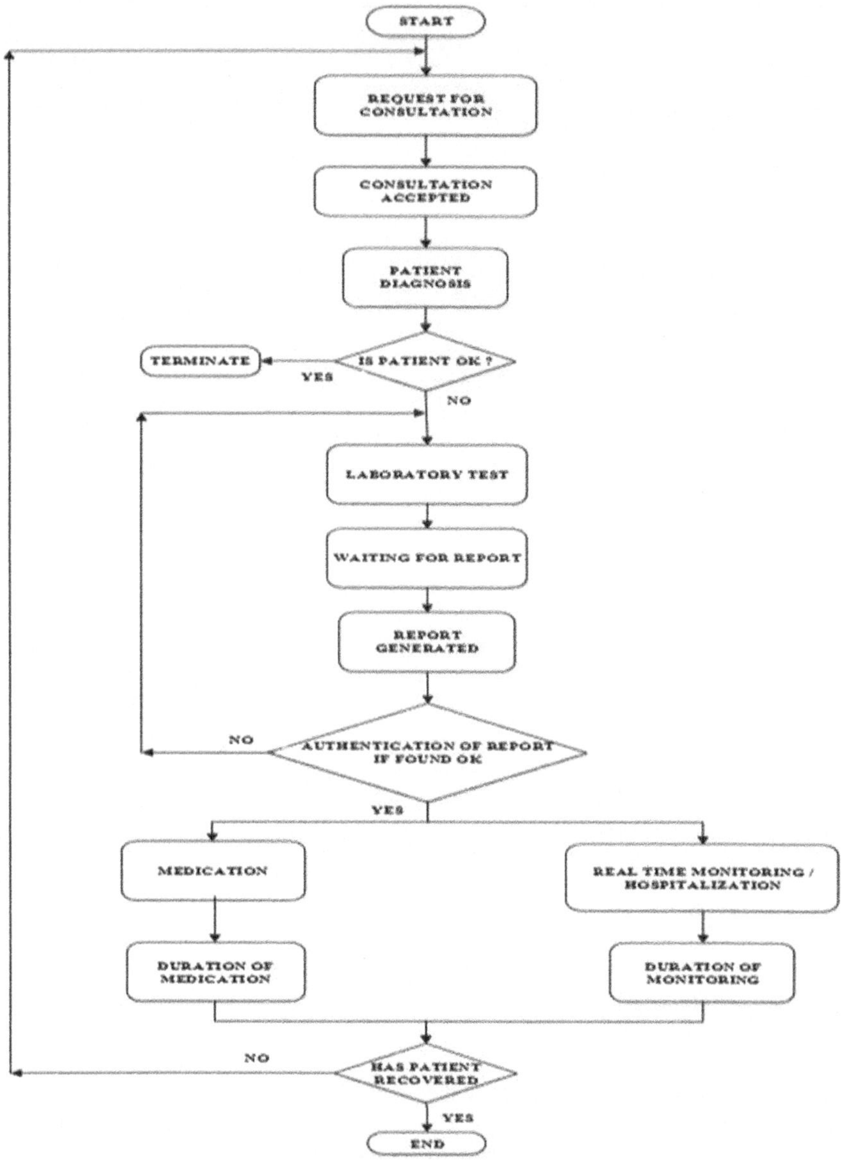

Fig. 3 Proposed flowchart

for carrying out various tests in the laboratory. Further authentication of the report is done, if reports found authentic then medication and real-time monitoring steps are followed. Otherwise, again diagnosis of the patient is done with laboratory tests. All the medication procedures are inspected by healthcare regulator. The entities and their roles are also describing in Table 3.

Table 3 Entity-role representation of proposed model

Entity	Role
Patient	He is the one who approaches the hospital authority for consultation. He starts the activity
Doctor	He is responsible for diagnosing the patient personally after the consultation is approved
Hospital authority	This authority is responsible for establishing the connection between patient and doctor. It takes action on the smart contract followed by the doctor's action
Healthcare regulator	This unit is responsible for the proper functioning of the hospital under the proper medical norms
Diagnostic department	This department is responsible for proper diagnosing of the patient by performing various laboratory tests and generation of authenticated reports

6 Scope of Improvement in Model

There are number of concerns of the healthcare industry that this model addresses but still there are areas which have to be properly dealt with to get more optimized results. Some of them are:

- **Mining**: Mining of blockchain includes linking transactions to the current blockchain ledger available among all blockchain users. Mining includes forming a hash of a transaction block which cannot be easily manipulated, preserving the credibility of the blockchain as a whole without the need for a central network. Mining is usually done on a specialized computer, need a fast processor, greater usage of power resources and more energy consumed than normal computer operations. The biggest reason for mining is that people opting to use a mining machine would be paid for doing so. Generally, IoT device like Raspberry Pi is a low-end device with quite limiting processing capacity. Unlike higher-end device are not a suitable choice for hashing.
- **Processing Time**: The time needed by the IoT system to perform the appropriate function will be longer, and this reaction period is having further scope to get increased.

7 Conclusion

In this chapter, a model for the healthcare industry is proposed by using both blockchain and IoT technology. IoT devices measure and collect real-time data which then is stored in the blockchain. Also, the various billing and insurance claim concerns could easily be addressed barring all fraudulent activities. The architecture we have proposed addresses these concerns as the blockchain technology is used as decentralized and distributed. The recorded transactions are permanently immutable, and the correspondence between the systems is encrypted. Along with healthcare model, we have tried to address various concerns of the healthcare industry. However, still there are areas which need improvement. As now the energy consumption by IoT devices is more; but with the advancement in technology, this can be lowered, thus allowing more people to use these services at lower costs. Also, the latency of adding a block in a blockchain can be shortened utilizing quicker, increasingly effective algorithms for encryption and faster processors.

References

1. Nakamoto S, Bitcoin A (2008) A peer-to-peer electronic cash system. Bitcoin. https://bitcoin.org/bitcoin. pdf
2. Gomber P (2017) Hinz-O. Nofer M. Schiereck D.,'Blockchain'. Springer 59(3):183–187
3. Bahga A, Madisetti VK (2016) Blockchain platform for industrial internet of things. J Software Eng Appl 9(10):533–546
4. Pongnumkul S, Siripanpornchana C, Thajchayapong S (2017) Performance analysis of private blockchain platforms in varying workloads. In 2017 26th International conference on computer communication and networks (ICCCN) (pp 1–6), IEEE
5. Ray PP (2018) A survey on Internet of Things architectures. J King Saud University-Comput Inf Sci 30(3):291–319
6. Hurrah NN, Parah SA, Sheikh JA, Al-Turjman F, Muhammad K (2019) Secure data transmission framework for confidentiality in IoTs. Ad Hoc Netw 95:101989
7. Kaw JA, Loan NA, Parah SA, Muhammad K, Sheikh JA, Bhat GM (2019) A reversible and secure patient information hiding system for IoT driven e-health. Int J Inf Manage 45:262–275
8. Alassaf N, Gutub A, Parah SA, Al Ghamdi M (2019) Enhancing speed of SIMON: a light-weight-cryptographic algorithm for IoT applications. Multimedia Tools and Applications 78(23):32633–32657
9. Rashid M, Parah SA, Wani AR, Gupta SK (2020) Securing E-Health IoT data on cloud systems using novel extended role based access control model. In: Internet of Things (IoT) (pp 473–489), Springer, Cham
10. Sheikh JA, Akhter S, Parah SA, Bhat GM (2018) Blind digital speech watermarking using filter bank multicarrier modulation for 5G and IoT driven networks. Int J Speech Technol 21(3):715–722
11. Ateniese G, Magri B, Venturi D, Andrade E (2017) Redactable blockchain–or–rewriting history in bitcoin and friends. In: 2017 IEEE European symposium on security and privacy (EuroS&P) (pp 111–126) IEEE
12. Taylor MB (2013) Bitcoin and the age of bespoke silicon. In: 2013 international conference on compilers, architecture and synthesis for embedded systems (CASES) (pp 1–10), IEEE
13. Böhme R, Christin N, Edelman B, Moore T (2015) Bitcoin: economics, technology, and governance. Journal of Economic Perspectives 29(2):213–238
14. Kosba A, Miller A, Shi E, Wen Z, Papamanthou C (2016) Hawk: the blockchain model of cryptography and privacy-preserving smart contracts. In: 2016 IEEE symposium on security and privacy (SP) (pp 839–858), IEEE
15. Kiayias A, Russell A, David B, Oliynykov R (2017) Ouroboros: a provably secure proof-of-stake blockchain protocol. In: Annual international cryptology conference (pp 357–388), Springer, Cham
16. Eyal I, Gencer AE, Sirer EG, Van Renesse R (2016) Bitcoin-ng: a scalable blockchain protocol. In: 13th {USENIX} symposium on networked systems design and implementation ({NSDI} 16) (pp 45–59)

17. Griggs KN, Ossipova O, Kohlios CP, Baccarini AN, Howson EA, Hayajneh T (2018) Healthcare blockchain system using smart contracts for secure automated remote patient monitoring. J Med Syst 42(7):130
18. Katuwal GJ, Pandey S, Hennessey M, Lamichhane B (2018) Applications of blockchain in healthcare: current landscape & challenges. arXiv preprint arXiv:1812.02776
19. Dhillon V, Metcalf D, Hooper M (2017) Blockchain in health care. In: Blockchain enabled applications (pp 125–138), Apress, Berkeley, CA
20. Drescher D (2017) Blockchain basics, vol 276. Apress, Berkeley, CA
21. Crosby M, Pattanayak P, Verma S, Kalyanaraman V (2016) Blockchain technology: Beyond bitcoin. Applied. Innovation 2(6–10):71
22. Miller D (2018) Blockchain and the internet of things in the industrial sector. IT Professional 20(3):15–18
23. Esposito C, De Santis A, Tortora G, Chang H, Choo KKR (2018) Blockchain: a panacea for healthcare cloud-based data security and privacy? IEEE Cloud Computing 5(1):31–37
24. Dey T, Jaiswal S, Sunderkrishnan S, Katre N (2017). HealthSense: a medical use case of Internet of Things and blockchain. In: 2017 International conference on intelligent sustainable systems (ICISS) (pp 486–491) IEEE
25. Hanley M, Tewari H (2018) Managing lifetime healthcare data on the blockchain. In: 2018 IEEE smartworld, ubiquitous intelligence & computing, advanced & trusted computing, scalable computing & communications, cloud & big data computing, internet of people and smart city innovation, pp 246–251, IEEE
26. Rashid M, Singh H, Goyal V (2019) Cloud storage privacy in health care systems based on IP and geo-location validation using K-Mean clustering technique. Int J E-Health Med Commun 10(4):54–65
27. Alhadhrami Z, Alghfeli S, Alghfeli M, Abedlla JA, Shuaib K (2017) Introducing blockchains for healthcare. In: 2017 international conference on electrical and computing technologies and applications (ICECTA) (pp 1–4) IEEE
28. Zyskind G, Nathan O (2015) Decentralizing privacy: using blockchain to protect personal data. In: 2015 IEEE security and privacy workshops (pp 180–184) IEEE
29. Samaniego M, Deters R (2016) Blockchain as a Service for IoT. In: 2016 IEEE International Conference on Internet of Things (iThings) and IEEE Green Computing and Communications (GreenCom) and IEEE Cyber, Physical and Social Computing (CPSCom) and IEEE Smart Data (SmartData) (pp 433–436) IEEE
30. Yuan Y, Wang FY (2016) Towards blockchain-based intelligent transportation systems. In: 2016 IEEE 19th international conference on intelligent transportation systems (ITSC) (pp 2663–2668) IEEE
31. Ali AA, El-Dessouky IA, Abdallah MM, Nabih AK (2017) The quest for fully smart autonomous business networks in IoT platforms. In: Proceedings of the 3rd Africa and Middle East conference on software engineering (pp 13–18)
32. Suliman A, Husain Z, Abououf M, Alblooshi M, Salah K (2018) Monetization of IoT data using smart contracts. IET Networks 8(1):32–37
33. Dorri A, Kanhere SS, Jurdak R (2017) Towards an optimized blockchain for IoT. In: 2017 IEEE/ACM second international conference on Internet-of-Things Design and Implementation (IoTDI) (pp 173–178) IEEE
34. Lindman J, Tuunainen VK, Rossi M (2017) Opportunities and risks of Blockchain Technologies–a research agenda

Chapter 13
Asset Security in Data of Internet of Things Using Blockchain Technology

Atul Banotra, Swastik Gupta, Sachin Kumar Gupta, and Mamoon Rashid

1 Introduction

Assets in terms of hardware and software are managed in the IT sector. These assets are manufactured at one end, brought to another end for selling, purchasing, configuring, and then installed at the customer's end [1]. It sounds very simple to trace such an asset that undergoes through various stages that have been mentioned above, but it is a very complex procedure. Tracing back these assets, as they move to start with one spot then onto the next is a significant difficult procedure. The previous technologies that have been developed so far were good, but there had not been any framework developed yet. As a result, Blockchain and IoT jointly came into consideration to solve the above issue.

A. Banotra · S. Gupta · S. K. Gupta
School of Electronics and Communication Engineering, Shri Mata Vaishno Devi University, Kakryal, Katra, Jammu and Kashmir 182320, India
e-mail: atulbanotraab@gmail.com

S. Gupta
e-mail: swastik.gupta@smvdu.ac.in

S. K. Gupta
e-mail: sachin.gupta@smvdu.ac.in

M. Rashid (✉)
School of Computer Science and Engineering, Lovely Professional University, Jalandhar 144411, India
e-mail: mamoon873@gmail.com

K. J. Giri et al. (eds.), *Multimedia Security*, Algorithms for Intelligent Systems,
https://doi.org/10.1007/978-981-15-8711-5_14

1.1 Blockchain Concept

Blockchain is a distributed, collaborative, and collective electronic registry that is used to document payments through multiple devices so that the database cannot be arbitrarily changed without modifying all corresponding frames and platform complicity [2]. This helps the stakeholders to review payments and to review them cheaply. The Blockchain platform is independently operated through a peer-to-peer infrastructure and a decentralized period molding database. They are independently verified by perceived self-interests operated by continuous coordination [3]. The effect is a stable process where there is minimal confusion about the stakeholder's consumer data or information protection. Blockchain use eliminates the endless methodological rigor attribute from digital currency. This guarantees that each variable of expenditure has been moved only once, addressing the enduring dual revenue problem. Blockchain was defined as a specification for the exchanging of values [4].

1.2 Internet of Things

Internet of things (IoT) is a model that is used for efficient computing and helps us in the functioning of different ways used in our daily lives. In IoT, interconnection of various devices is enabled to exchange data between them whenever interactions are done in these IoT systems. The most popular devices used in IoT are cellular phones, smart sensors, and coffee machines [5, 6]. As IoT systems help humans and various devices in easier connectivity, however it results in large volumes of data due to their high speeds and thus these systems get more complex in terms of data aspects.

In IoT systems, existing network infrastructure is used for the direction integration of remotely controlled physical objects and computer-based systems [7, 8]. This technology of IoT has resulted in minimum intervention of humans for performing any job with greater efficiency and accuracy. IoT systems have power of capturing data from different kinds of sensors as well. Various organizations are showing interests in IoT-based technology due to the lower costs of sensors and actuators that are used in this technology [9].

1.3 Blockchain in Internet of Things

Blockchain and the IoT are main innovations that are going to have a significant effect on manufacturing businesses during the next decades. This section explains how these two innovations increase performance, create incentives for new enterprise, tackle legal requirements standards, and enhance efficiency and exposure. The IoT facilitates the processing of telemetry data records in real time [10]. As the value of detectors and circuitry continues to drop, manufacturing industries can resolve cost barriers when embracing IoT architectures. Blockchain would allow the collaborative use of the decentralized, transparent, common registry accessible to stakeholders in the corporate channel to exchange vital required information collected from IoT.

Data security and confidentiality of the Internet of things (IoT), primarily because of the large and widespread existence of IoT channels, continue significant issues and problems [11]. Decentralization and protection of Blockchain solutions also require considerable resources, delays, and workload computing not appropriate for the majority of resource-controlled IoT machines. We have provided a small BC recompilation in our prior studies, specifically designed for IoT use with the removal of the POW and coin principle. Our methodology has been demonstrated in an intelligent home environment, consisting of three core levels—cloud infrastructure, extension, and clever house.

From its inception, the Internet of things (IoT) is completely developed and is part of the development Internet in the upcoming time. The ability to manage millions of computers globally is one of the technological barriers. Although IoT accessibility systems and processes do operate, they are focused on standardized frameworks that impose a new range of practical constraints for international administration [12].

Blockchain will in this way enable the sharing of key-relevant data got from the IoT using an appropriated, decentralized, shared record that is made available to the individuals in the business association for managing the advantage adequately and financially [1]. The framework worked by Blockchain and IoT is utilized to help the start to finish the procedure of asset management. This system focuses primarily on the functions like availability of assets, deployment of an asset, working of an asset, maintenance of an asset, proper handling of an asset, and replacement of asset in case it leads to the faulty situation. Blockchain innovation empowers the business to frame a record that organizes the sharing of at least one record [1]. Since the ledgers are distributed and decentralized, so rather than being controlled by a single organization, the ledger is controlled by the participants that share it. In this way, Blockchain technology enables us to maintain trust among organizations [3]. Our vision in the Blockchain-based technology is to develop trust among organizations so that they work together under one platform, to avoid manual action taken by the various people throughout the process. IoT devices will help in capturing real-time data to avoid manual actions.

1.4 Key Blockchain Features for Building Trust in Organizations

Blockchain will help to develop trust among organizations by its following key features as:

Blockchain Ledger as a Single Unit of Record: Transactions of the block which are appended to each other in chronological order will act as a source of trust for all those who are the part of it.

Permission: All the participants such as asset sellers, asset buyers, asset deployers, asset maintainers except the asset owner will need to have permission onto the Blockchain network.

Shared Data Model: Data that is common to all the participants or participating organizations will be maintained on the Blockchain network. Hence, it maintains transparency.

Smart Contract: Business protocols that govern the set of transactions regarding the asset will be maintained on to the Blockchain.

Proof of Existence: Proof received for transactions such as asset availability, asset deployment, asset maintenance will be digitally hashed and recorded.

The rest of the chapter is organized as follows: Sect. 2 discusses the existing literature survey that has been done in the related work and presents the research gap. Further, Sect. 3 includes architecture of asset management in IoT using Blockchain, and finally, Sect. 4 conducts a conclusion and future bearings.

2 Related Work

Blockchain innovation began with the mainstream cryptocurrency Bitcoin, and it enables advanced data to be appropriated without duplicated or modified. Many research scholars have presented various uses of the Blockchain technique integrating it either with IoT or with some other existing technology. A great deal of research work is going on in this field to make the platform more secure, more transparent, more reliable, more trustworthy, and more convenient.

According to Jun Wang et al. [14], Blockchain-based application can improve the present techniques of contract management, supply chain management, and hardware renting individually by solving two main issues. These issues are trusted among buyer–supplier in contract management, and transparency and traceability of products ought to be featured as the reason for further improvement. By using a public and private distributed ledger, users tried to solve these issues up to some extent. A brief description of the public and private ledger, blockchain enable supply chain management, blockchain enable equipment management and blockchain enable contract management is required.

In Zyskind et al. [15], the ongoing increment in revealed occurrences of reconnaissance and security ruptures trading off clients' protection raises doubt about the present model. The researcher makes use of the decentralization property of the Blockchain so that the user can control its data instead of the third party. He executes a convention that transforms a Blockchain into a computerized access-control administrator that does not require trust in an outsider. From a security point of view, analysts created different systems focusing on protection concerns concentrated on close to home information. Information anonymization techniques endeavor to secure identifiable data.

In Samaniego et al. [16], the main difficulty in orchestrating the Blockchain technology is that where should be hosted, the researcher assesses the utilization of cloud and fog as facilitating stages, i.e., hosting platforms. The researcher also pointed out that there is developing agreement on the ability of interfacing Blockchain advancement to IoT. The key question authors arise in this paper is that where the Blockchain be hosted? They also suggested that facilitating the Blockchain straightforwardly on asset obliged IoT devices are unadvisable due to lack of computing energy, lack of sufficient bandwidth, and need to maintain power. Therefore, they come up with the cloud and fog as a hosting platform.

According to Yuan et al. [17], Blockchain is one of the verified and confided models to structure the recently created Parallel Transportation Management System (PTMS). Blockchain can be used to set up a verified, trusted, and decentralized self-ruling Intelligent Transportation System (ITS) ecosystem. The researcher defines the relationship between the Blockchain Intelligent Transportation System (BITS) and PTMS which is used to establish a secure and prominent seven-layered Blockchain model. Moreover, for real-time monitoring of data in the supply chain model, a supply chain solution model has been proposed to accurately detect the asset in the supply chain. Apart from this, a similar investigation of the current Blockchain approach and IoT studies has been done in Table 1. A nearby investigation of the examinations as referenced in this area and Table 1 sets up a research gap that is expanding the use of integrated Blockchain and IoT in the asset management segment which has not been considered, up until now.

Table 1 Comparison of existing security approach of related surveys in the literature

Technology type	Objective	Inference
Blockchain and IoT [13]	To establish decentralized and distributed supply chain system using Blockchain and IoT	Efficient supply chain system Real-time monitoring of data
Blockchain [14]	To secure client information from unauthorized parties Also pays attention to usage of your personal data by the third party	Combines off-chain storage with Blockchain storage to secure personal data Third-party control of user's data is eliminated by using Blockchain concept
Blockchain and IoT [15]	A key test in the organization of Blockchain technology is the facilitating area (i.e., hosting location) To evaluate the use of cloud and fog as hosting platforms	Due to its distributed and decentralized association, Blockchain is being utilized inside IoT. For example; to manage device configuration, store sensor information and empower small-scale installments
Blockchain [16]	To define the relationship between BITS and PTMS Also, to provide seven layers of a conceptual model for Blockchain	Blockchain can be utilized to establish a secured, trusted, and decentralized autonomous ITS ecosystem Blockchain is one of the verified and confided designs to structure the recently created PTMS

(continued)

Table 1 (continued)

Technology type	Objective	Inference
IoT data with smart contract code of Blockchain [17]	To present a Blockchain solution and usage utilizing Ethereum smart contracts for adapting IoT information with mechanized installment including no mediator	Focused on the following: Architectural design Entity relations Interactions among participants Logic flow Implementation and testing of the overall system functionality
IoT Platform [18]	To present adaptable system layers, more devices, and stages depending on the IoT arrangements	Easy integration of IoT applications with executable business logic or smart contracts Integrating IoT stages with Blockchain; a new way for better straightforwardness, administrations devouring, and products that rely upon the group-based economy
IoT and Blockchain [19]	To solve the major issue of highly computational resources that includes high bandwidth overheads and processing time in Blockchain technology	Reduces the block validation processing time Statical assessment of the architecture under specific challenge model reinforces its usefulness in providing privacy and security Overhead processing is significantly reduced in contrast to that used in Bitcoin
Blockchain [10]	To expand the understanding of Blockchain technology in payment network	Research agenda suggests a few functional areas regarding the payment network such as operational issues, business environmental issues, and technology design flaws
Blockchain [20]	Providing peer-to-peer forms of digital money will make it possible for anonymous transfers to be sent directly from one group to another Do not need to bring the monetary entity in between for payment purposes	This eliminates the limitation of payment by excluding the attackers This secures the mode of payment It speeds up the payment process by removing mediators between the entities
Blockchain [21]	To provide a method for managing a collective agreement in the new digital world Provide for the use of Blockchain in both the monetary and non-monetary fields	This makes it possible for interested organizations to learn for certain that a physical event occurred by establishing an incontrovertible record in the public archive

3 Asset Management in IoT Using Blockchain

The basic idea behind using both Blockchain and IoT techniques for asset management over using IoT alone is that there are still three challenges, which IoT needs to be addressed: resource optimization (complex security method), privacy, and centralization. Blockchain is an appealing innovation for tending to the referenced security and protection challenges in IoT because of its key highlights, including decentralization, security, and self-ruling working of the device with the distributed ledger. Blockchain solution can offer a high level of privacy and security along with IoT

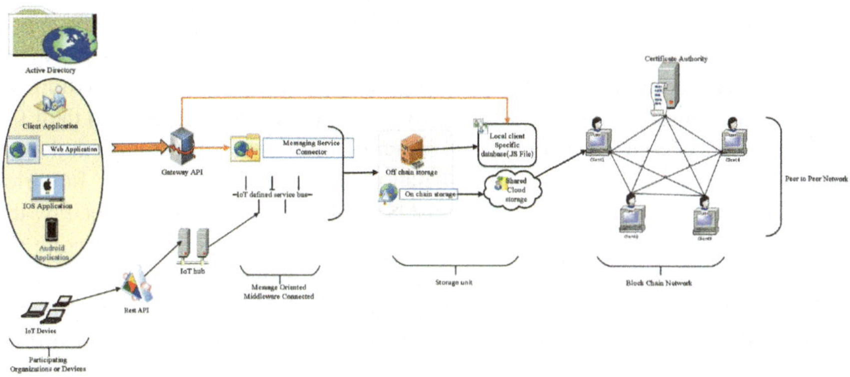

Fig. 1 Proposed architecture

hub. There are six "What" of asset management which needs to be addressed while deploying the integrated Blockchain and IoT network:

•What do you purchase?
•What is its worth?
•What is the defend maintenance?
•What is its condition?
•What is its remaining service life?
•What is the procedure to fix it?

The above questions are answered through the architecture model shown in Fig. 1.

Blockchain and IoT simplify asset management application by providing a solution using several components shown in the architecture. The Blockchain-integrated IoT architecture has been implemented in four phases as shown in Fig. 1.

In the first phase, there may be either an IoT device or a client-based application along with an active directory for identity and authentication. In an active directory, a group of organizations called consortium can federate themselves under a single platform or workbench. The provided platform will generate a new user account with the enterprise identity in the active directory. In the same directory, the specific role is provided to the signed user according to the smart contract code. The mapping of enterprise identity along with their specific role in the active directory will facilitate the validated login to client APIs and applications and therefore make use of the authentication policies of the organization. In case data is retrieved from the IoT sensor in a smart contract, based on the delivered data from IoT sensor smart contracts will take action. For example, during the logistic of an asset, if the asset is broken, then the sensor sends the data to the smart contract in the on-chain storage, and then the necessary action is taken. During live testing of an asset, the quantity of the asset while handling the other end will automatically be collected by the IoT sensor and delivered to the on-chain storage through a message-oriented middleware connector which is explained in the second phase.

In the second phase, the information of active directory and data collected by IoT sensor will route through message-oriented middleware connection which includes

GATEWAY-API. Blockchain incorporates a REST-based gateway service API. While doing an entry to a Blockchain ledger, the API produces and conveys messages to the message service connector which further accelerates the data to the storage unit. At the point when information is mentioned by the API, questions are sent to the LCS database. The LCS database contains an imitation of the on-chain storage and asset metadata that gives the setting and design data to bolstered smart contracts. Inquiries return the required information from the LCS database in a format informed by the asset metadata for the asset management contract. IoT sensors also accelerate the data to the REST API and interact with the IoT hub. After receiving the message, the IoT hub validates the data sent by the IoT device. After validation, the data that meets the contracted criteria will be sent to an IoT-defined service bus for further routing.

In the third phase, the routing information (asset information) is stored in the on-chain and off-chain storage. The on-chain storage stores the asset id, asset metadata, asset state, asset life cycle, and the actual data related to the asset. On-chain storage contains the asset metadata for the transaction to be composed of the Blockchain. It takes the data from the service bus and accelerates the hashed data to the cloud storage. Off-chain storage pushes the data to the LCS database in JSON format which is the replica of on-chain storage. This data can be queried, envisioned, and examined directly from the LCS database.

In the fourth phase, there is the Blockchain network which is the heart of architecture. It further contains CA, which is used for the enrollment of participants who to be a part of the shared ledger. Ledger is an instance of the database associated with each peer. It is a peer-to-peer communication, and every peer holds a record of transactions as a ledger or a copy of the ledger in this part of the architecture.

The above-integrated Blockchain and IoT architecture also provide us with the following features that individuals cannot be provided by and of these technologies. By going hand in hand, it makes certain things possible which are discussed below:

• It promotes the exchange of services and benefits that led to the establishment of a marketplace for services among devices [22, 23].

•It allows us to simplify many internal time-consuming workflows in a cryptographically verifiable manner.

• IoT and Blockchain effectively carry together more stable, trustworthy, and valuable insightful and collaborative flavors to our lives, allowing low-power wireless technology and embedded computing and intelligence to grow [24, 25].

• Blockchain and IoT are offering a new corporate network platform, communicating between network users regularly, maintaining data security, the transaction fees, and delivering accountability by peer-to-peer protocol [26].

• Blockchain and IoT are accredited to the confidence strap network clearly defining the guidelines on confidentiality and trustworthiness [27].

The proposed flowchart shown in Fig. 2 describes the role of entities in asset management. The asset management process starts with the query of the Purchaser (1) about the availability of the asset in the smart contract. Seller (2) takes action on the smart contract. If the asset is available, then the asset is released from the seller end. Otherwise, the process is terminated with raise and intend command.

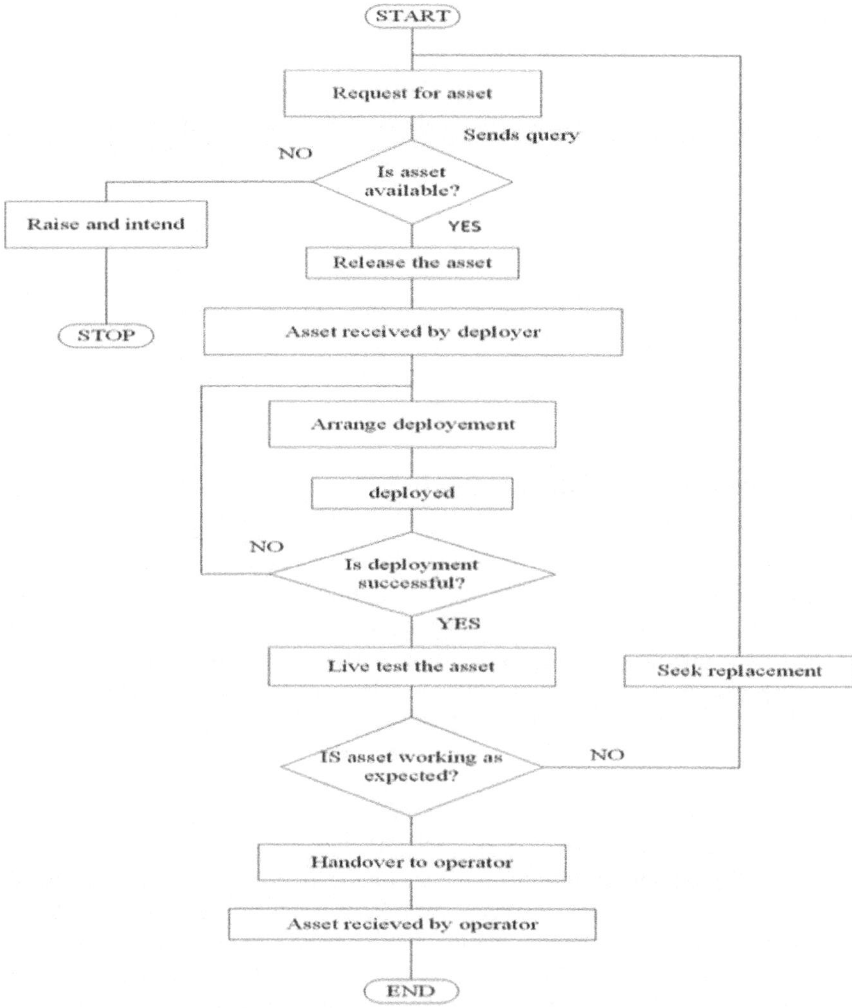

Fig. 2 Proposed flowchart

Now, the asset will be received by the Deployer (3) who will arrange, deploy, and install it properly and will generate the message about the successful deployment of the received asset. If the asset is not deployed properly, then it is being sent back to arrange deployment unit. In the next stage, the asset is live tested by the Inspection Unit (4). In case an asset is not working properly, it will go for seeking a replacement; otherwise, it is handed to the Operator (5). The role of different entities is shown in Table. 2.

Table 2 Role of the shown entities

S. no	Entity	Role
1	Purchaser	He is in charge of questioning about the availability of asset in the smart contract with the goal that he can buy it. He starts the activity
2	Seller	Seller takes action on the smart contract about the availability of the asset
3	Deployer	He is responsible for the arrangement and the deployment of the asset. He takes action on the smart contract followed by the seller's action
4	Inspection Unit	This unit is responsible for live testing the asset functioning, asset state, asset life cycle
5	Operator	Responsible for operating the received asset, and operator actions are picked by the purchaser

4 Future Trends of Blockchain in IoT

Apache Blockchain is usually utilized to guarantee regulatory compliance, boost supply chain management, and improve information protection, to name just a couple of cases. Blockchain is faster than many traditional centralized systems and works better than them by keeping the privacy and data record of all the transactions forever. In the future, it will be a technology being used by many MNCs, banks, government institutions, and persons also. This integration of Blockchain with IoT will open transparency and prevent corruption in many sectors, and trust will increase as this technology will boom and be incorporated by all the sectors. So, a list that how this technology will be adopted by the world and what changes we can see is given below:

Greater Customer Adoption: In the upcoming 10–20 years' time, we will see a huge demand in the IoT market as everybody wants privacy and security, and most importantly to save time. For this, IoT demand will increase and for this to work efficiently Blockchain is required. So, when the demand increases, we see a rise in the development of supply and companies adopting the new culture, i.e., future technology. The IoT architecture will be established, and the difficulties seen in transport, insurance, and agriculture will reduce vividly.

Increase in Smart House Demand: Everybody wants comfort, and this could be achieved through smart homes. Suppose you are coming home on a hot summer day and you want that your air conditioner gets on and the room should be chilly as you imagine, just push the button on your mobile phone and it will be done. These things are happening and will increase in time. Now here what Blockchain will do is that you pushed the button and the data was sent and received between you and your air conditioner; it seems normal but we do not want complete control and do not want our data to get tampered or shared. This is where Blockchain will come into effect, and suppose it happens, we can track the transaction, i.e., who did it. This is a simple example but at high levels, it is a risk and data privacy is important.

Use in Healthcare Departments: The healthcare departments are already following these, and there will be more increase in this technology. It is already being appraised all over the world. There area number of wearable devices, and patient's complete medical history could be stored in the cloud along with their images and pre-diagnostic conditions.

Integrating Blockchain and IoT with Artificial Intelligence: In the upcoming years, when the different sectors will start adopting the Blockchain and IoT model in their core services, then AI will come into effect. Currently, companies like Sentrian, Manna, Veros Systems, Neura, etc., are using AI with IoT; slowly Blockchain will be adopted, and the systems will perform far better. Currently, Blockchain runs on standard computers and a huge amount of power is required for processing a task completely. And integrating all these will be very much power and memory processing, and a steady increase will happen here as the architecture is the main problem here. But prediction problems, AI computing, security, and many more benefits could be sown if this technology is properly flourished.

Experts of Blockchain in IoT: The demand is increasing for the Blockchain experts especially the people having knowledge of IoT as well. Different industries have their own programs where they brush up their skills and make them industry-ready but all this process is too hectic for the industry and they have to bear the expenses for training individuals for this technology.

Energy and Resources Management with Blockchain and IoT: IoT devices equipped with Blockchain technology will become an efficient resource for the management to utilize energy properly. We can have IoT devices fitted in our homes and offices to maintain temperature, sprinklers, etc. With proper usage of these resources, an optimal quantity of energy will be required for processing data. Keeping a check on excessive use and all this data could be made transparent due to the Blockchain integrated with IoT. This could help us to prevent and keep a check on global warming, soil management, endangered species, and all problems regarding local industries, MNCs, etc., which are affecting our environment in any way.

Expansion of IoT Devices with Blockchain: According to Gartner, 95 percent of the companies will have IoT in their working devices. Now the companies are thinking about expanding these more and more. The need is arising for the devices to be more comprehensive and a way through which everyone can access this and have the benefit.

5 Conclusion

A model for asset management using Blockchain and IoT has been discussed and presented in this chapter. Blockchain holds a guarantee for protection and security in IoT devices. Blockchain is an appealing innovation for tending to the referenced security and protection. Our vision in the Blockchain and IoT-based innovation is to supplant many heritage approaches, manual moves made in asset management. This chapter additionally shows integrated Blockchain and IoT proposed design in four stages taking into consideration the middleware connection unit and storage unit. Here, the proposed four-layered architecture scheme also emphasizes the asset management segment that will be directed to survey the adequacy of the coordinated Blockchain and IoT framework. This may incorporate testing the coordinated Blockchain and IoT network in their real-time asset management projects.

References

1. Niranjanamurthy M, Nithya BN, Jagannatha S (2019) Analysis of blockchain technology: pros, cons and SWOT. Cluster Computing 22(6):14743–14757
2. Gomber P (2017) Hinz-O. Nofer M. Schiereck D.,'Blockchain'. Springer, 59(3):183–187
3. Bahga A, Madisetti VK (2016) Blockchain platform for industrial internet of things. J Softw Eng Appl 9(10):533–546
4. Pongnumkul S, Siripanpornchana C, Thajchayapong S (2017) Performance analysis of private blockchain platforms in varying workloads. In: 2017 26th international conference on computer communication and networks (ICCCN) (pp 1–6) IEEE
5. Ray PP (2018) A survey on Internet of Things architectures. J King Saud University-Computer Inf Sci 30(3):291–319
6. Hurrah NN, Parah SA, Sheikh JA, Al-Turjman F, Muhammad K (2019) Secure data transmission framework for confidentiality in IoTs. Ad Hoc Netw 95:101989
7. Kaw JA, Loan NA, Parah SA, Muhammad K, Sheikh JA, Bhat GM (2019) A reversible and secure patient information hiding system for IoT driven e-health. Int J Inf Manage 45:262–275
8. Alassaf N, Gutub A, Parah SA, Al Ghamdi M (2019) Enhancing speed of SIMON: a light-weight-cryptographic algorithm for IoT applications. Multimedia Tools Appl 78(23):32633–32657
9. Gubbi J, Buyya R, Marusic S, Palaniswami M (2013) Internet of Things (IoT): A vision, architectural elements, and future directions. Future Gener Comput Syst 29(7):1645–1660
10. Dorri A, Kanhere SS, Jurdak R (2017) Towards an optimized blockchain for IoT. In: 2017 IEEE/ACM Second International conference on internet-of-things design and implementation (IoTDI) (pp 173–178), IEEE
11. Rashid M, Singh H, Goyal V, Ahmad N, Mogla N (2020) Efficient Big Data-based storage and processing model in Internet of Things for improving accuracy fault detection in industrial processes. In: Security and privacy issues in sensor networks and IoT (pp 215–230), IGI Global
12. Singh P, Rashid E (2015) Smart home automation deployment on third party cloud using internet of things. Journal of Bioinformatics and Intelligent Control 4(1):31–34
13. Miller D (2018) Blockchain and the internet of things in the industrial sector. IT Professional 20(3):15–18
14. Wang J, Wu P, Wang X, Shou W (2017) The outlook of blockchain technology for construction engineering management. Front Eng Manage pp 67–75
15. Zyskind G, Nathan O (2015) Decentralizing privacy: Using blockchain to protect personal data. In: 2015 IEEE Security and privacy workshops (pp 180–184) IEEE
16. Samaniego M, Deters R (2016) Blockchain as a service for IoT. In: 2016 IEEE international conference on Internet of Things (iThings) and IEEE Green Computing and Communications (GreenCom) and IEEE Cyber, Physical and Social Computing (CPSCom) and IEEE Smart Data (SmartData) (pp 433–436). IEEE
17. Yuan Y, Wang FY (2016) Towards blockchain-based intelligent transportation systems. In: 2016 IEEE 19th International conference on intelligent transportation systems (ITSC) (pp 2663–2668), IEEE

18. Ali AA, El-Dessouky IA, Abdallah MM, Nabih AK (2017) The quest for fully smart autonomous business networks in IoT platforms. In: Proceedings of the 3rd Africa and middle east conference on software engineering (pp 13–18)
19. Suliman A, Husain Z, Abououf M, Alblooshi M, Salah K (2018) Monetization of IoT data using smart contracts. IET Networks 8(1):32–37
20. Lindman J, Tuunainen VK, Rossi M (2017) Opportunities and risks of Blockchain Technologies–a research agenda. In: 50th Hawaii international conference on system sciences, HICSS/IEEE Computer Society, Waikoloa, USA, pp. 1533–1542
21. Nakamoto S (2019) Bitcoin: a peer-to-peer electronic cash system. Manubot. Posted on Cryptography Mailing List at metzdowd. com. Retrieved from https://www.gartner.com/it-glossary/internet-of-things/
22. Crosby M, Pattanayak P, Verma S, Kalyanaraman V (2016) Blockchain technology: beyond Bitcoin. Appl. Innovation 2(6–10):71
23. Ali MS, Vecchio M, Pincheira M, Dolui K, Antonelli F, Rehmani MH (2018) Applications of blockchains in the Internet of Things: a comprehensive survey. IEEE Commun Surveys Tutorials 21(2):1676–1717
24. Christidis K, Devetsikiotis M (2016) Blockchains and smart contracts for the internet of things. Ieee Access 4:2292–2303
25. Miorandi D, Sicari S, De Pellegrini F, Chlamtac I (2012) Internet of things: Vision, applications and research challenges. Ad Hoc Netw 10(7):1497–1516
26. Atzori L, Iera A, Morabito G (2010) The internet of things: a survey comput. Netw. 54(15):2787–2805
27. Brunner A, Abderrahmane N, Muralidharan A, Halfpap P, Sume O, Zimprich S (2017) Trade finance disrupted: a blockchain use case. J Fin Transf 45:41–48

Chapter 14
Review of Secure Distributed Range-Free Hop-Based Localization Algorithms in the Wireless Sensor Networks

Deepak Prashar and Nishant Jha

1 Introduction

The function of locating sensor coordinates where it is situated in a particular target region is called localization [1]. Localization in WSN is a very critical activity, because we install nodes that are called sensors in a particular target region, and these nodes transmit the generated data to a user who ultimately takes necessary action. Currently, if the position of the node is not specified, the user would not be able to take suitable action at a given location. Localization in WSN is therefore a major challenge. Sensor node forms in WSN [2] are anchor node and node unknown. Anchors are nodes which have awareness of their common target-area location. These nodes know their position because they are manually deployed, they are attached to global positioning system (GPS) [3]. Such nodes have a big effect on localization as more anchors provide more specificity in deciding the position. Unknown are the nodes which have no awareness of their specific target-area location. These nodes use the information regarding the anchor node to determine their position.

Most of the localization algorithms are dependent on the nodal density that is to be monitored and localized [4]. However, the approaches which depend on counting the hops or estimating the distance are not vitally dependent on node density. Also with a significant increase in nodes, the cost trade-off can also become an important factor. High density can also lead to larger transmission cost and propagation delay. So a trade-off of the number of nodes versus cost and total latency must be considered while construction of a localization algorithm. The presence of obstacles or an irregular deployment area also affects the localization process. The main

D. Prashar (✉)
Department of Computer Science, Lovely Professional University, Punjab, India
e-mail: Deepak.prashar@lpu.co.in

N. Jha
Department of Computer Science, Lovely Professional University, Punjab, India
e-mail: nishant.11702196@lpu.in

K. J. Giri et al. (eds.), *Multimedia Security*, Algorithms for Intelligent Systems,
https://doi.org/10.1007/978-981-15-8711-5_15

assumption in WSNs is to have a line-of-sight communication within the nodes. Obstacles may cause diffraction and deflection of signals and thus incorrect estimation leading to an inaccurate result. Initially, the sensor networks were designed with nodes being static or stationary. But with the requirement of new applications such as IoT, the network became mobile. The localization algorithm must be adaptable and deal with the mobility of nodes either anchors or targets or both. Mobility in WSN can be modeled using various methods such as random way point (RWP) [5] in which each sensor node moves independently by choosing a random destination and a random speed. Localization has become one of the most prominent tasks in WSNs in the present time. Applications that are position-centric, location estimation are the major requirement, and this leads to the emergence of various localization techniques. The localization methods are generally categorized as range-based and free range. Along with the position estimation, security of the position computation process is the prime concern as the wrong position estimation through the presence of adversaries in the environment may compromise the whole process. Moreover, with the advancement of the multimedia applications, the risks that are associated with its security have also been raised to a large extent. When it is related to the applications that use the localization, there is a requirement of various techniques like steganography, watermarking, and encryption.

The main highlights of this chapter are:

1. It describes the localizations process and its types in a detailed manner along with the issues related to the localization process.
2. It provides the detail literature work pertaining to the issue of security for the range-free hop-based localization schemes available.
3. It also covers the multimedia security and various authentication methods pertaining to the localization.
4. It also points out the issues that are of main concern on which the further research work can be done in the times to come.

The whole chapter is divided into various sections like in Sect. 1 the introduction of the localization process and its major concerns are discussed. Section 2 covers the localization process followed by its types as covered in Sect. 3. The issues pertaining to the localization process are covered in Sect. 4. The need of secure localization and the requirements are mentioned in Sect. 5. Literature work related to the security is covered in Sects. 6. Multimedia security and its associated techniques are discussed in Sect. 7. Section 8 covers the gaps on which the further research can be carried out followed by the conclusion as defined in Sect. 9 of the chapter.

2 Localization Process

In WSN, localization procedure for the unknown nodes gets divided into three parts [1] and the procedure is also shown with Fig. 1.

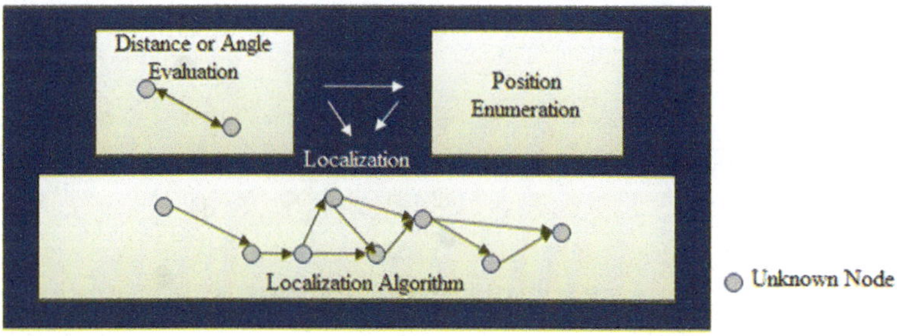

Fig. 1 Description of localization process

- **Distance or Angle Evaluation**: First of all, the distance or angle measurement is carried out in the middle of two nodes, and then this detail will help in the localization. There are various methods for determining the distance or angle between the nodes, but all methods have some advantages and some drawbacks which are addressed later in this chapter.
- **Position Enumeration**: The position of the unknown node is calculated through distance or angle estimation. Various methods for position enumeration are trilateration, multilateration, and triangulation.

  - **Trilateration**: The position is estimated as shown in Fig. 2 using the intersection point of the three circles in this method. Here the coordinates of the anchors are shown in the notation of $x1$ and $y1$ and so on, and x, y represents unknown node coordinates, d defines the distances.
  - **Multilateration**: Multilateration is the same as trilateration, but the difference is that it uses the unknown node's position enumeration as represented in Fig. 3 other than three anchor nodes.
  - **Triangulation**: In triangulation instead of using distances, angles are used to estimate the location of the unknown node. Two cases are considered like the first case is shown in Fig. 4 in which two anchor nodes calculates the angle of

Fig. 2 Working of trilateration method

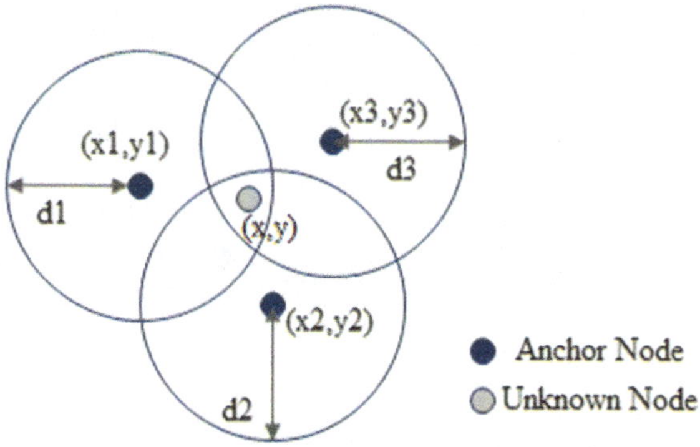

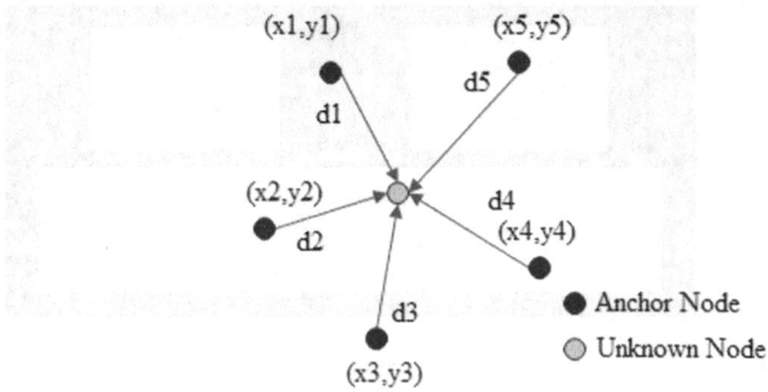

Fig. 3 Working of multilateration method

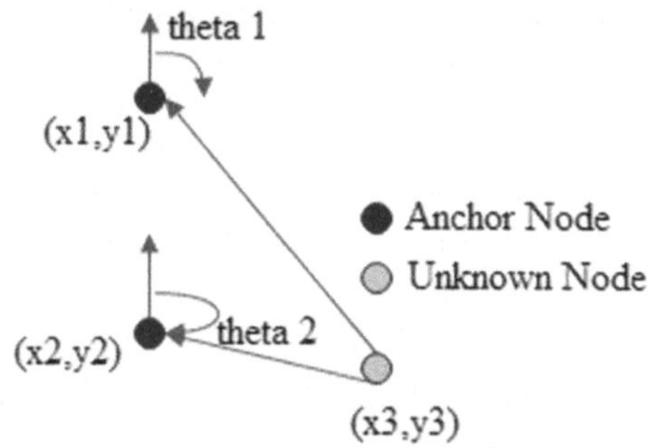

Fig. 4 Triangulation with two anchors

arrival (AoA) and then enumerate the position of the unknown node, second case as shown in Fig. 5 that requires at least three anchor nodes.

– **Localization Algorithm**: Algorithm helps the other nodes estimate their position with the details available. There are various types of algorithms designed to enumerate the location of the unknown nodes based on their specialization and environmental requirement.

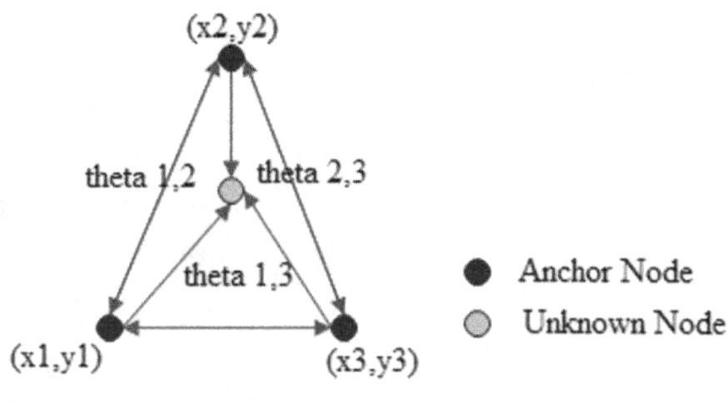

Fig. 5 Triangulation with three anchors

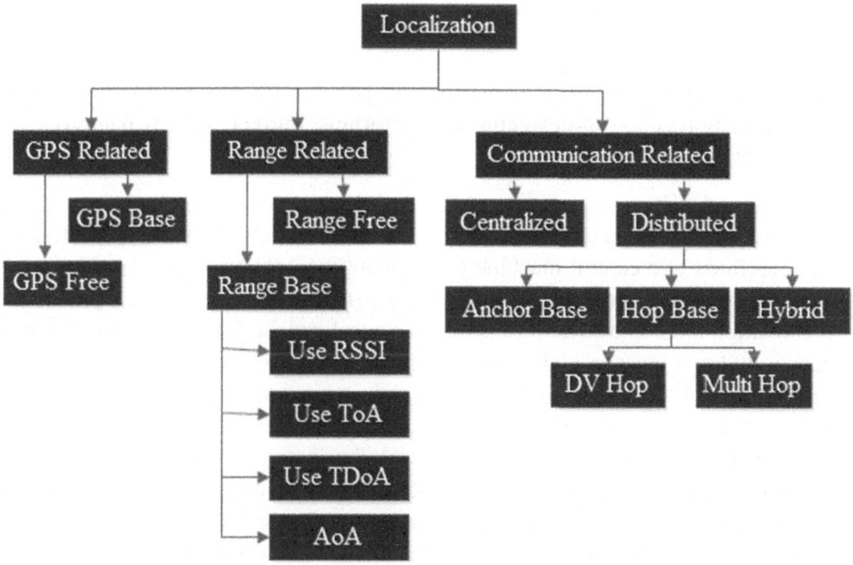

Fig. 6 Classification of localization techniques in WSN

3 Localization Types

Various localization schemes are available, that can be divided into various categories as shown below in Fig. 6. The localization schemes are outline into four categories. Each category has its own benefits and pitfalls.

3.1 GPS Related

GPS-related approach is also categorized into two types: one is GPS based and another is GPS-free or without GPS, both have own benefits and pitfalls

- **GPS-Based**: In this approach, all of the sensor nodes are getting furnished with GPS. With the GPS localization accuracy is very high, but the problem is GPS communicates in the line of sight (LOS); but in dense area, many obstacles can be there, that becomes the problem in communication and also equipping all the sensor nodes with GPS makes this approach very costly.
- **GPS-Free**: To solve the limitations of GPS-based approach, instead of using GPS with the entire sensor node; connect the limited number of sensor nodes with GPS. Those nodes become an anchor node. Then unknown nodes uses the anchor nodes to localize them using any localization algorithm either range-based or range-free one.

3.2 Communication-Based

It outline whether it is the central or distributed algorithm. Central and distributed both are different approaches for designing the algorithms. Further, these two approaches also have own benefits and pitfalls.

- **Centralized**: They need different estimation methods and calculations to be perform on a central station. Each sensing node collects measurement information, and it is to a central machine to calculate the unknown node location. One challenge is that if the main station fails for whatever reason the entire process of localization fails. The advantage is that in centralized the burden on the individual sensor nodes decreases, and it gives the greater accuracy.
- **Distributed**: Each sensor node can perform independent calculation at its end, making the sensor network more robust and fault tolerant. In addition, the cost is divided equally across all nodes. Various distributed algorithms are available, and those provide good accuracy for the node localization. The disadvantage is in distribution every individual node performs the calculation so the chances of error increases in distributed schemes. The advantage is that this approach makes the network more fault tolerable and robust.

3.3 Anchor Related

This defines whether the localization algorithm is using the anchor node or not. Anchor-based algorithms require anchor nodes for determining the positions of unknown nodes. Generally, all the algorithms fall under this category. Accuracy is high when the anchor ratio is more; but on another hand, it makes the network costly. While there is no need for anchor nodes in anchor-free localization algorithms, but more complex computation is required for anchor-free algorithms.

3.4 Range Related

Range-oriented schemes are based on distance between nodes, requirement of angle, and estimation of speed. The methods like triangulation, trilateration, or multilateration [6] are used while determining the location using the range-oriented scheme. The different ranging methods that localization uses for the estimation of the distances or angle is received signal strength indicator (RSSI) [7], time difference of arrival (TDoA), time of arrival (ToA) [8], and AoA [9].

Table 1 describes the tabular comparison of various range-based techniques. RSSI is the simplest amongst all three as it requires no additional hardware, and it is easily scalable. But this gives low accuracy. AoA has better accuracy but it requires antenna

Table 1 Comparison of range-based techniques

Ranging approach	Computation metric	Benefits	Pitfalls
RSSI [7]	Signal power	• No additional hardware required • Cheaper • No need for synchronization • Scalable • Less overhead	• Less flexibility • Less accuracy • Susceptible to noise and attenuation • Memory cost
ToA [8]	Signal perceiving time	• More accuracy • Less overhead	• Rigid measurement
AoA [9]	The angle subtended by anchor node on the node	• Yields good accuracy as compared to RSSI • Only two anchor nodes are required	• A need for the antenna array • Hardware limitations

arrays and has hardware limitations. ToA has high accuracy but it involves strict computations.

3.5 *Range-Free Schemes*

Range-free methods mitigate the issues of range-based methods [10]. This approach is categorized into anchor-based and anchor-free categories [11]. The hop count shows how near or far the sensing nodes are from each other. If they come under the transmission range of each another, then they are adjacent to each other and the hop-count value is taken as one. Most of the range-free localization schemes make use of nodes known as anchor nodes. Various range-free algorithms developed in the recent past are Centroid, Approximate Point in Triangulation (APIT), DV-Hop, Convex Position Estimation (CPE), and Multi-Dimensional Scaling (MDS). They have been discussed in detail in Chap. 3 of this thesis work.

Table 2 shows a tabular comparison of primitive range-free localization techniques. Centroid scheme has the least localization accuracy followed by APIT. In terms of power consumption, VN-APIT shows enhancements over APIT, but the accuracy is similar to APIT. DV-Hop has better accuracy than APIT and eliminated the need of having at least three anchors for estimating the unknown node location. CPE had the best of all accuracy but it was somewhat complex and more overhead experience.

Table 2 Comparison of range-free techniques

Algorithm	Type	Benefits	Pitfalls
Centroid	Decentralized	• Low overhead • Simple computations • Scalable	• Low accuracy • Nodes must be static • At least 3 anchors required
APIT	Decentralized	• No ideal assumptions required • Simple computations	• RSSI required • High power consumption • Low accuracy
VN-APIT	Decentralized	• Independent of node deployment • More rational and logical than APIT	• Better or equal accuracy than APIT • Fewer computations than APIT
DV-Hop	Decentralized	• Localization even with two anchors • Better accuracy than APIT • Provides connectivity information of nodes	• More overhead • Less scalable • Mobility makes complexity large • More memory requirement
CPE	Centralized	• Higher accuracy	• Single point of failure

4 Issues in Localization

The accuracy of the localization is the most important concern in WSN localization because if in localization accuracy is not achieved, then WSN is not effective. But there are some more issues, and those have some significant effects in WSN. Various issues in WSN are described below in detail.

4.1 Accuracy

Accuracy in WSN is very important for most of the applications in localization [12]. In range-oriented category, the accuracy is better as compared to the range-free category because the range-oriented uses the measurement process to find the Euclidean distance among the pair of the nodes.

4.2 Energy Consumption

WSN is resource limited network, so energy consumption is an important issue in WSN. As, WSN is made up of small sensors that contain the restricted energy or not replaceable batteries that become an issue in network when batteries get discharged [13]. The more consumption of energy occurs in data transmission. Even when the sensors are idle, energy is required. So there is a need to design the effective algorithm so that they can use less energy.

4.3 Overhead

The overhead in localization schemes concerns with communication, processing, and hardware cost required by the sensor nodes. The process of localization in a 3D network requires a large efficient data-structure; modeling techniques are needed to make 3D computing more compared to 2D computing [14].

4.4 Optimization

Optimization plays its part in every sector of human activity, and in all computer science engineering research activities that depend on it [15]. An important step is the selection of a perfect optimization algorithm for optimization. Specific methods of localization are carried out by integrating biologically motivated computational approaches, such as artificial bee colony (ABC) [16], genetic algorithm (GA) [17], particle swarm optimization (PSO) [18], differential evolution (DE) [19], and several others.

4.5 Mobility

Similar to WSN, mobile wireless sensor network (MWSN) is becoming most popular in the network those require mobility. BS can be mobile who get the information from all the static nodes or the mobile nodes that sense the environmental conditions in the network. Localization in MWSN is also important task. In MWSN, localization process contains three parts [20].

4.6 3D Deployment

The classic meaning of the localization is to locate the 2D WSN sensor node; but it is in 3D plane in different areas such as surveillance, underwater ecosystems, space monitoring and exploration, and environment monitoring [21].

4.7 *Security*

When WSN is deployed in the aggressive environmental area, there is a risk of security. Different types of attacks are there in WSN such as black hole, sybil, and sink-hole; those can affect the localization in WSN. Localization security can be considered into two ways [22]. First node attacks. Second, it is an information attack. Attacks mainly divided into two parts that are internal attack and external attack.

One of the most extensively researched algorithms is DV-Hop as it can be implemented in an effective manner. Also, there are various amendments that have been done on it and on account of those new variants of DV-Hop algorithms have come out. All the review has been done in context to DV-Hop in Sect. 6 of this chapter.

5 Need of Secure Localization

Security is also a major issue in WSN. With localization, one important point that also needs major attention in WSN is security. As without security, localization has no meaning in WSN because if there is no security, then there will be no surety about the true location estimation. As when the gathered data is sent by the sensors to the BS, there is a chance that any unauthorized entity may intercept the data or may modify the data. So whenever a receiver gets the data, authentication of the source of the data should be there and also data should be in the encrypted form in order to reduce the effect of any malicious activity. There are some security requirements that need to be addressed by the algorithm designed for the purpose of secure localization. Figure 7 represents the attacker model on the smart cities enterprise where there are some sensor nodes, anchor nodes, and the attacker nodes. Here the attacker tries to execute various types of attacks on the smart network.

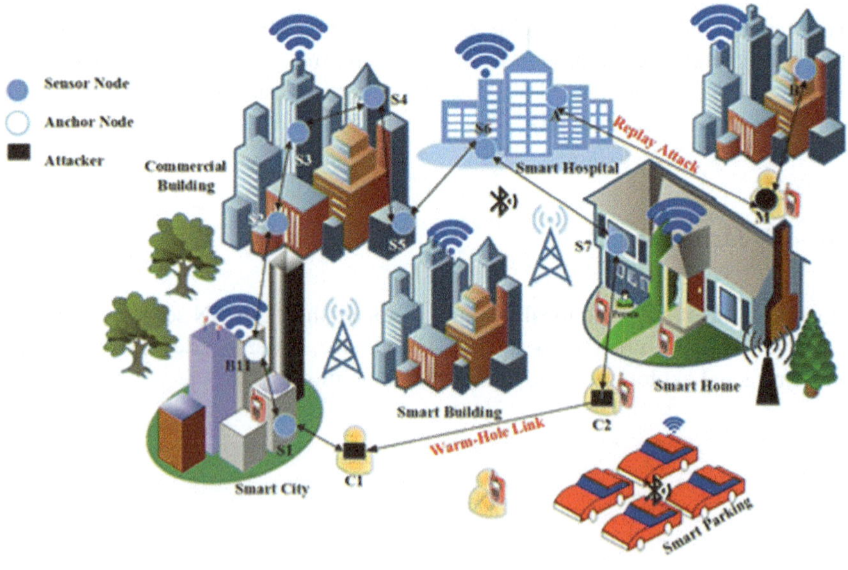

Fig. 7 Representation of attacker model in smart city network

5.1 Security Requirements

- **Authentication**: In localization, each sensor node should be able to ensure whether the sender node is authenticated one or not.
- **Integrity**: During the data transfer in the middle of the sender and receiver data should not be changed until the receiver not used it for the localization.
- **Availability**: The information required by the sensor nodes should be available whenever necessary to find out the location.
- **Non-Repudiation**: The sender and receiver nodes that exchange the information during the localization process cannot at the later stage deny the exchange of information.
- **Privacy**: Make use of cryptography to secure the data while communication is going on among the nodes.

5.2 Various Attacks in Localization

- **Wormhole Attack**: In the wormhole attack, attacker sniffs the data packet from one location in the network and transfers them through a fake link to another position where the attacker is present. Afterward, it broadcasts the packets. This process affects the localization and routing. The working of wormhole attack is shown in Fig. 8.
- **Blackhole Attack**: This attack in WSN is a simple routing attack. In this attack, the node setup by the attacker in the network that is the malicious node will receive all the messages coming to it and does not further forward those messages in a network just similar to the black hole.
- **Replay Attack**: Replay attack is easy to deploy so most of the time attacker perform this attack. Particularly when the attacker has not enough capability, then an attacker cannot compromise more than one sensor node in a network. In the

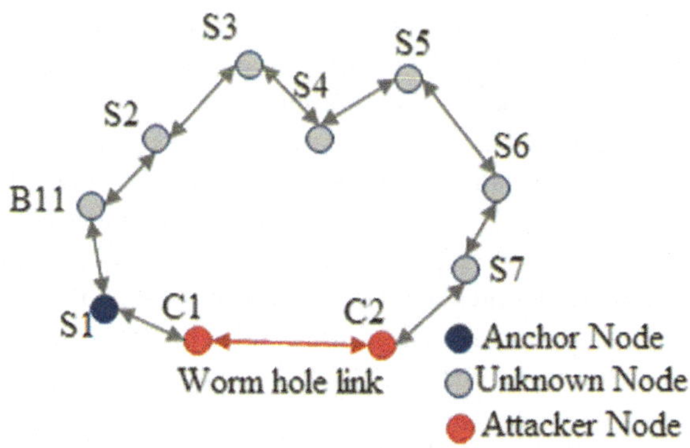

Fig. 8 Working of wormhole attack

replay attack attacker jams, the sender–receiver communication sends the same message or replays the message to the recipient pretending to be a sender.

- **Flooding Attack**: This attacker continuously sends the connection establishment requests to the particular target node. When the target node hits to request, the resources would be allocated to the attacker. The result of this attack is the memory and the energy resources of the node would be a deluge.
- **Sybil Attack**: The node holds the multiple identities in the network. In the localization, a localizing node receives the more than one location-reference from one node leads to incorrect location estimation.
- **Selective Forwarding**: A compromised node acts as the black hole in this attack and does not further send the important messages or drops those messages. It is not easy to detect, as the attacker drop only selected messages instead of dropping every message.
- **Alteration Attack**: The attack is carried out in this attack on the information being transferred in the midst of the unknown node and anchor node. Attacker can change hop-count between the nodes or can change the time.
- **Distance Fraud Attack**: The malicious node tries to pretend the closer node to the legitimate node in this method.
- **Spoofing Attack**: The malicious node in the network is set up in this attack by pretending to be a legitimate node in the network to give wrong details of localization.
- **Overshadowing**: The original message appears as noise in the strong signal of the attacker.

6 Secure Localization-Based Related Work

Security always remains one of the main concerns for the applications based on localization as the wrong position estimation will hamper the overall localization process. Here, the threats for the localization and the various localization initiatives are described. Also, the security aspects in DV-Hop localization are also discussed.

In [23], the authors provide a survey of secure localization in WSN. At last, the author also provided a review of various secure localization algorithms designed by the other authors. In [24], authors discuss the different attacks and also stresses that the security scheme must add security through cryptography, misbehavior-detection, and robust position estimation (RPE). In [25], the author has provided a survey for the security of WSN. In this, author discuss the security issues in WSN, those were identified by the researchers and also various attacks are also explained. In [26], the author narrates the various security susceptibility in WSN. In [27], the authors present the survey related to security in WSN. In this, author defines the various susceptibilities related to WSN and also defines the prevention method on the behalf of the network-protocol-layer.

In [28], authors develop secure localization algorithm in WSN. This scheme is decentralized and range-free and known as secure range-independent localization

(SeRLoc). In this approach, the authors provide the security by using the encryption and verification of the anchor *id*. A simulation result narrates that this approach yields the best accuracy for the localization uses the fewer anchors and communication cost is also less. In [29], authors propose one new approach for secure localization in WSN named as high-resolution range-independent localization (HiRLoc). After getting the signals from the anchor's unknown node finds the location in the form of region of intersection (ROI) by every sector. In SeRLoc, sensor know their location only with the one message communicated by every anchor. Whereas, this approach reduces the ROI by changing the orientation of the antenna and by changing the transmission range by power-control. HirLoc provides the security by using the encryption at the transmission time of the messages by anchors and by the authentication method for message transmission. In [30], authors propose techniques for localization those works well in the presence of the attack specifically when the anchor node in the network becomes compromised node. One approach to find the malicious node by collating the estimated distance and calculated distance. Another approach uses the round trip time (RTT), where this method filters the repeated anchor signal based on the examination of the anchor signal includes the extra-delay. In [31], authors suggest a hybrid approach that provides powerful location computing without central organization and jamming vulnerability known as robust position estimation (ROPE) that restricts malicious to spoof sensor node position.

In [32], authors propose a scheme known a distributed reputation-based beacon trust system (DRBTS). This approach uses the idea of a reputation to ban the nodes from participating in localization. In [33], the authors propose a new algorithm based on hop-count known as secure hop count-based localization (SHOLOC). Results reflect that this scheme is not feasible to defend the wormhole attack. In [34], authors propose one new secure localization approach for WSN known as secure localization with elliptic curve cryptography (ECC). This approach uses the ECC with ToA. ECC has chosen over the Rivest–Shamir–Adelman (RSA) because it has lower key length, lower bandwidth, higher computing capability, and less hardware requirement. In [35], one robust WSN localization algorithm is presented by the authors. The author proposes the distributed algorithm for known as bilateration that employs a combined way to deal with every type of attack. In [36], three algorithms are proposed by authors to estimate the position in the presence of the neighboring node that communicate the false details of their position in the network. In [37], authors propose secure localization schemes against wormhole attack with conflicting sets. In this, initial author considers the two classes of the wormhole attack. In [38], authors develop one new approach for secure localization is known as efficient gradient descent (EGD) in resource restricted WSN. The author combines the iterative gradient descent with selective trimming of irregular measurement in this approach. In [39], a new approach based on detection and classification of attacks in WSN is provided. In this, BS does not trust the nodes involved in localization, and thus, new approach is used. In [40], authors propose a node localization scheme for reputation-based security (RBS) in WSN. The author has proposed a reputation model for security risks in this approach. The author narrates that in the future, they will enhance the security for other malicious attack without an increase in cost and overhead.

In [41], the authors propose a new approach that secures the basic DV-Hop localization from the wormhole attack in WSN. In this, the author specifies the effect of the wormhole attack on basic DV-Hop following the development of a new approach called label-based (LB) DV-Hop. In [42], the authors propose one new approach for secure localization in WSN. Here, this approach divides the secure localization into two steps and reduces the effect of WSN attacks. The first step is to find the malicious node. The next step is to use the least square Taylor series (LSTS) method to find the unknown node position. A simulation result specifies that this approach is efficient and robust for node localization in WSN. In [43], authors develop one new approach for secure localization in WSN based on basic localization algorithm DV-Hop. In this approach, the author assumes all the anchors in the network send their detail to the BS confidentially, and after that, BS broadcast this detail in the network by authentication scheme and a sink node in this scheme contain some intrusion detection schemes. In [44], the authors propose a secure DV-Hop scheme known as SDV-Hop which effectively removes the errors resulting from the wormhole attack and increases the accuracy. In [45], authors propose secure localization that is known as temporal spatial consistent-based detection (TSCD).

Some of the latest research that has been done in this regard is described. For achieving the features like integrity and confidentiality, a cryptographic approach is defined [46]. In this approach, the authors have used hash-based message authentication code (HMAC) along with the elliptic curve and then the position computation is done using trilateration. This approach does not address the impact of malicious nodes in the localization process. Authors develop an approach for security for wormhole attack in WSN. The effect of radio model and the malicious nodes is done by the author on the basic DV-Hop, correction-based DV-Hop (CDV-Hop), and improved DV-Hop (IDV-Hop) [47]. Mutual authentication-based approach for security in localization is given by the author, in whom two major issues are addressed: one is the random deployment and another one is mobility of the nodes [48]. It uses the certificate values that are provided by the BS to each one of the nodes involved in the localization process. Similarly, another approach based on vector refinement and outlier elimination is given by the author that improves the localization process further [49]. In [50], authors propose a secure localization approach in WSN based on node relationship list for DV-Hop. In this scheme, there is a marking method which is utilized to differentiate the trusted node from malicious node. Trust-based localization is given by the author [51]. In this approach, they have implemented the secure algorithm in underwater WSN.

7 Multimedia Security and Authentication Techniques-Related Work

Information is one of the main assets for applications that are based on data processing and when they are multimedia-based applications, it becomes more important to secure them on account of various threats like replication and manipulation of data. The need of security is more crucial when these applications are based on localization aspects. Hence, there are various methods and approaches like cryptography, watermarking, and steganography that make the applications secure to a large extent. Also, the integration of watermarking with encryption techniques provides additional security and more robustness. Object localization is very important in wireless multimedia sensor network (WMSN) which is based on range-free localization along with machine learning algorithms.

In [52], authors propose localization system focused on historical data analysis (HDA-RTL) and in real time. They undergo two types of analysis for multimedia applications. One is real-time location based on neighborhood location and another one is real-time positioning based on mobile anchor node. In [53], authors develop a bacterial foraging optimization (BFO), an effective form of optimization that has been commonly utilized in a number of research and engineering applications, in order to increase the localization accuracy in WMSN. Within specific network environment, they perform detailed simulations; the simulation results reveal that the proposed algorithm achieves substantially higher positioning accuracy than the standard DV-Hop algorithm. In [54], authors propose a model for WMSNs based on energy-efficient and reliable routing protocol because the quality of service is critical for traffic-intensive multimedia data, such as photographs and videos. A multi-sink WMSN requires cluster heads (CHs) to transfer the generated data to the nearest sink, thereby minimizing the overhead transmission. They propose a novel evolutionary game-based routing (EGR) protocol for WMSNs with multiple sinks, exploiting the evolutionary game theory to select CHs. In [55], authors propose fingerprint localization method for indoor localization. Because of the indoor environment's complexity and variability, it is difficult for traditional indoor localization algorithms to achieve better accuracy and stability in the localization. They propose a high-performance random forest-based fingerprint localization algorithm (HPFLRF), which has higher precision and stability.

Apart from this, many researchers have worked in the domain of multimedia security and authentication methods that can also be used in the applications pertaining to localization. Some of the notable work has been carried by them has been highlighted [56–65].

8 Research Gaps

There are many research areas that still need to explore as inferred from the extensive literature work that is executed above as mentioned below.

1. Localization of node in WSN is more important because without detail of the position of the event occurrence its detection is not possible. So, localization is an important task in WSN [66].

2. Accuracy is based on whether the algorithm is range-based or range-free. In range-based localization, accuracy is high compared to range-free. But using GPS will make the system more expensive. Hence, there is a demand for a range-free techniques that are more efficient and less costly [67, 68].
3. Distributed and centralized nature of localization makes a huge impact in WSN. Decentralized localization in WSN is more appreciable as compared to the centralized on account of single point failure. But, the design of distributed localization algorithm is also a very challenging task [69].
4. Since many algorithms have been designed by the authors for the localization, the localization in the unsecured environment is a very challenging task.[70–72].
5. Localization in a secure manner with higher accuracy is not an easy task. So, there is a need for the design of secure and more accurate algorithm algorithms [73, 74].
6. Since the DV-Hop algorithm is widely used algorithm for localization, there are issues related to its accuracy. Many amendments have been done in the DV-Hop but still, there is a scope for more improvement in the form of the robust algorithm [75].

9 Conclusion

In WSN, localization has become one of the main paradigms for the applications that demand the position estimation of the process in hand. Localization process is mainly divided into range-based or range-free. Range-free-based algorithms do not require the Global Positioning System (GPS) for the computation of the unknown node position and thus are more in demand. In the current scenario, DV-Hop is the most extensively used hop-based localization algorithm. In this chapter, we have covered the detailed literature study pertaining to the range-free localization algorithms and schemes based on their working in terms of security. All the secure algorithms must ensure the prime security requirements like integrity, authentication and privacy for making the localization process more robust. There is a demand for secure range-free localization schemes with less complexity to mitigate all possible attacks. Apart from this with the advancements in the multimedia applications, the need of security has arisen and there are many techniques like watermarking, steganography, and encryption that are used in WMSN. The study of mobility and integration of secure multimedia approaches in localization will be the future work.

References

1. Bachrach J, Taylor C (2005) Localization in sensor networks. Handbook of sensor networks: Algorithms and architectures, pp 277–310

2. Yong-Min L, Shu-Ci W, Xiao-Hong N (2009) The architecture and characteristics of wireless sensor network. In: International conference on computer technology and development, vol 1, pp 561–565
3. Cheng B, Du R, Yang B, Yu W, Chen C, Guan X (2011) An accurate GPS-based localization in wireless sensor networks: a GM-WLS method. In: 2011 40th International conference on parallel processing workshops, pp 33–41
4. Yick J, Mukherjee B, Ghosal D (2008) Wireless sensor network survey. Comput Netw 52(12):2292–2330
5. Han G, Jiang J, Zhang C, Duong TQ, Guizani M, Karagiannidis GK (2016) A survey on mobile anchor node assisted localization in wireless sensor networks. IEEE Commun Surv Tutor 18(3):2220–2243
6. Boukerche A, Oliveira HA, Nakamura EF, Loureiro AA (2007) Localization systems for wireless sensor networks. IEEE Wirel Commun 14(6):6–12
7. Zanca G, Zorzi F, Zanella A, Zorzi M (2008) Experimental comparison of RSSI-based localization algorithms for indoor wireless sensor networks. In: Proceedings of the workshop on real-world wireless sensor networks, pp 1–5
8. Yu K, Guo YJ, Hedley M (2009) TOA-based distributed localization with unknown internal delays and clock frequency offsets in wireless sensor networks. IET Signal Proc 3(2):106–118
9. Rong P, Sichitiu ML (2006) Angle of arrival localization for wireless sensor networks. In: 2006 3rd Annual IEEE communications society on sensor and ad hoc communications and networks, vol 1, pp 374–382
10. He T, Huang C, Blum BM, Stankovic JA, Abdelzaher TF (2005) Range-free localization and its impact on large scale sensor networks. ACM Trans Embedded Comput Sys (TECS) 4(4):877–906
11. Pei Z, Deng Z, Xu S, Xu X (2009) Anchor-free localization method for mobile targets in coal mine wireless sensor networks. Sensors 9(4):2836–2850
12. Suo H, Wan J, Huang L, Zou C (2012) Issues and challenges of wireless sensor networks localization in emerging applications. In: International conference on computer science and electronics engineering, vol 3, pp 447–451
13. Cardei M, Thai MT, LiY, Wu W (2005) Energy-efficient target coverage in wireless sensor networks. In: Proceedings IEEE 24th annual joint conference of the IEEE computer and communications societies, vol 3, pp 1976–1984
14. Yadav V, Mishra MK, Sngh AK, Gore MM (2009) Localization scheme for three dimensional wireless sensor networks using GPS enabled mobile sensor nodes. Int J Next-Gener Netw (IJNGN) 1(1):60–72
15. Aioffi W, Mateus G, Quintao F (2007) Optimization issues and algorithms for wireless sensor networks with mobile sink. In: International network optimization conference, pp. 1–6
16. Karaboga D, Okdem S, Ozturk C (2012) Cluster based wireless sensor network routing using artificial bee colony algorithm. Wireless Netw 18(7):847–860
17. Bhondekar AP, Vig R, Singla ML, Ghanshyam C, Kapur P (2009) Genetic algorithm based node placement methodology for wireless sensor networks. In: Proceedings of the international multiconference of engineers and computer scientists, vol 1, pp 18–20
18. Kulkarni RV, Venayagamoorthy GK (2011) Particle swarm optimization in wireless-sensor networks: a brief survey. IEEEE Trans Syst Man Cybern Part C (Appl. Rev.) 41(2):262–267
19. Kuila P, Jana PK (2014) A novel differential evolution based clustering algorithm for wireless sensor networks. Appl Soft Comput 25(C):414–425
20. Bao H, Zhang B, Li C, Yao Z (2012) Mobile anchor assisted particle swarm optimization (PSO) based localization algorithms for wireless sensor networks. Wirel Commun Mobile Comput 12(15):1313–1325
21. Samanta S, Tembhare PU, Pote CR (2013) A survey on 3d localization in wireless sensor networks. Int J Comput Eng Res 3(1):90–94
22. Zhang Y, Liu W, Fang Y, Wu D (2006) Secure localization and authentication in ultra-wideband sensor networks, IEEE J Sel Areas Commun 24(4):829–835

23. Srinivasan A, Wu J (2007) A survey on secure localization in wireless sensor networks. Encycl Wirel Mobile Commun 126:1–26
24. Boukerche A, Oliveira HA, Nakamura EF, Loureiro AA (2008) Secure localization algorithms for wireless sensor networks. IEEE Commun Mag 46(4):96–101
25. Zhou Y, Fang Y, Zhang Y (2008) Securing wireless sensor networks: a survey. IEEE Commun Surv Tutor 10(3):6–28
26. Kavitha T, Sridharan D (2010) Security vulnerabilities in wireless sensor networks: a survey. J Inf Assur Secur 5(1):31–44
27. Chen X, Makki K, Yen K, Pissinou N (2009) Sensor network security: a survey. IEEE Commun Surv Tutor 11(2):52–73
28. Lazos L, Poovendran R (2004) SeRLoc: secure range-independent localization for wireless sensor networks. In: Proceedings of the 3rd ACM workshop on Wireless security, pp 21–30
29. Lazos L, Poovendran R (2006) HiRLoc: high-resolution robust localization for wireless sensor networks. IEEE J Sel Areas Commun 24(2):233–246
30. Liu D, Ning P, Du W (2005) Detecting malicious beacon nodes for secure location discovery in wireless sensor networks. In: Distributed computing systems, 2005. ICDCS 2005. Proceedings. 25th IEEE international conference, pp 609–619
31. Lazos L, Poovendran R, Čapkun S (2005) ROPE: robust position estimation in wireless sensor networks. In: Proceedings of the 4th international symposium on Information processing in sensor networks, pp 43–49
32. Srinivasan A, Teitelbaum J, Wu J (2006) DRBTS: distributed reputation-based beacon trust system. In: 2006 2nd IEEE international symposium on dependable, autonomic and secure computing, pp 277–283
33. Zeng Y, Zhang S, Guo S, Li X (2007) Secure hop-count based localization in wireless sensor networks. In: 2007 International conference on computational intelligence and security (CIS 2007), pp 907–911
34. Vijayalakshmi V, Palanivelu TG (2008) Secure localization using elliptic curve cryptography in wireless sensor networks. Int J Comput Sci Netw Secur 8(6):255–261
35. Li X, Hua B, Shang Y, Xiong Y (2008) A robust localization algorithm in wireless sensor networks. Front Comput Sci China 2(4):438–450
36. Alfaro JG, Barbeau M, Kranakis E (2009) Secure localization of nodes in wireless sensor networks with limited number of truth tellers. In: 2009 Seventh annual communication networks and services research conference, pp 86–93
37. Chen H, Lou W, Wang Z (2010) Secure localization against wormhole attacks using conflicting sets. In: International performance computing and communications conference, pp 25–33
38. Garg R, Varna AL, Wu M (2012) An efficient gradient descent approach to secure localization in resource constrained wireless sensor networks. IEEE Trans Inf Forensics Secur 7(2):717–730
39. Zeng Y, Cao J, Hong J, Zhang S, Xie L (2013) Secure localization and location verification in wireless sensor networks: a survey. J Supercomputing 64(3):685–701
40. He J, Xu J, Zhu X, Zhang Y, Zhang T, Fu W (2014) Reputation-based secure sensor localization in wireless sensor networks. Sci World J. https://doi.org/10.1155/2014/308341
41. Chen H, Lou W, Wang Z, Wu J, Wang Z, Xia A (2015) Securing DV-Hop localization against wormhole attacks in wireless sensor networks. Pervasive Mob Comput 16:22–35
42. Mazinani SM, Safari M (2015) Secure localization approach in wireless sensor network. Int J Mach Learn Comput 5(6):458–462
43. Liu X, Yang R, Cui Q (2015) An efficient secure DV-Hop localization for wireless sensor network. Int. J. Secur. Appl 9:275–284
44. Wang H, Feng L, Li R, Zhang Y (2016) The secure localization algorithm of SDV-HOP in wireless sensor networks. Telkomnika 14(3):65–74
45. Chen H, Lou W, Ma J, Wang Z (2008) TSCD: a novel secure localization approach for wireless sensor networks. In: 2008 Second international conference on sensor technologies and applications (sensorcomm 2008), pp 661–666
46. Zhang T, He J, Li X, Wei Q (2012) A signcryption-based secure localization scheme in wireless sensor networks. Phys. Procedia 33:258–264

47. Ran X, Shu L, Mukherjee M, Wu Y, Chen Y, Sun Z (2016) Impact of irregular radio and faulty nodes on localization in industrial WSNs. In: International wireless internet conference, pp 36–48
48. Kumar G, Rai MK, Kim HJ, Saha R (2017) A secure localization approach using mutual authentication and insider node validation in wireless sensor networks. Mob Inf Syst. https://doi.org/10.1155/2017/3243570
49. Li X, Yan L, Pan W, Luo B (2017) Secure and robust DV-Hop localization based on the vector refinement feedback method for wireless sensor networks. Comput J 60(6):810–821
50. Li J, Wang D, Wang Y (2017) Security DV-hop localisation algorithm against wormhole attack in wireless sensor network. IET Wirel Sens Syst 8(2):68–75
51. Han G, Liu L, Jiang J, Shu L, Rodrigues J (2016) A collaborative secure localization algorithm based on trust model in underwater wireless sensor networks. Sensors 16(2). https://doi.org/10.3390/s16020229
52. Zhao Y, Xu J, Xu F, Song S, Zhang Y (2018) Real-time localization in wireless sensor network with multimedia applications. Multimedia Tools Appl 77(16):21791–21801
53. Zhou C, Yang Y, Wang Y (2019) DV-Hop localization algorithm based on bacterial foraging optimization for wireless multimedia sensor networks. Multimedia Tools Appl 78(4):4299–4309
54. Habib MA, Moh S (2019) Robust evolutionary-game-based routing for wireless multimedia sensor networks. Sensors 19(16):35–44
55. Huang P, Zhao H, Wang W (2019) HPFLRF: a high performance fingerprint localization algorithm based on random forest. In: Advanced multimedia and ubiquitous engineering, pp 358–363
56. Hurrah NN, Parah SA, Loan NA, Sheikh JA, Elhoseny M, Muhammad K (2019) Dual watermarking framework for privacy protection and content authentication of multimedia. Future Gener Comput Syst 94:654–673
57. Parah SA, Sheikh JA, Bhat GM (2014) Fragility evaluation of intermediate significant bit embedding (ISBE) based digital image watermarking scheme for content authentication. In: 2014 International conference on advances in electronics computers and communications, pp. 1–6
58. Gull S, Loan NA, Parah SA, Sheikh JA, Bhat GM (2018) An efficient watermarking technique for tamper detection and localization of medical images. J. Amb. Intell. Hum. Comput. 1–10
59. Gull S, Parah SA (2019) Color image authentication using dual watermarks. In: 2019 Fifth international conference on image information processing (ICIIP), pp 240–245
60. Parah SA, Sheikh JA, Loan NA, Ahad F, Bhat GM (2018) Utilizing neighborhood coefficient correlation: a new image watermarking technique robust to singular and hybrid attacks. Multidimens Syst Signal Process 29(3):1095–1117
61. Giri KJ, Peer MA, Nagabhushan P (2014) A channel wise color image watermarking scheme based on discrete wavelet transformation. In: 2014 International conference on computing for sustainable global development (INDIACom), pp 758–762
62. Giri KJ, Peer MA, Nagabhushan P (2015) A robust color image watermarking scheme using discrete wavelet transformation. IJ Image, Graphics and Signal Processing 1:47–52
63. Giri KJ, Bashir R (2017) Digital watermarking: a potential solution for multimedia authentication. In: Intelligent techniques in signal processing for multimedia security, pp 93–112
64. Giri KJ, Bashir R (2018) A block based watermarking approach for color images using discrete wavelet transformation. Int J Inf Technol 10(2):139–146
65. Giri KJ, Bashir R, Bhat JI (2019) A discrete wavelet based watermarking scheme for authentication of medical images. Int J of E-Health Med Commun (IJEHMC) 10(4):30–38
66. Bangash JI, Abdullah AH, Khan AW (2014) Issues and challenges in localization of wireless sensor networks. Sci Int 26(2)
67. Brito LM, Peralta LMR (2008) An analysis of localization problems and solutions in wireless sensor networks. Tékhne-Revista de Estudos Politécnicos 9:46–172
68. Shi W, Barbeau M, Corriveau JP, Garcia-Alfaro J, Yao M (2017) Secure localization in the presence of colluders in WSNs. Sensors 17(8). https://doi.org/10.3390/s17081892

69. Zia T, Zomaya A (2006) Security issues in wireless sensor networks. In: 2006 International Conference on Systems and Networks Communications (ICSNC'06), pp 40–40
70. Pal A (2010) Localization algorithms in wireless sensor networks: current approaches and future challenges. Netw Protocols Algor 2(1):45–73
71. Jiang J, Han G, Zhu C, Dong Y, Zhang N (2011) Secure localization in wireless sensor networks: a survey. J. Commun 6(6):460–470
72. Ammar W, ElDawy A, Youssef M (2010) Secure localization in wireless sensor networks: a survey. arXiv:1004.3164
73. Halder S, Ghosal A (2016) A survey on mobility-assisted localization techniques in wireless sensor networks. J Netw Comput Appl 60(C):82–94
74. Amundson I, Koutsoukos XD (2009) A survey on localization for mobile wireless sensor networks. In: International workshop on mobile entity localization and tracking in GPS-less environments, pp 235–254
75. Gupta V, Singh B (2018) Performance evaluation of DV-HOP localization algorithm in wireless sensor networks. Soft Comput: Theories Appl 584:433–440

The manufacturer's authorised representative in the EU is Springer Nature Customer Service Centre GmbH, Europaplatz 3, 69115 Heidelberg, Germany. If you have any concerns regarding our products, please contact ProductSafety@springernature.com

Printed and bound by CPI Group (UK) Ltd, Croydon, CR0 4YY
30/09/2026
02225291-0010